MAY 1 9 1988
UNIVERSITY OF CALIFORNIA

Computer-Aided Design of Polymers and Composites

PLASTICS ENGINEERING

Series Editor

Donald E. Hudgin

Princeton Polymer Laboratories
Plainsboro, New Jersey

1. Plastics Waste: Recovery of Economic Value, *Jacob Leidner*
2. Polyester Molding Compounds, *Robert Burns*
3. Carbon Black-Polymer Composites: The Physics of Electrically Conducting Composites, *Edited by Enid Keil Sichel*
4. The Strength and Stiffness of Polymers, *Edited by Anagnostis E. Zachariades and Roger S. Porter*
5. Selecting Thermoplastics for Engineering Applications, *Charles P. MacDermott*
6. Engineering with Rigid PVC: Processability and Applications, *Edited by I. Luis Gomez*
7. Computer-Aided Design of Polymers and Composites, *D. H. Kaelble*

Other Volumes in Preparation

Computer-Aided Design of Polymers and Composites

D. H. KAELBLE

Arroyo Computer Center
Thousand Oaks, California

MARCEL DEKKER, INC. New York and Basel

Library of Congress Cataloging in Publication Data

Kaelble, D.H., (date)
Computer-aided design of polymers and composites.

(Plastics engineering ; 7)
Includes index.
1. Polymers and polymerization--Data processing.
I. Title. II. Series: Plastics engineering (Marcel Dekker, Inc.) ; 7.
QD381.9.E4K34 1985 547.7'028'54 84-23046
ISBN 0-8247-7288-1

MARCEL DEKKER, INC.
270 Madison Avenue, New York, New York 10016

Current printing (last digit):
10 9 8 7 6 5 4 3 2

PRINTED IN THE UNITED STATES OF AMERICA

Preface

This book on computer-aided design of polymers and composites introduces and discusses the subject from the viewpoint of atomic and molecular models. Thus the origins of stiffness, strength, extensibility, and fracture toughness in composite materials can be analyzed directly in terms of chemical composition and molecular structure.

This discussion is introduced by a review of current concepts of composite reliability (Chapter 1) and composite characterization techniques (Chapter 2). The central section (Chapters 3 and 4) of this discussion introduces the computer models which describe the relations between chemical, physical and mechanical responses of polymer composites. Six simplified CAD/CAM (computer-aided design and computer-aided manufacturing) models are demonstrated in the final sections (Chapters 5 and 6). The personal computer program listings for these models are provided in an Appendix. These programs are written in BASIC for the TRS80 T.M. Model 1, and can easily be translated for use on other microcomputers (see Section 5.1).

The detailed effects of temperature, stress, strain, time and environmental exposure upon composite stiffness, strength and fracture toughness can be generated by mathematical simulation from atomic and molecular properties. By establishing clearly defined mathematical links between chemical and mechanical responses, the controlling influence of chemistry and characterization on composite mechanical reliability is clarified. The models also provide useful new training aids in material science. These model calculations are illustrative in nature and provide a basis for numeric estimation and prediction of the simpler properties of composite laminates. The form in which the models are developed permits easy alteration and combination to meet specific requirements of the reader. This is the author's first effort in CAD/CAM discussion, so the reader should be aware of a certain unpolished enthusiasm for this new and exciting subject area.

The author gratefully acknowledges financial support for the research relating to this discussion to the U.S. Army Research Office through Grant No. DAAG 29-80-C-0137 and the U.S. Department of Energy under JPL Subcontract 954739. The author's colleagues and management at the Rockwell International Science Center provided both financial support and much technical consultation in the development of this discussion. Above all, my kind and helpful wife Elaine deserves the most gratitude and thanks. Since completing this book I have taken early retirement from the Rockwell International Science Center and am now independently engaged in research and consulting on CAD/CAM for polymers and composites.

D. H. Kaelble

Introduction

The contents and organization of this book require a brief introduction and explanation. This text is designed to be used in conjunction with the six computer models listed in the Appendix. When used with an appropriate microcomputer this text is capable of electronically generating an exhaustive and detailed set of tables which are typical of a handbook. The use of the computer and computer modeling is the central theme of this text. The programs are written in BASIC for the TRS80$^{T.M.}$ Model 1 microcomputer, and can be easily translated for use on other microcomputers (see Section 5.1).

The term CAD/CAM (computer-aided design and computer-aided manufacturing) has come to be associated with various fields of engineering. This book introduces the CAD/CAM concept to the molecular engineering of polymers and polymer composites. In this sense, this book is a new experiment and a special challenge to the reader. Proper use of the book demands an interactive use of the computer models presented in the Appendix and discussed in the text. Nothing less will suffice since the text is, in fact, a brief introduction to the computer models (Chapters 3 and 4) followed by fairly detailed demonstrations of their interactive character with extensive verification with experimental studies (Chapters 5 and 6).

The concept of molecular design and engineering of chemical properties is well advanced in pharmaceutical chemistry and genetic engineering, but appears to have lagged behind in polymer science. The properties of polymers are defined by this author in *Physical Chemistry of Adhesion* (Wiley-Interscience, New York, 1971) as resulting from (1) chemical composition, (2) free volume, and (3) macromolecular network structure. The viscoelastic theory and mechanical property models of that earlier text are extensively utilized in the computer models of this book. Only brief reference and introduction to the detailed derivations of the earlier text are presented here, and therefore the interested reader who intends to explore the foundations of the computer models is referred to my earlier detailed analysis of polymer structure and properties.

The definitions and measurement of mechanical strength of both polymers and polymer composites are still controversial. In this text, the definitions of strength are replaced by statistical concepts of design for reliability, hence, reliability is emphasized and quantitatively defined in Chapter 1 and extensively applied to experimental analysis throughout the text. In the author's view, there now exists a substantial basis for the manufacturing science of polymers and polymer composites and this subject is introduced in Chapter 2 and discussed in more detail in Chapter 4.

The structure of the text is organized around six CAD/CAM computer models, which are identified as follows:

Model Number	Type (CAD or CAM)	Title
1	CAM	Atomic to Molecular Properties
2	CAD	Polymer Chemistry to Physical Properties
3	CAM	Polymerization & Crosslinking
4	CAD	Polymer Chemistry to Mechanical Properties
5	CAM	Composite Fracture Energy & Strength
6	CAD	Peel Mechanics

In proceeding down the above listing, one notes that the manufacturing (CAM) and design (CAD) models alternate to provide an interleafing and overlapping structure to the computer-based analysis. The second important feature in the model development is the three level transition from chemistry (Models 1 and 2) to chemorheology (Models 3 and 4), to composite fracture mechanics (Models 5 and 6).

Each of the computer models in this text is provided with a brief introduction and practical examples of sample calculations. Model calculations provide an immediate insight into the multiple effects that changes in chemical composition or composite morphology are likely to have upon both performance and reliability. This result is difficult to achieve by more traditional research methods. In this sense, this book is pioneering a new area of research.

The user of this book should have an understanding of polymer physical chemistry as well as some understanding of viscoelasticity. Without this background, the user may find it difficult to design the proper strategies for adapting the simplified computer programs presented here to the user's own problems. The models presented here should also provide instructional tools in an introductory polymer or composites course. A short list of suggested background reading is listed below.

This is based upon, and is an extension of, the molecular models of polymer response derived in the first reference of this suggested reading list.

Suggested Introductory Reading

D. H. Kaelble, "Physical Chemistry of Adhesion," Wiley-Interscience, New York, 1971.

I. M. Ward, "Mechanical Properties of Solid Polymers," Wiley-Interscience, New York, 1971.

N. G. McCrum, B. E. Read, and G. Williams, "Anelastic and Dielectric Effects in Polymer Solids," Wiley, New York, 1967.

J. D. Ferry, "Viscoelastic Properties of Polymers," 3rd Ed., Wiley, New York, 1980.

J. Delmonte, "Technology of Carbon and Graphite Composites," Van Nostrand, New York, 1981.

R. M. Jones, "Mechanics of Composite Materials," McGraw-Hill, New York, 1975.

Composite Reliability, ASTM Spec. Tech. Pub. STP 580, American Society of Testing & Materials, Philadelphia, 1975.

Contents

List of Tables

List of Figures

List of Figures

1
Polymer Composite Reliability

1-1. Current Concepts

The structural performance, reliability and durability of polymer composites can now be correlated with three generic classes of internal defects. The first generic class of chemical structure defects (size 1 - 100Å) that control critical design properties such as glass transition T_g, moisture absorption, and dimensional changes can be controlled by chemical analysis of raw materials prior to manufacture. A second generic class of manufacturing defects (size greater than 10 μm) include inclusions, voids and debonds which are related to manufacturing process control and recognized by ultrasonics, optical scanning and other techniques sensitive to interfacial imperfections. The interaction of these two classes of intrinsic defects with environmental and mechanical stresses produces a third class of macroscopic fatigue defects such as interconnected microcracks and macroscopic crack growth which can be detected by visual inspection and ultrasonic emission.

The recognition of intrinsic structural defects, and their contributions to polymer composite reliability, represents an important extension in the analytic modeling and reliability predictions for structural polymers, adhesively bonded metals and high strength fiber reinforced composites in which the physical chemistry parameters appear as primary control variables. This discussion introduces and discusses combined deterministic/statistical models for polymer composite reliability. The molecular process which determines the relation between environmental condition and macroscopic structural effect is detailed within such models and provides important criteria for chemical and manufacturing optimization of polymer composite reliability. Experimental data of aging effects on the statistical strength distributions of structure polymers, metal-to-metal joints and reinforced composites are examined and compared with model predictions.

Concepts of structural design for reliability are old and well used. A classical expression of these concepts is well illustrated in an excerpt of a poem by Oliver Wendell Holmes as described in Table 1-1. The design of high performance, reliability, and durability into mobile structures such as spacecraft, airplanes, cars, electronic-circuitry or the one-horse shay described in Table 1-1 is a special branch of structural engineering in which weight minimization is an important design constraint. The rapidly evolving polymer com-

TABLE 1-1 THE DEACON'S MASTERPIECE: Or the Wonderful "One-Hoss-Shay."*
A Logical Story

Have you heard of the wonderful one-hoss-shay,
That was built in such a logical way
It ran a hundred years to a day,
And the, of a sudden - ah, but stay,
I'll tell you what happened without delay,

At age one hundred years to the day
There are traces of age in a one-hoss-shay
A general flavor of mild decay
But nothing local, as one may say.
There couldn't be, - for the Deacon's art
Had made it so like in very part
That there wasn't a chance for one to start.
And yet, as a whole, it is past a doubt
In another hour it will be worn out!

This morning the parson takes a drive.
All at once the horse stood still,
Close by the meet'n'-house on the hill.
-First a shiver, and then a thrill,
Then something decidedly like a spill,-
And the parson was sitting upon a rock,

-What do you think the parson found,
When he got up and stared around?
The poor old chaise in a heap or mound,
You see, of course, if you're not a dunce,
How it went to pieces at once,-
All at once, and nothing first,-
Just as bubbles do when they burst.

*Exerpts from a poem by Oliver Wendell Holmes, in "The Autocrat of the Breakfast Table," pp. 252-256, The Riverside Press, Cambridge, Mass. (1895) relating to "Structural design for reliability."

posite technology represents new inputs which make exotic advanced structures such as the man-powered airplane "Gondor," and the Space Shuttle functional within their design constraints.

One specific objective of this discussion is to review current reliability concepts and identify new appraoches, particularly relevant to polymer composite materials. A second objective is to indicate the important role that predictive models and computer aided design and manufacture (CAD/CAM) can play in achieving polymer composite reliability. A final objective is to illustrate the utility of determining the molecular processes which control macroscopic reliability and define environmental aging and nondestructive evaluation (NDE) directly in terms of the molecular process.

Current concepts of engineering design to achieve structural reliability are detailed in an excellent review edited by Swedlow, Cruse, and Halpin.[2] Several of the important current definitions of reliability provide valuable insight into design concepts. A widely used and generally accepted definition of reliability is described by the so-called "bathtub curve" of Fig. 1-1, and the following definition of reliability R by the following equation:

$$R = \exp - \int_0^t \lambda(t)\ dt \tag{1-1}$$

where $\lambda(t)$ is defined as the failure rate at time t, of the "bathtub curve" and Eq. (1-1) defines a break in period in which manufacturing defects are covered by warranty and repaired without charge. Following this break in period, the failure rate is minimized and dominated by circumstantial failure until a period where the wearout process characteristic of environmental aging begins to dominate. As pointed out by Heller,[3] human existance is modeled by high infant mortality (burn-in) followed by primarily accidental mortality during early maturity and terminating in wear out by aging as shown in Fig. 1-1.

With ceramic materials where time effects are nondominant and failure is primarily determined by stress initiated failure modes, the concept of struc-

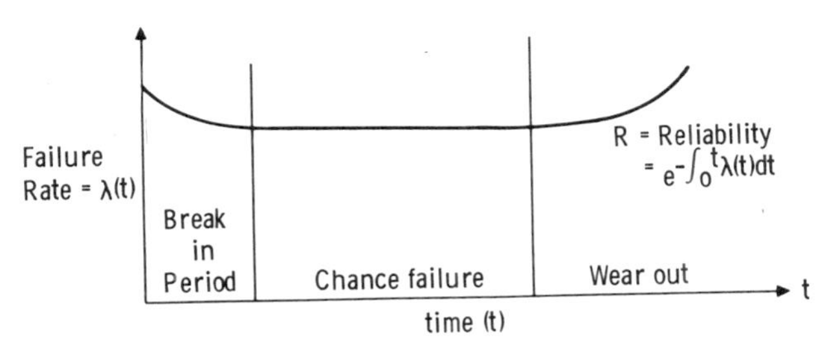

FIG. 1-1 Failure rate criteria for reliability.

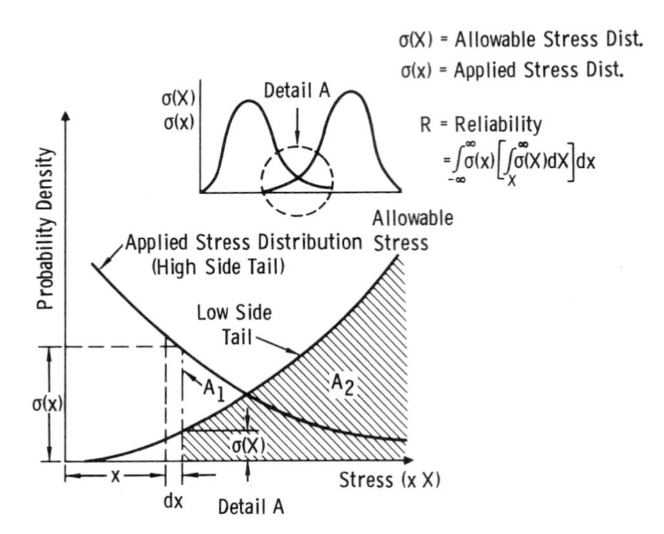

FIG. 1-2 Applied and allowable stress distribution analysis of reliability.

tural reliability can be defined by applied and allowable stress distributions as shown by the curves of Fig. 1-2, and the following definition of reliability R as given in the following relation:

$$R = \int_{-\infty}^{\infty} \sigma(x) \int_{X}^{\infty} \sigma(X)dX \; dx \tag{1-2}$$

where:

$$\int_{-\infty}^{\infty} \sigma(x) \; dx = 1.0$$

$$\int_{-\infty}^{\infty} \sigma(X)dX = 1.0$$

describe the areas beneath the stress distribution curves of Fig. 1-2. If the distribution of applied stresses $\sigma(x)$ is less than the distribution of allowable stresses $\sigma(X)$, the value R = 1.0, the greater the interpenetrations of these two distributions as shown by detail A of Fig. 1-2, the lower the structural reliability R.

In metal structures which exhibit ductility and slow crack growth under fatigue loading, a fracture mechanics approach to failure prediction is employed in the design of fracture resistant structures. The procedures to ensure reliability outlined by Tiffany[5] and graphed in Fig. 1-3 include defect characterization (left view) combined with experimental studies of crack growth rates (right view) and the determination of critical crack size where rapid fracture

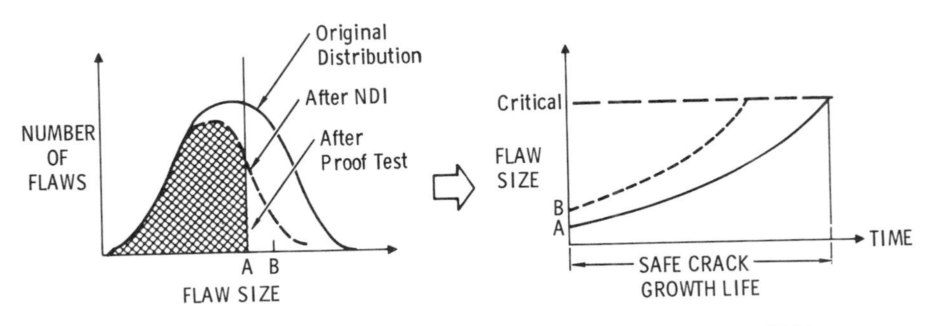

FIG. 1-3 Fracture mechanics criteria for structure reliability.

occurs. This design strategy for structure reliability assumes that pre-existing defects are either inherent in the material or produced during fabrication. Flaw detection by nondestructive inspection (NDI) becomes of primary importance in eliminating detectable flaws. Sacrificial proof testing may be applied to eliminate all flaws above a given size. The result of both NDI with rejection at flaw size B or proof testing with sacrificial failure at flaw size A is to establish the determined safe crack growth curves in the right view of Fig. 1-3.

An intrinsic problem in designing large structures involves the extrapolation of strength and reliability data obtained from small coupon specimens to values appropriate to the large structure. In practice this problem is further complicated by the fact that manufacturing defects in the large structure may differ from those in small coupon specimens manufactured separately. Materials which are not particularly sensitive to flaws, such as metals exhibit far less scatter in strength than brittle materials which are flaw sensitive. At the micro-mechanical response level the scatter in strength is due to the distribution of flaws. As discussed by Jerima and Halpin,[6] the extreme value (or Weibull[7]) statistical distribution function has been widely applied in materials science and engineering.

The Weibull distribution function defines reliability R by the following relation:[6-8]

$$R = 1 - F = \exp\left[-V_i K_i N_i \left(\frac{\sigma}{\beta}\right)^m\right] \tag{1-3}$$

where F is failure probability, V_i, K_i, N_i are respectively related to size, shape, and numeric complexity effects upon the mean strength β at reliability level $R = 1/e \simeq 0.368$. The Weibull distribution shape factor m is determined from analysis of experimental data where the measured strengths are arranged serially, $j = 1,2,3 \ldots\ldots N$ in decreasing order of σ and the reliability is defined as follows:[7]

$$R = 1 - F = \frac{j - 0.50}{N} \tag{1-4}$$

By taking logarithms of Eq. (1-3) we obtain the following linear relation:

$$\ell n \, \ell n \, R^{-1} = \ell n \, (- \ell n R) = m(\ell n \, \sigma - \ell n \, \beta) + \ell n \, V_i K_i N_i$$

where the ordinate becomes $\ell n \, (-\ell n R)$, the slope is m and the intercept at $\ell n \, (-\ell n R) = 0$ is a reference mean strength β_0 for the reference condition $V_i K_i V_i = V_0 K_0 V_0 = 1.0$.

The several implications inherent in the Weibull criteria for structural reliability are graphically summarized in Fig. 1-4 and discussed in greater detail by Jerima and Halpin.(6) The left views of Fig. 1-4 trace the progressive effects of increasing size, shape with a hole stress concentration, and numeric complexity by N holes. In the upper right of Fig. 4, the predicted lowering of mean strength from β_0 for the reference coupon specimen to β_1, β_2, and β_3 is defined by the following relation:(6)

$$\frac{\beta_j}{\beta_i} = \left(\frac{V_i K_i N_i}{V_j K_j N_j}\right)^{1/m} \tag{1-5}$$

which assumes that the Weibull slope m remains constant with varied V, K, and N. Both Eq. (1-5) and the lower right view of Fig. 1-4 point out the fundamental importance of the Weibull slope m in structural design. As shown in the lower right curves of Fig. 1-4, if the Weibull slope factor m displays a high value m = 30, the increase in size or number complexity, such as $V_j/V_0 = 100$ or $N_j/N_0 = 100$, only slightly lowers the mean strength with $\beta_j/\beta_0 = 0.86$. However, where the Weibull shape factor m is lowered to represent typical structural materials such as m = 4 for aluminum, the prediction is that with $V_j/V_0 = 100$ or $N_j/N_0 = 100$, a major lowering of mean strength with $\beta_j/\beta_0 = 0.32$ results. Thus, where m represents a material property relating to the distribution of microflaws, the translation of this material property to large scale structural reliability R is well delineated in Weibull definitions.

This section has provided a brief review of statistical definitions of structural reliability which form the basis of current design practice in composite structures. The Weibull distribution provides an entry point for translatinmg microscopic material responses, as defined by m and β_0 into predictions of large scale structural reliability.

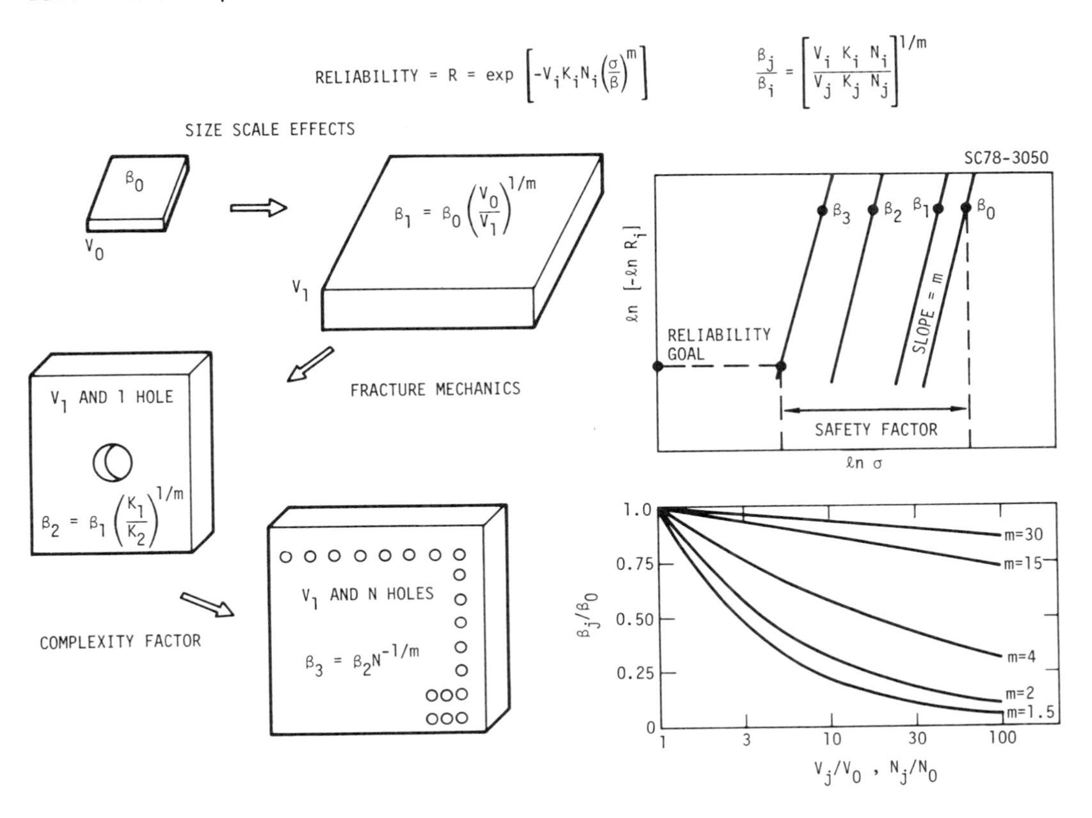

FIG. 1-4 Weibull criteria for structure reliability.

1-2. Dual Path Concept

The structural performance, reliability, and durability of polymer composites can now be correlated with three generic classes of defects. The first class of generic defects (size 1 - 100Å) that control critical design properties such as glass transition T_g, moisture absorption, and dimensional changes can be controlled by chemical analysis and selection of raw materials prior to manufacture.(9-12) A second generic class of manufacturing defects (size greater than 10 μm) include inclusions, voids, and debonds which results from non-optimum process control in fabrication and manufacture. This second class of defects are detected by ultrasonics, optical scanning and other methods sensitive to interfacial imperfections.

The interaction of these two classes of intrinsic defects with environmental and mechanical stresses produces a third class of macroscopic fatigue defects such as networks of interconnected microcracks as well as singular macroscopic crack growth. These fatigue defects can be detected by ultrasonic emission, moisture diffusion analysis, and optical inspection.

Recognition that intrinsic chemical and manufacturing defects may, in large part, determine polymer composite reliability represents an important extension of analytical modeling in which physical chemistry parameters appear as primary control variables. A preferred dual path approach for correlating environmental (plus mechanical) aging with macroscopic strength is shown in Fig. 1-5. In addition to the statistical correlations for structure reliability discussed in the preceeding section, the dual path approach adds detailed spectroscopic analysis to define the molecular process of aging and strength change. As shown in Fig. 5, the proposition of process scale-up characterization to define and control manufacturing defects is inherent in the dual path approach. Utilization of the dual path approach usefully combines both deterministic and statistical models for polymer composite reliability. A highly evolved flow chart for polymer composite reliability analysis is shown by the block diagram of Fig. 1-6.[12]

The left column of Fig. 1-6 deals with studies of composite system response and develops the important statistical correlation between environmental aging and composite reliability. The central column of Fig. 1-6 defines special studies of interfacial bonding, which range from surface chemistry to macroscopic characterization of surface roughness. These studies are dominantly related to isolation and control of manufacturing defects introduced by improper surface treatment and process conditions for bonding and curing.

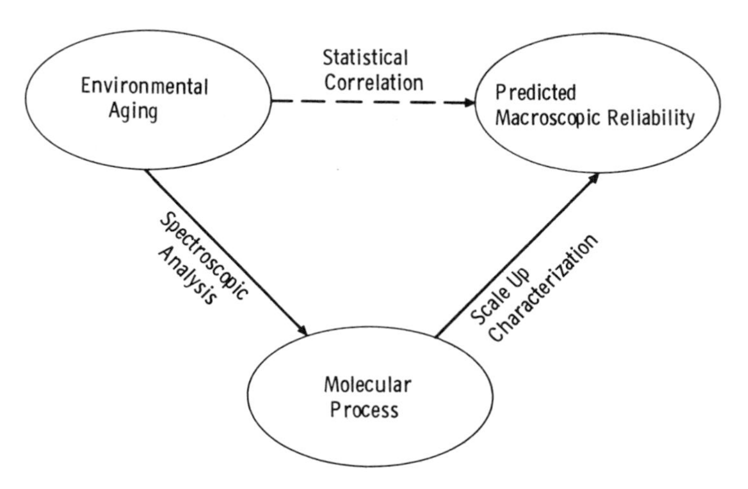

FIG. 1-5 Preferred dual path for correlating environmental aging with macroscopic strength.

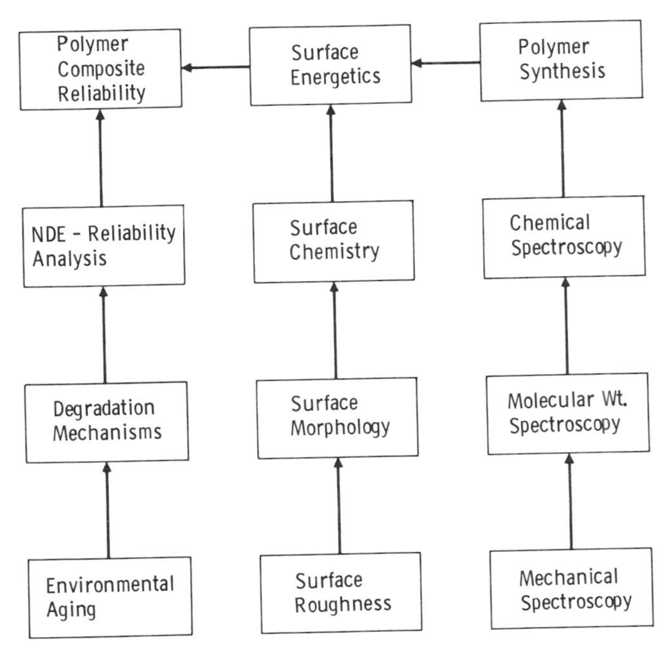

FIG. 1-6 Technical approach to polymer composite reliability.

The right column of Fig. 1-6 defines essential studies in polymer properties which include chemical analysis, polymer synthesis, molecular weight characterization and mechanical spectrum analysis. In polymer studies, these four areas of study are presently closely interconnected. Failure of polymer cohesive response in composite applications is readily identified in molecular terms and corrective action involving polymer chemistry and chemical analysis becomes available to composite reliability studies.

The present problem with regard to full implementation of the technical approach shown in Fig. 1-6 is the failure of material scientists to convert their detailed physical data into parameters directly useful to the design engineer for large scale structures. This problem has been recognized and a semi-formal approach for resolution has been proposed by Kelley and Williams(15) in the form of a morphological scheme termed the "Interaction Matrix." The interaction matrix, as shown in Table 1-2, lists mechanical property requirements as defined by the design engineer as column headings. A list of molecular properties as defined by the material scientist are listed in descending order as row headings. Within the box, defined by a design requirement and a molecular property, the degree of correlation is assigned on a numeric scale such as shown in Table 1-2. By summing these numbers across both the rows and columns as

TABLE 1-2 Interaction Matrix Between Molecular Property and Mechanical Requirement; 3 = Strong Interaction, 2 = Medium, 1 = Negligible, - = Unknown, Σ = Sum of Interactions

	Mechanical Requirement*					
Molecular Property	T_g	E_e	τ_0	n	E_g	Σ
Volume Fraction Plasticizer	3	3	3	1	1	11
Volume Fraction Filler	2	3	2	3	1	11
Degree of Crystallinity	1	3	3	3	1	11
Molecular Weight	3	3	1	1	1	9
Crosslink Density	1	3	1	2	1	8
Chain Stiffness	3	1	0	2	1	7
Monomeric Friction Coefficient	3	1	3	0	0	7
Heterogeneity Index	2	1	2	1	1	7
Entanglement Molecular Wt	1	3	1	1	1	7
Solubility Parameter	3	1	0	0	2	6
Σ	22	22	16	14	10	

*T_g = glass temp; Modulus (E) vs time (t) = $E(t) = E_e + [E_g - E_e][1 + t/\tau_0]^{-n}$ where E_e = elastomeric modulus, E_g = glass modulus, τ_0 = glass to rubber relaxation time, n = exponent.

shown by the Σ values in Table 1-2, the position of both rows and columns can be rearranged to maximize the strong correlations within a matrix as shown in the upper left section of Table 1-2. This summing and rearrangement of the interaction matrix tends to focus communciations and research toward the most fruitful sector of defined engineering need and molecular response.

The development of quantitative and preferably deterministic relations between engineering design parameters and molecular response variables is, of course, the desired end point in the dual path approach. An illustrative reliability and durability model for the polymer subphase in composite response is detailed in the following set of six relations:

GLASS TRANSITION RESPONSE

$$T_{g\infty} = \frac{2}{R_o} \left[\frac{\Sigma U_c}{\Sigma h}\right] + C(t) \tag{1-6}$$

$$T_g = T_{g\infty} + \left(\frac{\partial T_g}{\partial \sigma}\right)\sigma + \left(\frac{\partial T_g}{\partial C}\right) C_{H_2O} + \left(\frac{\partial T_g}{\partial M^{-1}}\right) M^{-1} \tag{1-7}$$

RHEOLOGICAL RESPONSE

$$\log (1/a_T) = \frac{17.4\ (T - T_g)}{51.6 + T - T_g} \tag{1-8}$$

$$M_i = \frac{M_o - M_\infty}{\left(1 + \frac{t}{a_t \tau_1}\right)} n + M_\infty \exp -\left(\frac{t}{\tau_m a_T}\right) \tag{1-9}$$

RELIABILITY-FAILURE RESPONSE

$$R_W = \exp -\left(\frac{\sigma}{\sigma_b}\right)^{m(\sigma)} \exp -\left(\frac{\varepsilon}{\varepsilon_b}\right)^{m(\varepsilon)} \exp -\left(\frac{t}{t_b a_T}\right)^{m(t)} \tag{1-10}$$

$$R_i = R_\infty + (1 - R_\infty)\ R_W \tag{1-11}$$

The details of symbols and parameter definitions for this model are summarized in Table 1-3.

This model starts by introducing chemical structure and measurement time contributions to define a reference $T_{g\infty}$ in Eq. (1-6). This relation was empirically developed by Hayes[17] and correlated with glass transition theory by Kaelble.[18] Monomer sequence distribution as discussed by Johnston[19] and semicrystallinity as discussed by Boyer[20] complicate the use of Eq. (1-6) for some polymers. The important concept embodied in the relation is that functional group properties of the monomeric unit determine an important thermal transition relating to environmental durability.

TABLE 1-3 Nomenclature for Polymer Reliability Relations

Symbol	Meaning
$T_{g\infty}$	Reference glass transition defined by monomer composition.
ΣU_c	Summation of molecular molar cohesion.
Σh	Summation of molecular degrees of freedom.
$C(t)$	Time scale correction factor $C(t) \approx 25°C$.
T_g	Nominal T_g as affected by mechanical (tensile) stress σ, moisture concentration C_{H_2O}, and U.V. radiation effects on polymer reciprocal molecular weight (M^{-1}, number average.
a_T	Time shift factor for rheological response.
T	Test temperature.
M_i	Time dependent modulus.
M_0	Glass (solid) state modulus.
M_∞	Rubbery state modulus.
t,n	Test time and exponent.
τ_1	Relaxation time for glass to rubber transition.
τ_2	Terminal time for rubber to flow transition.
R_i	Reliability ($\equiv$ survival probability).
R_∞	Residual reliability at infinite time.
τ_0	Relaxation time for Weibull failure process.
σ_0	Stress (tensile) for Weibull failure process.
ε_0	Strain (tensile) for Weibull failure process.
$m(t)$, $m(\sigma)$, $m(\varepsilon)$	Weibull distribution shape factors for time (t), stress (σ), and strain (ε) dominated failure.

In Eq. (1-7) the effects of mechanical stress σ, bulk moisture concentration C_{H_2O} and U.V. radiation effects on reciprocal polymer molecular weight M^{-1} are identified. Common values for the partial derivatives are as follows:

$$\left(\frac{\partial T_g}{\partial \sigma}\right) \approx -0.1 \ \frac{{}^oC \ cm^2}{Kg} \text{ (for polystyrene, Ref. 18)}$$

$$\left(\frac{\partial T_g}{\partial C_{H_2O}}\right) \approx -6.0^oC/wt\% \ H_2O \text{ (for epoxy, Ref. 21)}$$

$$\left(\frac{\partial T_g}{\partial M^{-1}}\right) \approx -0.06^oC/M^{-1} \text{ (general, Refs. 18, 22)}$$

The fundamental issue which is directly expressed by Eq. (1-7) is the direct description of local or macroscopic changes in nominal values of glass temperature T_g due to separate or combined effects of mechanical stress, moisture and U.V. exposure effects on M^{-1}. In general, one must assume effective local concentration effects in σ, C_{H_2O}, and M^{-1}. As examples, stress concentrations at crack tips, or high surface moisture concentrations will locally reduce T_g.

The familiar WLF equation[23] as presented in Eg. (1-8) is a nearly universal relation for calculating the time shift factor a_T at test temperature T which characterizes all rheology dominated responses of the polymer interfacial subphase. Since T_g in this relation is previously defined by Eq. (1-7), all the comments presented above for spatially localized T_g states are, of course, transformed by Eq. (1-8) into spatially localized time response states in terms of a_T within both the bulk polymer and at the interface. It is particularly evident that a polymer composite at temneratures near its reference $T_{g\infty}$ will display a potentially wide distribution of a_T values with local variations in T_g.

The values of a_T generated from Eq. (1-8) appear as time t shift factors in both Eq. (1-9), which defines subphase modulus M_i and Eq. (1-10), which defines the Weibull type reliability R_i. A modified power law relation suggested by Kelley and Williams[15] for approximate fitting of viscoelastic data in polymeric materials over extremes of time response is presented in Eq. (1-8). An example calculation of M_i response over 30 decades of reduced time (t/a_T) for a typical amorphous polymer such as polystyrene is shown in the solid curve of Fig. 1-7.

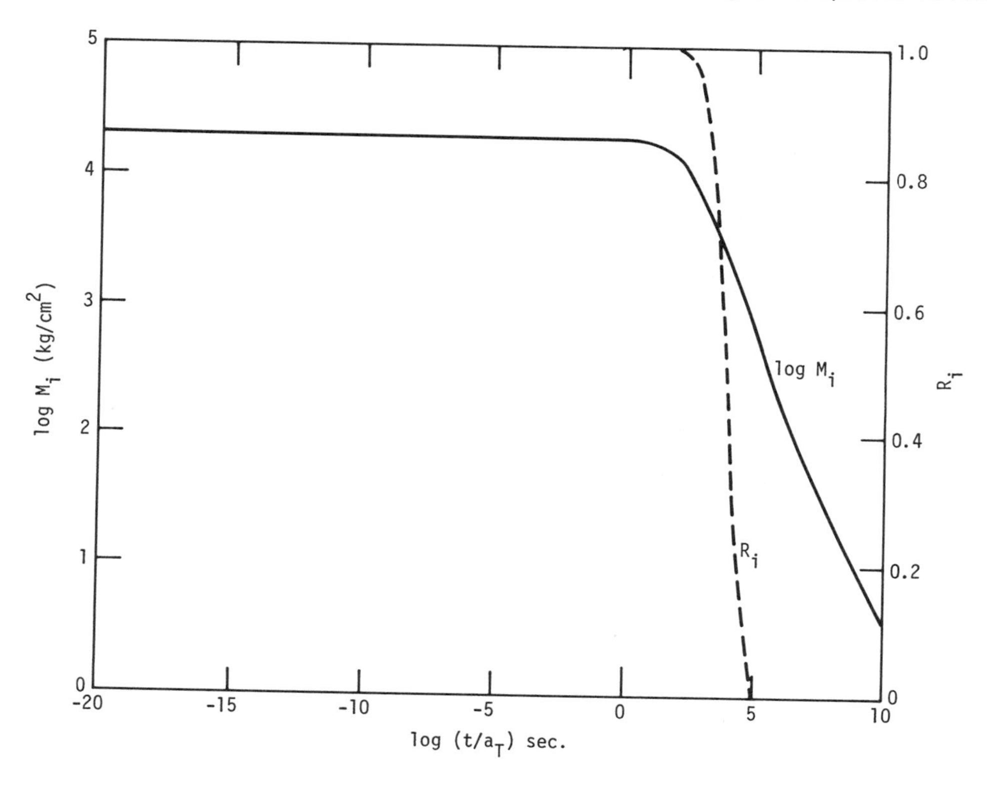

FIG. 1-7 Calculated function of M_i and R_i for M_∞ = 20,000 kg/cm^2, M_∞ = 2.0 kg/cm^2, τ = 100 s, R_∞ = 0, τ_0 = 10^4 s, m = 1.0.

In Eq. (1-10), which defines a reduced time function for reliability (equivalent to survival probability), the a_T parameter appears as a time reduction factor. The complexity in time scales of failure for uniform failure conditions is presented in physical terms by the Halpin and Polley[24] model for fracture. The Halpin-Polley model describes a statistical distribution of flaws as present in all polymeric materials and mathematically connects these defects to the well known Weibull frequency distribution[6-8] which assumes that failure is initiated at the weakest link. Gardon[25] has shown that time fluctuations in steady state peel fracture appear to fit a Gaussian frequency distribution which is essentially identical to the distribution defined by Eq. (1-3) when the Weibull shape factor exponent takes the value m = 3.67. The dashed curve of Fig. 1-7 presents a plot of Eq. (1-10) with m(t) = 1.0 which is relatively low for typical polymer failure.

A simple calculation is introduced to demonstrate the fundamental propositions in terms of stress-strain behavior and failure response of a polymeric subphase. First, assume the M_i and R_i properties of this subphase are described by the curves in Fig. 1-7. Also assume the subphase is tested in tension somewhat below the subphase glass temperature with:

$$T - T_g = -30^oC$$

$$\left(\frac{\partial T_g}{\partial \sigma}\right) = -0.10^oC \frac{cm^2}{Kg}$$

and with

$$C_{H_2O} = M^{-1} = 0$$

By use of Eq. (1-7) and Eq. (1-8) the curves shown in Fig. 1-8 can be calculated to correlate stress magnitude σ with log t/a_T. The curves in Fig. 1-8 illustrate the WLF prediction of stress effects on the time shift factor and show

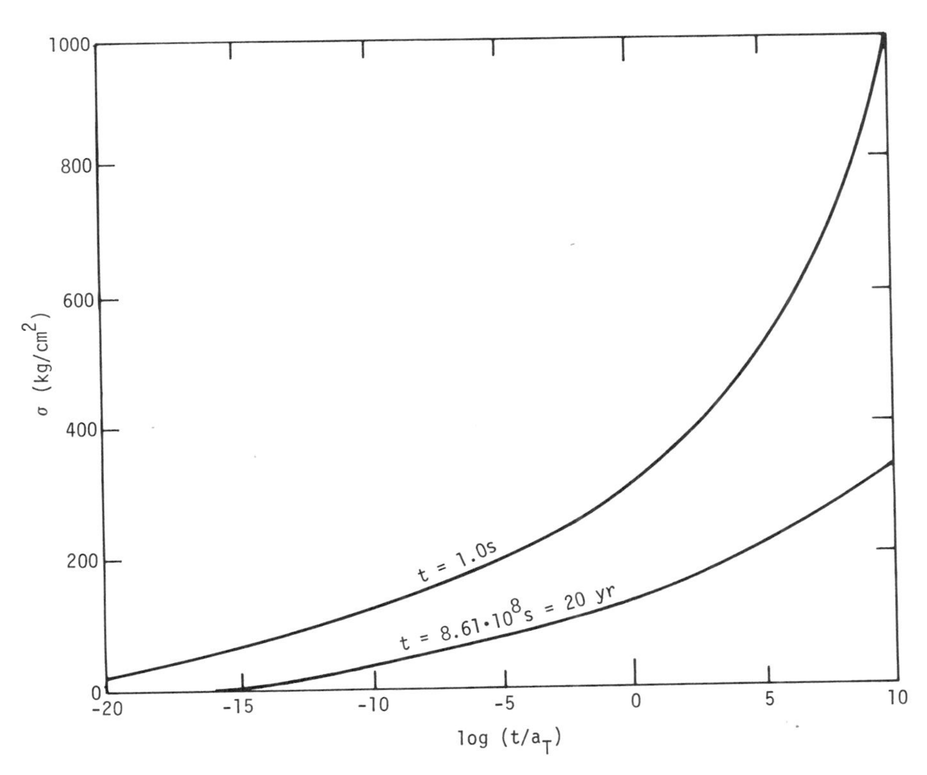

FIG. 1-8 Illustrative relations between tensile stress σ and time shift factor log (t/a_T).

that tensile stresses readily supported by the cohesive or interfacial strength of a glassy polymer can change rheological response time of the material over many decades and, in fact, produce rubbery state response when $\sigma \geqslant 300$ Kg/cm^2 for the example shown. For a known stress-strain response for the material, such as defined by simple time dependent Hookian response:

$$\sigma(t) = \varepsilon \cdot M_i(t)$$

It follows that stress-strain curves in tensile creep can be calculated. This calculation applies for the case of instant loading to constant stress σ, and maintaining this stress constant to a designated constant time t, which we set at t = 1.0 sec and t = $6.31 \cdot 10^8$ sec = 20 yr. Isochronal creep stress vs strain curves can be constructed from the curves of Fig. 1-7 and Fig. 1-8.

The 20 year life of the polymer composite reflected by tensile creep curves for t = 20 yr = $6.31 \cdot 10^8$ sec shown in lower Fig. 1-9 can be readily generated in the same fashion as the t = 1 sec response curves shown in upper Fig. 1-9. In summary, the intent of this section and the above illustrative examples is to show that the right vertical arrows which connect chemical synthesis to mechanical spectroscopy in Fig. 1-6 are detailed by structure-property relations. In Eq. (1-8) the time shift factor a_T can be reset by changes in temperature T, stress σ, moisture concentration C_{H_2O}, and reciprocal polymer molecular weight M^{-1} to accelerate natural degradation processes which determine long term durability and reliability.

The data summary of Table 1-4 reports experimentally determined Weibull strength distributions of structural adhesives, fibrous reinforced composites, and adhesive bonded metal joints under different conditions of environmental aging. These strength distributions were obtained in conjunction with physical chemical characterizations which clarify the shift in the Weibull values of mean strength σ_0 and distribution shape factor $m(\sigma)$ reported in Table 1-4 will, perhaps, provide practical insights into the proposal dual path conept sketched in Fig. 1-5.

1-3. Multiphase Structural Adhesives

Epoxy structural adhesives, toughened by a rubbery subphase, are now extensively used in aircraft structures.[28] Current adhesively bonded primary structures such as Air Force PABST (primary adhesive bonded structures technology) include carboxy terminated butadiene acrylonitrile (CTBN) rubber as a chemically combined constituent of the epoxy structural adhesive.[29] In CTBN modified epoxy adhesives, the rubber subphase precipitates during curing of the

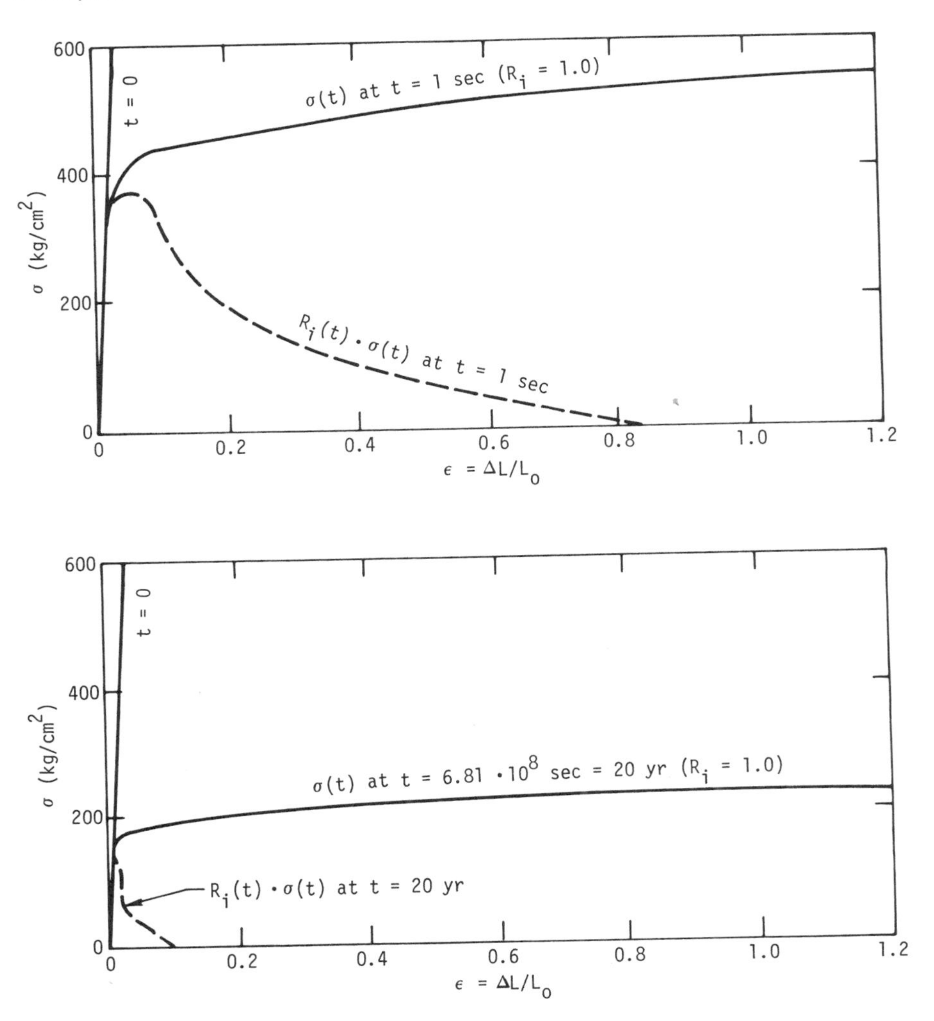

FIG. 1-9 Calculated tensile creep stress $\sigma(t)$ vs strain $\varepsilon(t)$ (solid curves) and reliability $R_i(t)$ reduced stress $R_i(t)*\sigma_i(t)$ vs strain $\varepsilon(t)$ (dashed curves) at t = 1 s (upper view) and t = 20 yr (lower view).

adhesively bonded joint. This produces the multiphase morphology and microstructure that leads to a strongly enhanced fracture toughness[30-34] where G_{IC} = 2.0 to 4.0 kJ/m^2 with 15% CTBN modifier, compared to 0.1 to 0.2 kJ/m^2 for unmodified epoxy resins.[35]

Several detailed studies of the fracture properties of CTBN modified epoxy adhesives now show that their high fracture toughness is related to cavitation and crazing (stress-whitening) due to dilational strains in the triaxial stress field of the crack tip.[34,35] Since extensive microstructure degradation precedes final failure in CTBN toughened epoxy, an important ques-

TABLE 1-4 Weibull Strength Distributions

Composite Polymer		Test	Strength Distribution $R = \exp -(\sigma_b/\sigma_0)^{m(\sigma)}$	
EPON 828/CTBN % CTBN	T(°C)	Tensile	σ_0 (Kg/cm)	$m(\sigma)$
0	-150	N = 15	812	7.64
17	-150	14	679	9.78
50	-150	14	1274	15.5
0	100	15	95.6	6.82
17	100	15	42.1	8.33
50	100	15	26.6	5.44
Uniaxial Graphite/Epoxy		Interlaminar		
Herc. AS/3501-5		Shear	$\sigma_0(Kg/cm^2)$	$m(\sigma)$
23°C air + 232°C spike		N = 18	1054	7.60
100°C water + 232°C spike		16	601	2.20
Metal-Adhesive Joint AT2024T3-HT424 Epoxy		Single Lap		
SET (hr)	BET (hr)	Shear	$\sigma_0(Kg/cm^2)$	$m(\sigma)$
0	0	N = 12	232	14.5
0	165, 449	12	184	15.4
0	808, 1023	12	165	10.0
21	0	12	208	15.0
20	669, 983	12	160	18.1
Ti-6Al-4V-HT424 Epoxy				
SET (hr)	BET (hr)		$\sigma_0(Kg/cm^2)$	$m(\sigma)$
0	0	N = 12	270	7.65
0	(670, 1016)	12	182	6.22
21	0	12	272	7.65
21	(591, 997)	12	202	6.35

SET = surface exposure time
BET = bond exposure time at 54°C and 195% relative humidity.

tion is whether the statistical distribution of cohesive strengths is adversely modified.

A special study was undertaken to clarify the multiphase morphology effects upon the statistical distributions of cohesive strengths using the materials summarized in Table 1-5. Microtensile test specimens (ASTM Method 1708-66) were die cut from th cured epoxy films while they were heated to rubbery state response at 120°C. Tensile tests were conducted at a strain rate ε of 0.09 min^{-1}. Single specimen tests, at temperatures from -200°C to 200°C as shown in Fig. 1-10a, were conducted to determine the temperature dependence of nominal tensile strength σ_b. At -150°C and 100°C, which represent lower and

TABLE 1-5 Co-reactants for Three-Dimensional Epoxy-Nitrile Rubber Block Copolymers

1. Epoxy: DGEBA (Epon 828, Shell Chemical Company), 100 pbw (parts by weight), $M_n \simeq$ 380 gm/mole.

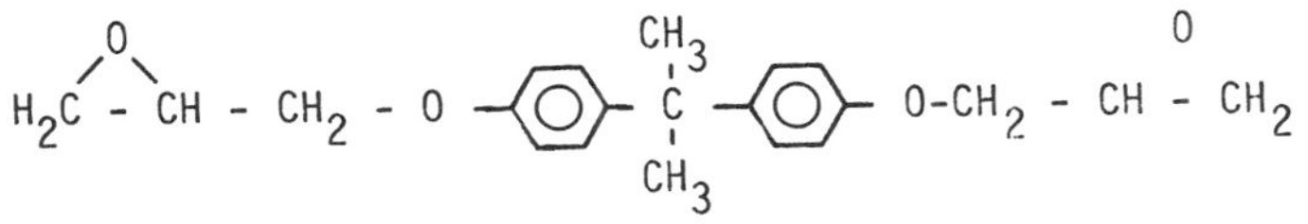

$$H_2\overset{O}{C - CH} - CH_2 - O - C_6H_4 - \underset{CH_3}{\overset{CH_3}{C}} - C_6H_4 - O{-}CH_2 - \overset{O}{CH - CH_2}$$

2. Catalyst: Piperidine - 5 pbw

3. Carboxy terminated nitrile rubber (HYCAR CTBN, B.F. Goodrich Chemical Company) - 0, 17, 29, 39, 50% by weight based on 100 pbw Epoxy + 5 pbw piperidine.

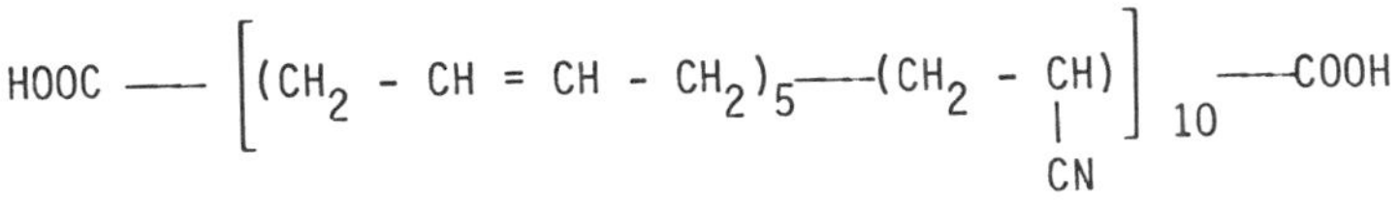

$$HOOC - \left[(CH_2 - CH = CH - CH_2)_5 - (CH_2 - \underset{CN}{CH})\right]_{10} - COOH$$

$M_n \simeq$ 3300 - 3500 gm/mole

4. Mix items (1), (2), (3), above, degas, and cure for 16 hours at 120°C under dry N_2.

upper service temperatures in structural applications, larger groups of specimens were tested to determine the statistical distribution of σ_b. Six groups of specimens, as summarized in upper Table 1-3, describe the bounds of strength variation indicated in Fig. 1-10a by the extreme high and low tensile strengths indicated by the bracketed error bars at -150°C and 100°C for each strength curve.

In Fig. 1-10b, the strength data points are arranged serially j = 1, 2,...N in decreasing order of σ_b, and the survival probability ($\equiv$ reliability) is defined by Eq. (1-4) where N is the number of observations and F is the failure probability. Six groups of data show the cumulative distribution of survival probability. The ratio of extreme strength values (Table 1-4) for pure epoxy is not dramatically modified by either extremes of test temperature or composition. The study thus supports the view that the statistical distribution of cohesive strengths is not strongly or adversely modified by addition of CTBN. Inspection of Fig. 1-10 provides the following conclusions in relation to fracture integrity:

1. The effect of temperature change, from -150°C to 100°C, dominantly influences the cohesive strengths of CTBN toughened epoxy.

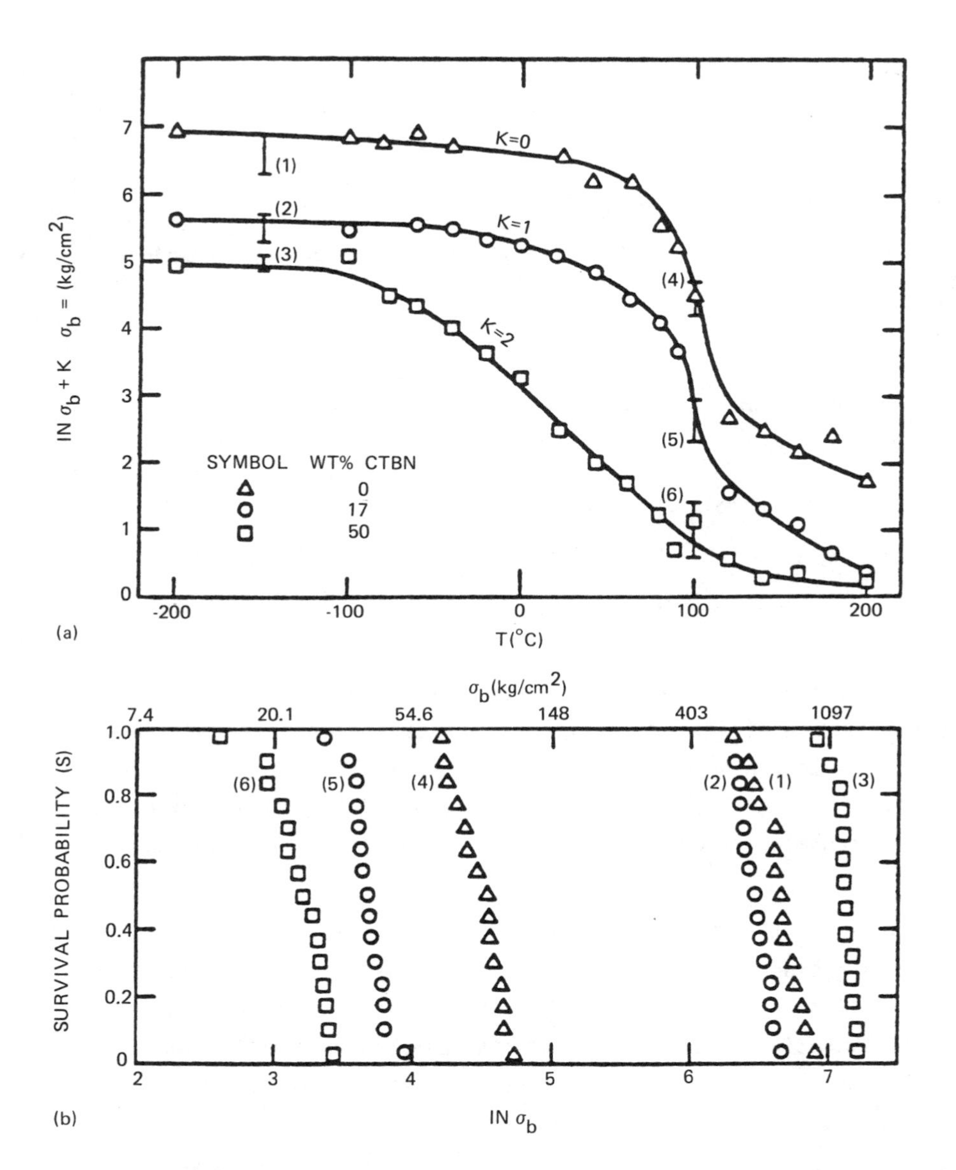

FIG. 1-10 (a) The temperature and composition dependence of the tensile strength for a rubber modified epoxy. (b) The stress dependence of the survival probability for rubber modified epoxy at -150°C and 100°C.

2. Chemical composition changes, from 0 - 50% by weight CTBN epoxy modifier, is the second most dominant strength determinant.

3. Statistical strength variation, for a survival probability range from 0.05 to 0.95, is a minor strength determinant.

It thus appears that the 15 - 40 fold enhancement of fracture toughness by inclusion of approximately 15% CTBN in structural epoxy resins displays its principal adverse effect in shifting the entire distribution of high temperature tensile strengths to lower values, without change in shape. This adverse effect on high temperature strength is physically related to the lower crosslinking density for the three-dimensional network formed between CTBN and epoxy coreactants.

1-4. Fibrous Composite Reliability

Hydrothermal exposure (combined high moisture and temperature) of graphite reinforced epoxy matrix composites can produce irreversible reduction in shear strength and modifies the Weibull distribution of survival probabilities as shown in studies by Kaelble and Dynes.[36] Comparative chemical analysis of two catalytically cured graphite-epoxy composites and an aromatic amine cured epoxy matrix composite shows that curing mechanisms and epoxy network structure influence both thermal response and environmental durability. The results of chemical analysis of the three epoxy prepregs are summarized in Table 1-6. The chemical analysis follows standard procedures outlined in previous reports.[10,12] The data of Table 1-6 shows that a BF_3 type boron is present as a catalytic curing agent in 3501-5 epoxy (Hercules) and 934 epoxy (Fiberite) and present only as a trace constituent in 5208 epoxy (NARMCO). This BF_3 catalyst decomposes to initiate homopolymerization of epoxy groups at lower tempertures than co-reaction between epoxy and DDS (diaminodiphenyl-sulfone) curative which is common to all three systems. The higher level of free DDS curative in 5208 epoxy correlates with the lower degree of cure as indicated by liquid chromatography and higher heat of cure indicated by differential scanning calorimetry (DSC). The numeric information of Table 1-6 also shows the importance of differentiating between total DDS curative as measured by IR spectroscopy and free amine as measured by quantitative molecular separation using liquid chromatography. The chemical analysis data of Table 1-5 forms part of the materials and processes approach to chemical defects definition as outlined in Fig. 1-6.

A comprehensive environmental durability characterization has been carried out on the three composites described in Table 1-6. Some results of this study serve to highlight the direct importance of chemical analysis in composite reliability and durbility predictions.

TABLE 1-6 Chemical Characterization of Graphite-Epoxy Prepreg Materials

	This Study		Reference System
1) Epoxy Marix	Hercules 3501-5	Fiberite 934	NARMCO 5208
2) Graphite Fiber	Hercules Type AS	U. Carbide T300	U. Carbide T300
3) % Total DDS Curative by IR Spectroscopy	29.2	27.8	22.1
4) % Free DDS Curative by Liquid Chromatography	18.1	14.5	17.8
5) Epoxide Equivalent	205	227	173
6) Wt% BF_3 Type Boron	0.047	0.022	0.0005
7) Relative Degree of Cure by Liquid Chromatography	22	27	6.9
8) Heat of Polymerization by DSC (cal/g polymer)	107	107	140

In the DSC thermograms of Fig. 1-11, the rate of chemical curing correlates with the amplitude of dH/dt, the heat release rate. The low temperature initiation of curing in 934 and 3501-5 epoxy correlates with the detected presence of the BF_3 catalyst as compared to uncatalyzed 5208 which requires much higher temperatures to complete the curing process.

The use of catalytically assisted curing clearly correlates with easier processability as shown in the lower temperature cure cycles shown in Fig. 1-11. The thermal scans of dynamic damping response in the fully cured uniaxial composites of these materials, as shown in Fig. 1-12, reveal that catalytic curing at lower tempertures produces a crosslinked network with substantially lower T_g. As shown in Fig. 1-12, the 934 epoxy displays onset of T_g response by an initial increase in tanδ at T = 200°C while cured 3501-5 epoxy shows initial rise in tan δ at 225°C and 5208 epoxy at T = 250°C for dry fully cured composite.

The effects of prior moisture exposure to full saturation is shown in the damping curves of Fig. 1-13 to lower the T_g related initial increase in tan δ response to about T = 140°C for 934 epoxy and T = 160°C for 3501-5, while the 5208 resin maintains low tan δ response to above 200°C.

The prediction offered by the dynamic damping response as measured by Rheovibron is that cured 3501-5 resin will display glass state response at 232°C

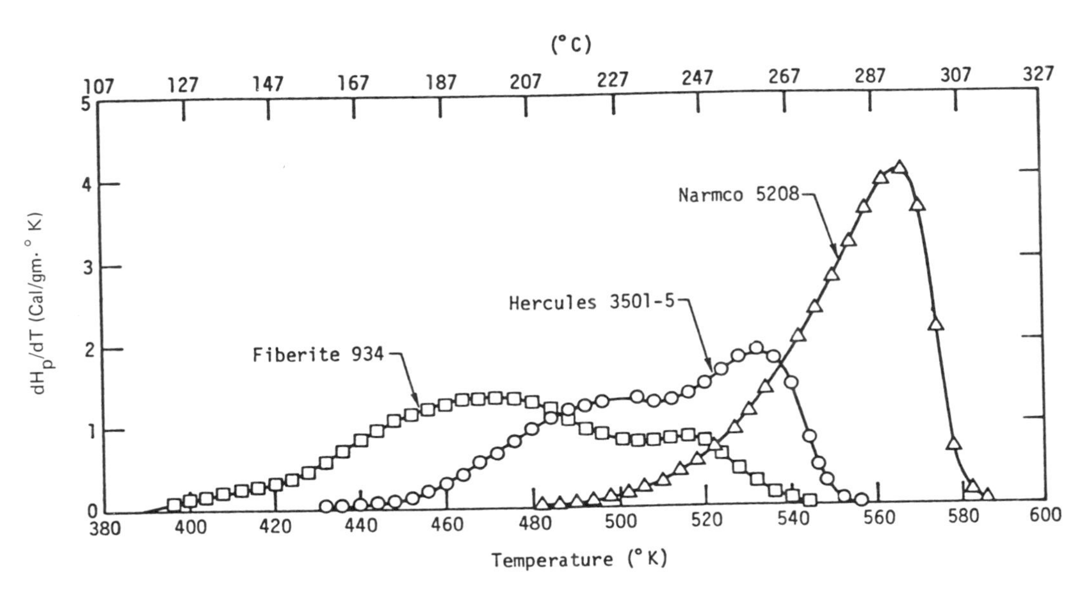

FIG. 1-11 DSC thermograms for curing reactions of commercial epoxy matrix materials extracted from prepreg (DSC scan rate ϕ = 20°C/min).

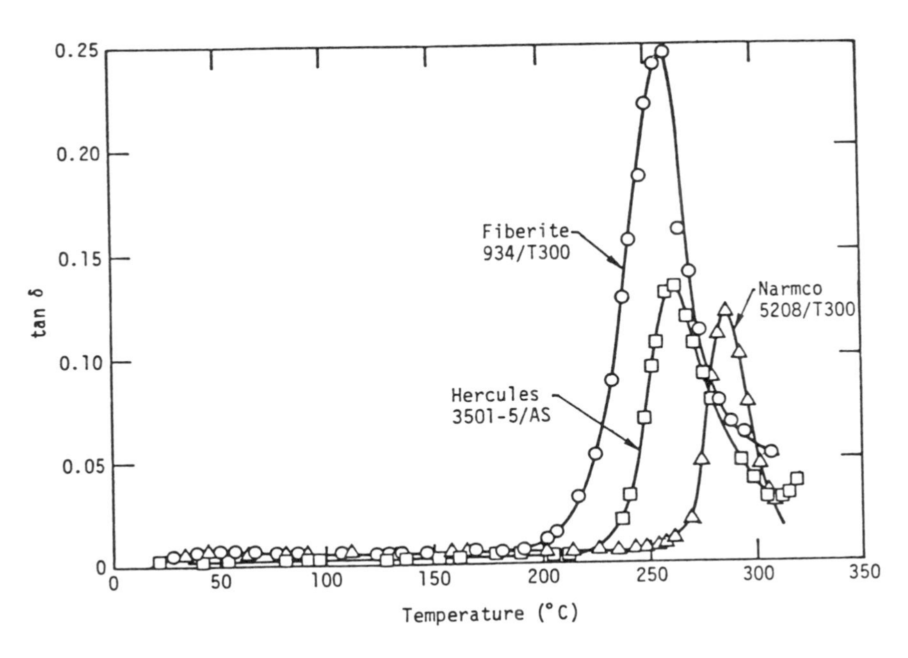

FIG. 1-12 Rheovibron thermal scans for flexural damping in cured reinforced graphite-epoxy composite in the dry unaged condition.

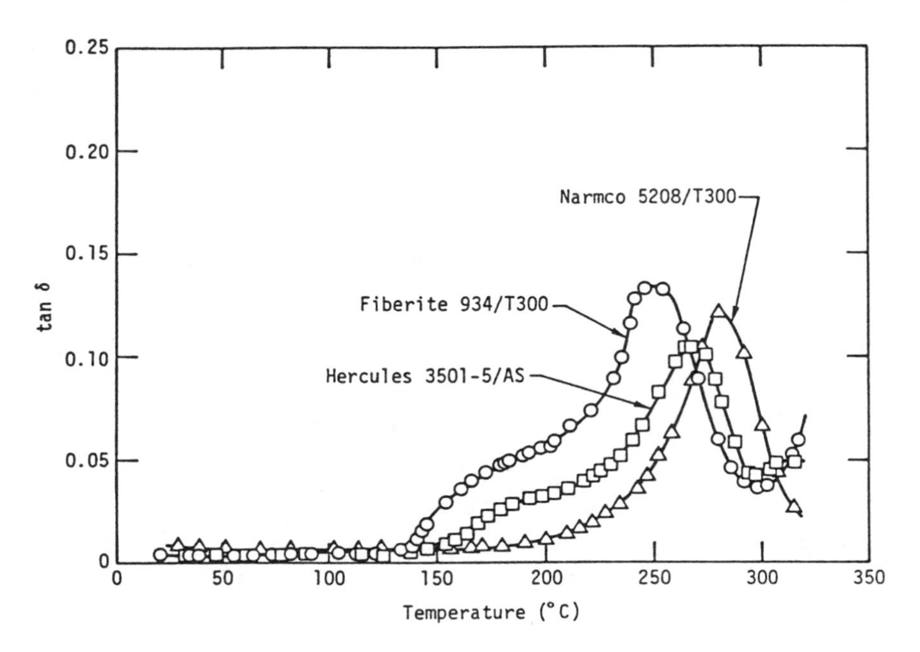

FIG. 1-13 Rheovibron thermal scans for flexural damping in cured uniaxial reinforced graphite epoxy composite in the wet-aged condition.

in the dry state, and low strength rubbery response due to water plastization in the moisture saturated state at 232°C. The curves of Fig. 1-14 show the clearly separated interlaminar shear strength distributions for uniaxial 3501-5 epoxy matrix composite where dry and moisture exposed specimens were subjected to an equivalent 232°C thermal spike prior to strength measurement at 23°C. As predicted, the interaction of high moisture and high temperature produced internal damage which lowered both mean shear strength λ_0 and also substantially lowered the Weibull distribution shape factor m.

1-5. Metal Joint Reliability

All metals, except gold, are chemically reactive with oxygen and moisture at room temperature and tend to form oxide or hydrated oxide surface layers.[37] In addition to oxidative chemical reaction, these high energy surface films are subject to further physical adsorption by water vapor and volatile organic contaminants. In structural metal to metal joints, the reliability of the bond may be directly related to thre chemical stability of the interfacial bond. Weibull statistical analysis has recently been applied to investigate the effects of both surface exposure time (SET) and bond exposure time (BET) on the distributions of single layer shear bonds of aluminum alloys.[38] Smith and Kaelble[39] have recently conducted a detailed study of

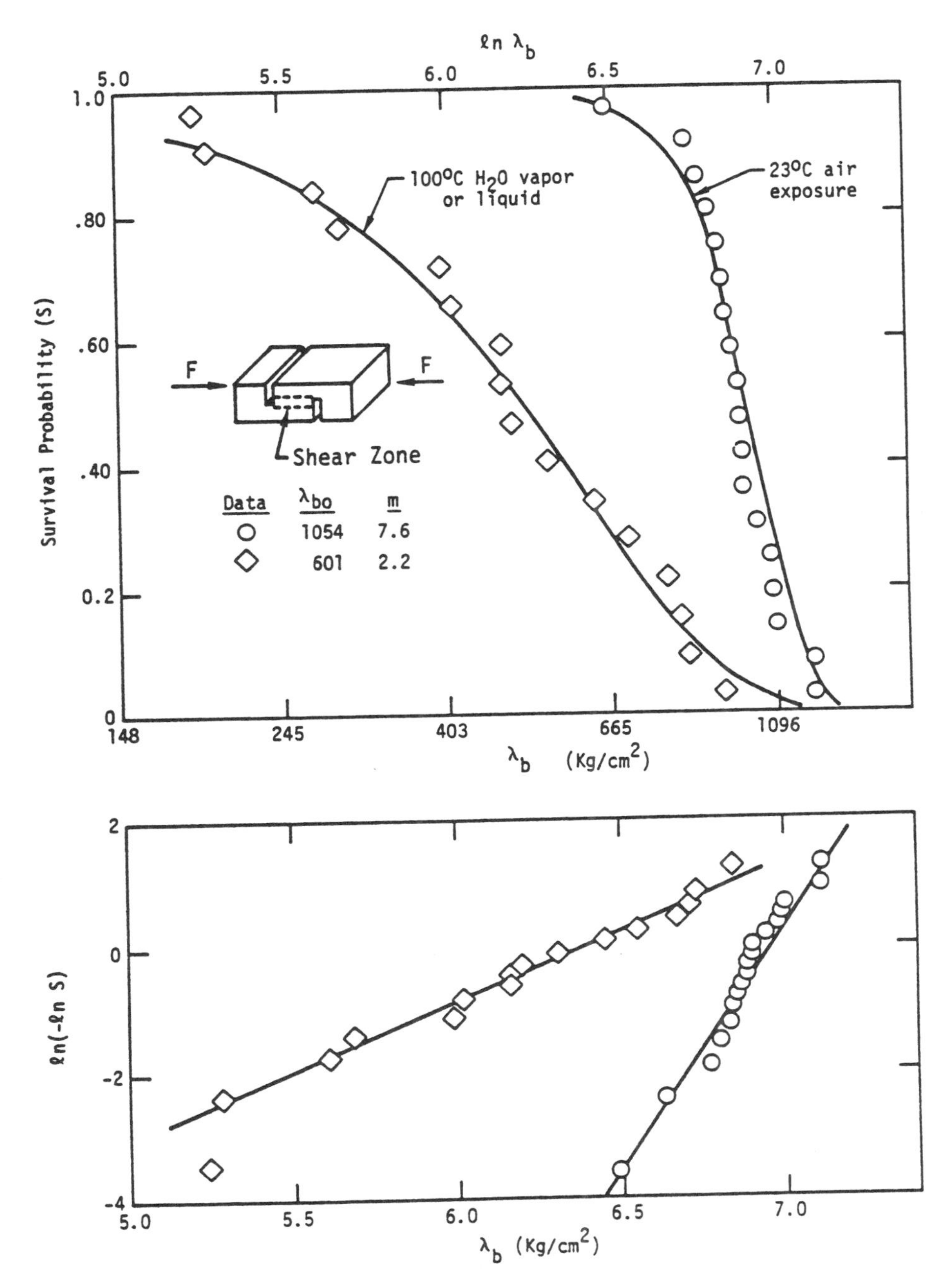

FIG. 1-14 Cumulative distribution function of survival probability.

TABLE 1-7 Metal Joint Reliability Studies

1. Metal Adherends: Unclad 2024-T3 aluminum alloy surface treated by standard FPL sulfuric chromate etch and T8-6Al-4V titanium alloy treated by standard phosphate fluoride cleaning process. Coupon size 0.063 in. thick, 1 in. wide, and 4 in. long.

2. Adhesive: HT 424 epoxy-phenolic film adhesive (from American Cyanamid) with glass fiber carrier and standard weight 0.0135 ± 0.005 lb/sq. ft. Unfilled HT 424 primer with parts A and B used with adhesive.

3. Bonding Process: Treated metal coupons spray primed with 0.001 in. thickness HT 424 primer solution using clean dry argon carrier gas. Primer layers dried 30 min ambient 23°C and 60 min at 66°C. An adhesive film is placed in the 1.000 in. × 0.500 in. overlap between two meal adherends. Six such joints are aligned in a bonding jig with the glass carrier acting to provide constant glue line thickness 0.008 in. Cure cycles with 60 min temperature rise to 171°C and 60 min cure cycle at 171° followed by cooling to room temperature.

4. Tensile Lap Shear Testing: 1.5 in. × 1.0 in. × 0.063 in. aluminum alignment shims bonded to eliminate offset. Tests at 23°C using 0.01 in./min Instron crosshead rate and 4.5 in. jaw separation.

combined SET and BET aging in normal (50% R.H., 23°C), and high moisture (95% R.H., 54°C). The experimental methods of the study by Smith and Kaelble are outlined in Table 1-7 with full details contained in the published report.

In this study the surface chemistry and related wettability of both adherends and adhesive were analyzed prior to bonding. The results of this surface energy analysis can be plotted on a surface energy map where the ordinant α and abscissa β as shown in Fig. 1-15 respectively refer to dispersion (nonpolar) and polar components of surface energy and interfacial bonding mechanisms. The theory of interfacial adhesion experimentally verified in this analysis defines the thermodynamic work of adhesion W_a by the following relation:

$$W_a = W_{13} = 2(\alpha_1 \alpha_3 + \beta_1 \beta_3) \quad (1\text{-}12)$$

where α_1, β_1 define the dispersion and polar surface properties of adhesive and α_3, β_3, those of the metal adherend.

As shown in Fig. 1-15, and further documented in detailed kinetic studies by Kaelble and Dynes,[40] the α_3 and β_3 properties of aluminum alloy change dramatically with surface aging time after FPL etch. The α_1, β_1 surface properties of HT 424 adhesive are shown to lie below the curve for Al 2024-T3 at

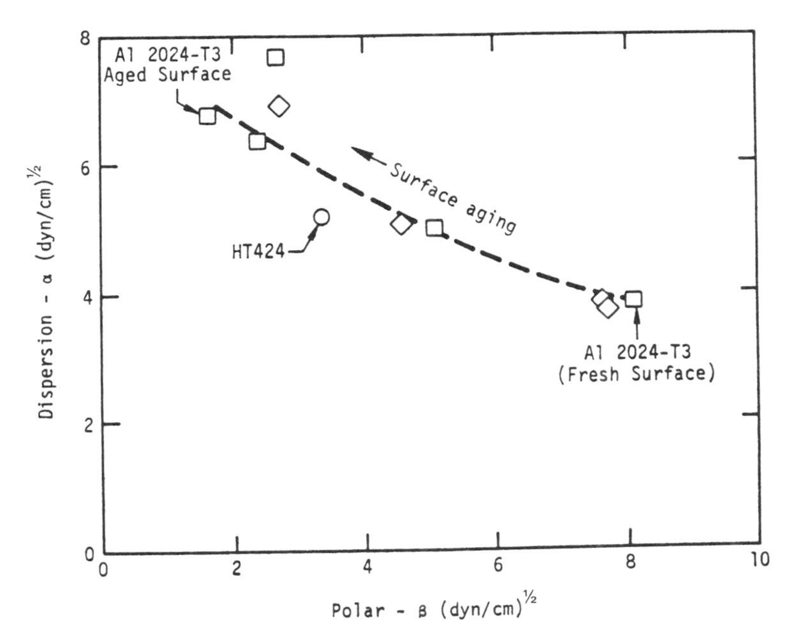

FIG. 1-15 Dispersion (α) and polar (β) components of the solid-vapor surface tension $\gamma_{SV} = \alpha^2 + \beta^2$ for HT424 primer (Phase 1) and Al 2024-T3 adherend (Phase 3).

all stages of surface aging with predicts proper bonding between adhesive and adherend in air.

The interfacial work of adhesion W_a as defined by Eq. (1-12) will decrease with surface aging time (SET) as shown by the upper curve of Fig. 1-16. As shown in the lower curves of Fig. 1-16, the lap shear bond strength varies with SET in a fashion that correlates closely with the predictions from work of adhesion calculations. A simple but now widely demonstrated correlation between surface energtics and fracture mechanics is available in the following relations for critical stress σ_c for Griffith type crack initiation under normal stress loading:[41,42]

$$\sigma_c = \left(\frac{2E}{\pi C}\right)^{1/2} \left(R^2 - R_o^2\right)^{1/2} \geqslant 0 \tag{1-13}$$

where E, and C are a characteristic modulus and crack length which are assumed constant and the surface energy parameters R and R_o are defined by the following relations:[41,42]

$$R_o = 0.25\ (\alpha_1 - \alpha_3)^2 + (\beta_1 - \beta_3)^2 \tag{1-14}$$

$$R^2 = (\alpha_2 - H)^2 + (\beta_2 - K)^2 \quad (1\text{-}15)$$

$$H = 0.5\ (\alpha_1 + \alpha_3) \quad (1\text{-}16)$$

$$K = 0.5\ (\beta_1 + \beta_3) \quad (1\text{-}17)$$

In Eq. (1-15) two new surface energy parameters α_2 and β_2 define the environment (phase 2) at the crack tip. The model for critical stress defined by Eq. (1-13) can be presented on surface energy coordinates as shown in Fig. (1-17).

As the adhesive joint changes from dry air immersion with $\alpha_2 = \beta_2 = 0$ to equilibrium respone with water immersion with $\alpha_2 = 4.67\ (\text{dyn/cm})^{1/2}$ and β_2 =

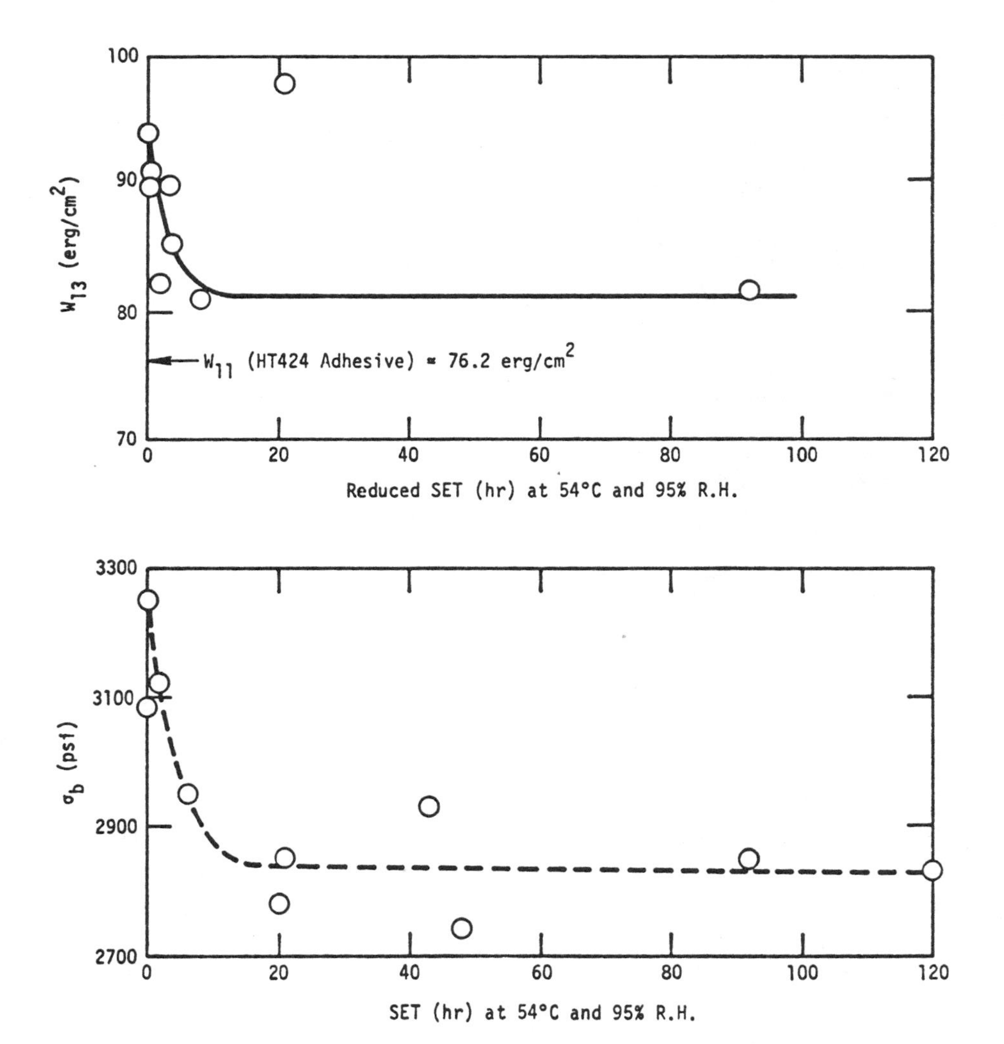

FIG. 1-16 Dependence of interfacial work of adhesion W_{13} (upper curve) and lap shear bond strength σ_b (lower curve) at varied SET.

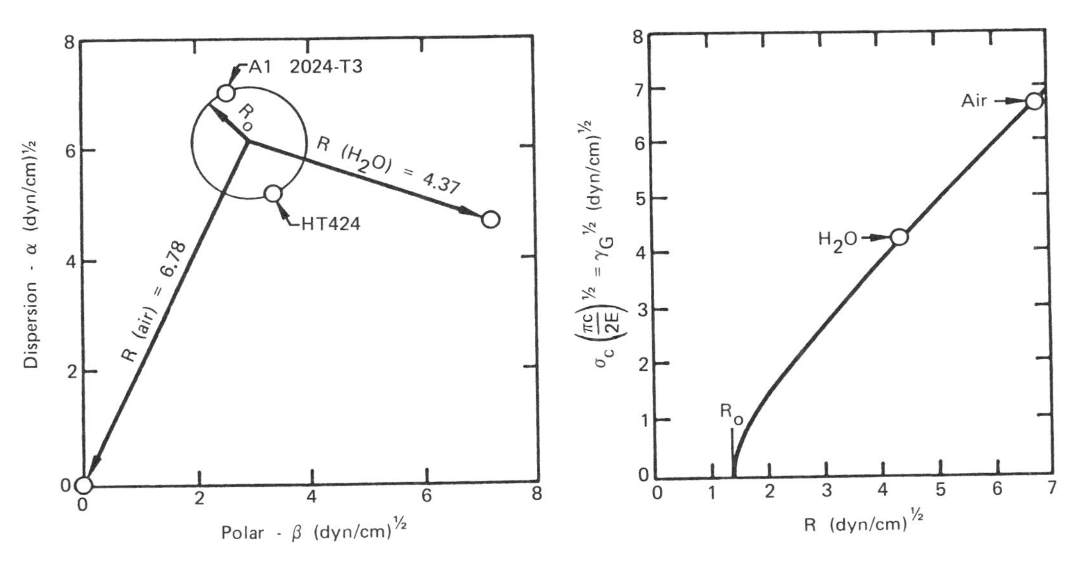

FIG. 1-17 Modified Griffith analysis of the effect of H_2O immersion in reducing the critical interfacial stress σ_I for interfacial failure between HT424 and etched Al 2024-T3 ($\phi_I = 1 - \phi_C = 1.0$).

7.14 $(\text{dyn/cm})^{1/2}$, the predicted decrease in critical stress σ_c of the HT 424 to Al 2024-T3 interface is:

$$\frac{\sigma_c(H_2O)}{\sigma_c(\text{air})} = 0.644 \qquad (1\text{-}18)$$

Extensive joint strength testing of this system was completed to determine the response surface of lap shear strength vs both SET and BET under high moisture (95% R.H., 54°C). These results are summarized on the response surface of Fig. 1-18 where each point represents the average of six tests. Comparing joint strengths for fully aged (20 hr SET, 1000 hr BET) and unaged (0 hr SET, 0 hr BET) provides (see Table 1-3) the following expeimental ratio:

$$\frac{\sigma_b(\text{aged, wet})}{\sigma_b\ (\text{unaged, dry})} = \frac{2275\ \text{psi}}{3300\ \text{psi}} = 0.69 \qquad (1\text{-}19)$$

which is in close agreement with the prediction of Eq. (1-18).

An essentially parallel detailed study of surface aging showed a shift in α_3 and β_3 for phosphate-fluoride cleaned titanium alloy similar to that detailed in Fig. 1-15. Application of the modified Griffith analysis as shown in Fig. 1-19 provides the following predicted moisture degradation of bond strength at the HT 424 to Ti-6Al-4V interface:

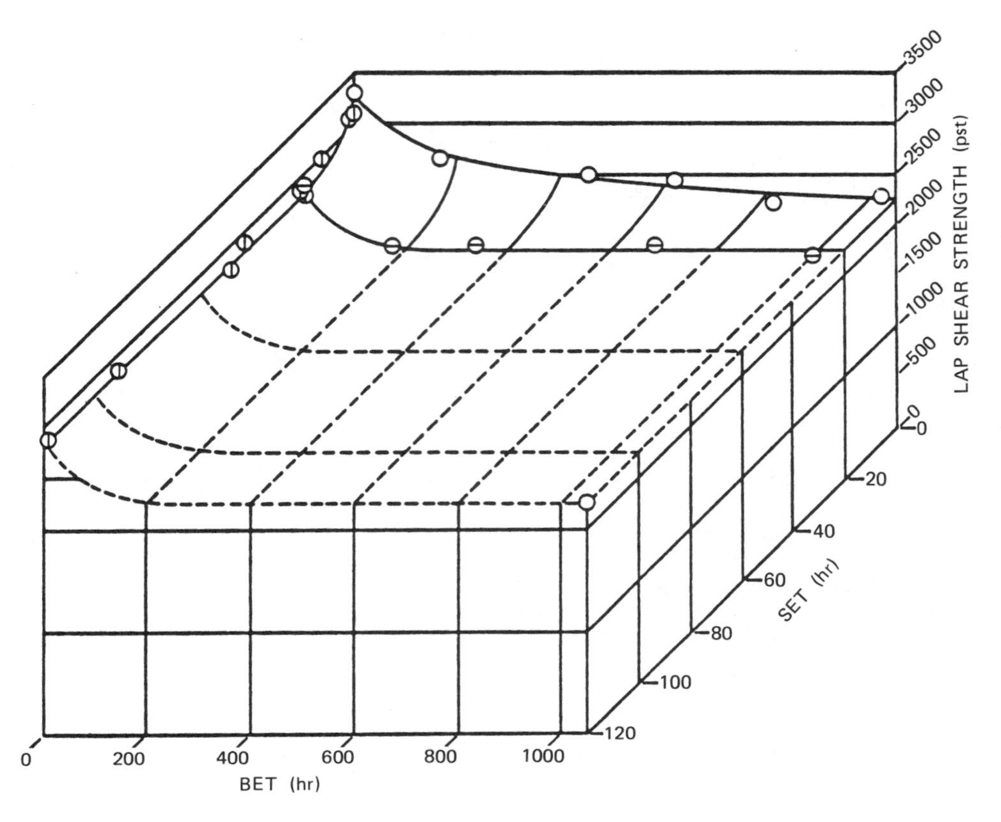

FIG. 1-18 SET and BET response surface for lap shear bond strength for Al 2024-T3 - HT424.

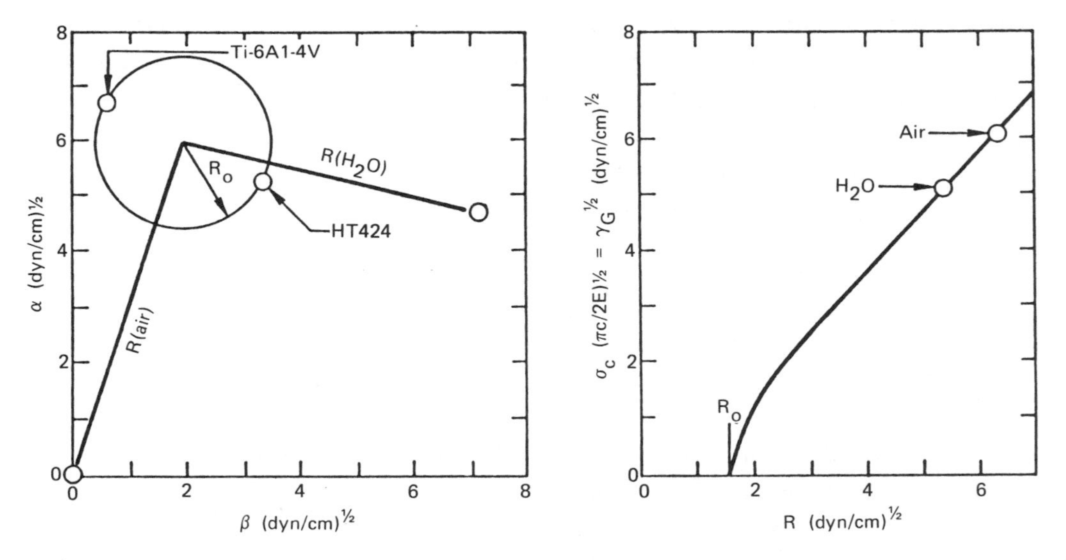

FIG. 1-19 Modified Griffith analysis of the effect of H_2O immersion in reducing critical failure stress σ_I for interfacial failure between HT424 and phosphate-fluoride treated Ti-6Al-4V ($\phi_I = 1 - \phi_c = 1.0$).

$$\frac{\sigma_c\,(H_2O)}{\sigma_c\,(\text{dry air})} = 0.84 \tag{1-20}$$

An equivalent calculation for moisture degradation of the HT 424 cohesive bond produces the following prediction:

$$\frac{\sigma_c(H_2O)}{\sigma_c\,(\text{dry air})} = 0.63 \tag{1-21}$$

Joint strength testing of the HT 424 to titanium alloy was carried out under conditions of separate and combined SET and BET in high moisture (95% R.H., 54°C). The SET vs BET response surface of shear bond strength σ_b is shown in Fig. 20 where each point is an average of six strength tests. As shown in Fig. 20, the shear bond strength reaches an equilibrium value under extended moisture aging. Comparing joint strengths for fully aged (22 hr SET, 1000 hr

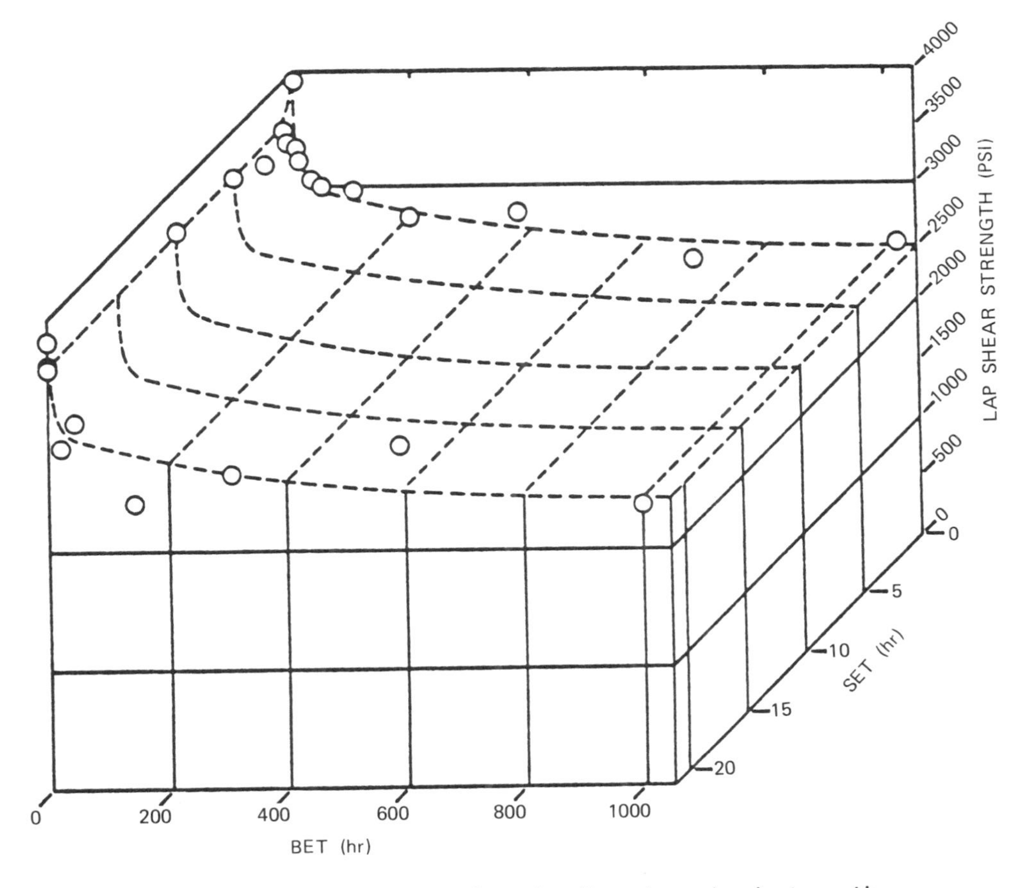

FIG. 1-20 SET vs BET response surface for lap shear bond strength for Ti-6Al-4V - HT424.

BET) and unaged (0 hr SET, 0 hr BET) for Ti-6Al-4V to HT 424 bonds (see Table 1-4) provides the following experimental ratio:

$$\frac{\sigma_b(\text{aged, wet})}{\sigma_b\ (\text{unaged, dry})} = \frac{2873\ \text{psi}}{3840\ \text{psi}} = 0.75 \tag{1-22}$$

which lies intermediate between the cohesive failure prediction of Eq. (1-21) and the interface prediction of Eq. (1-20). Microscopic visual inspection of the fracture surfaces for the HT 424 to titanium lap shear points shows predominant (above 50%) cohesive failure for lap shear bonds described in Fig. 1-20.

The Weibull plots of Fig. 1-21 show shear bond strength distributions for unaged and fully aged aluminum (upper view) and titanium (lower view) joints. The titanium bonds show lower Weibull m values in both unaged and aged states as evidenced in Fig. 1-21 and the data summary of Fig. 1-4. A design requirement of high reliability shear strength where R = 0.98 or ln (-lnR) = -4 is shown by Fig. 1-21 to predict higher performance for the aluminum alloy joints

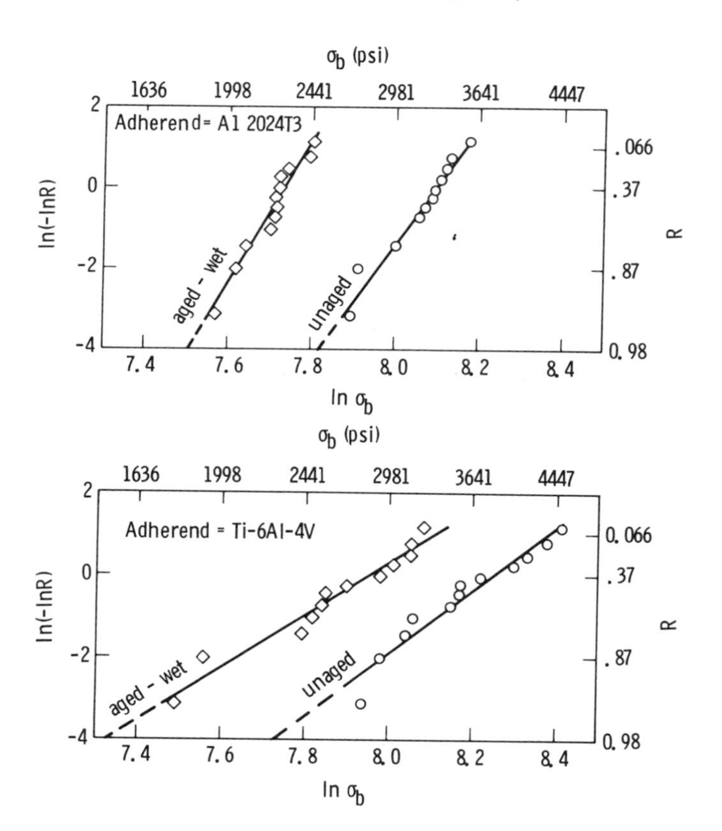

FIG. 1-21 Comparison of Weibull shear strength distributions for aluminum (upper view) and titanium (lower view) adherends.

in both unaged and aged-wet states. Conversely, if mean strength with R = 0.37 and ln (-lnR) = 0 is applied as a design criteria, the curves show titanium alloy joints to display higher unaged and aged strengths.

1-6. Detailed Test Plans

Previous sections of this discussion have presented and discussed the importance of recognizing the molecular processes which influence polymer composite reliability. A need for further development of the molecular theory of polymer reliability is also made evident in these several examples from structural polymers, fibrous composites, and bonded metal joints. The simple design concept of structural strength becomes replaced by a more sophisticated and holistic (or encompassing) design concept for reliability and durability by use of a combined molecular and mathematical modeling. If molecular theory of structure reliability were complete, the chemical analysis test plan outlined in Fig. 1-22 would provide sufficient design data to provide structural reliability predictions. Lacking a complete molecular theory requires that detailed experimental test programs for physical and mechanical analysis as outlined in Fig. 1-23 be employed in conjunction with chemical analysis. The detailed test programs outlined in Fig. 1-22 and Fig. 1-23 require advanced instrumentation and computer aided data processing.

Executing a complete chemical analysis of a polymer composite as outlined in Fig.1-22 generally provides sufficient detailed information to replicate a commercial polymer adhesive or coating. In order to improve the reliability and durability of adhesives or coatings, essentially all aspects of the detailed physical and mechanical test program of Fig. 1-23 need to be employed. The upper rank experiments in Fig. 1-23 involving DSC, surface energetics, and thermal mechanical analysis (TMA) simulate manufacturing steps for surface treating, bonding, and curing polymer composites. The central tier of experiments in Fig. 1-23 incorporate NDE tests, adhesive joint tests in conjunction with study of polymer material response. The lower portions of Fig. 1-23 outline phases of study involving aging and failure mechanisms. The lower extremity of Fig. 1-23 describes a data analysis with correlation of molecular process and composite response. The detailed test plans outlined by Fig. 1-22 and Fig. 1-23 are presently utilized to determine and improve polymer reliability. The challenge to both theoretical and experimental analysis is, of course, to lower the cost and increase the reliability of currently functioning test programs as detailed in Fig. 1-22 and Fig. 1-23.

In summary, we may return to the mythical "one-hoss-shay" of Holmes poem (see Table 1-1) which presents the nearly perfect example of design for

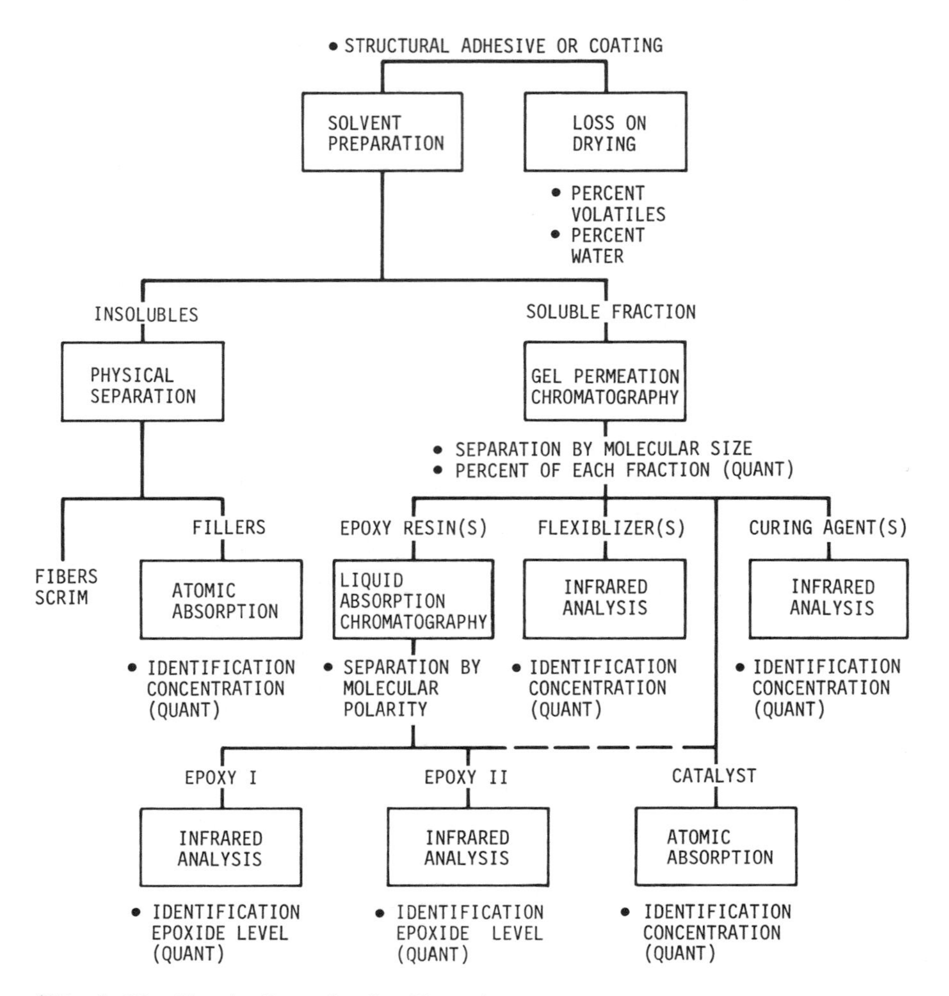

FIG. 1-22 Chemical analysis flow chart.

structural reliability and durability. An important point presented and hopefully well illustrated in this discussion is the importance of adding the analysis of molecular processes to present statistical theory of structure reliability. Implementation of the interaction matrix (see Table 1-2) establishes a direct avenue for communication between the specialist in engineering design and the materials scientist. The dual path approach (see Figs. 1-5 and 1-6) with combined deterministic/statistical testing and analysis is validated by extensive studies of which several are briefly reviewed in this discussion. A more general development and application of molecular theory of polymer reliability can lower the cost and increase the efficiency in present detailed test programs (see Figs. 1-22 and 1-23) for analyzing chemical, physical, and mechanical aspects of polymer composite durability.

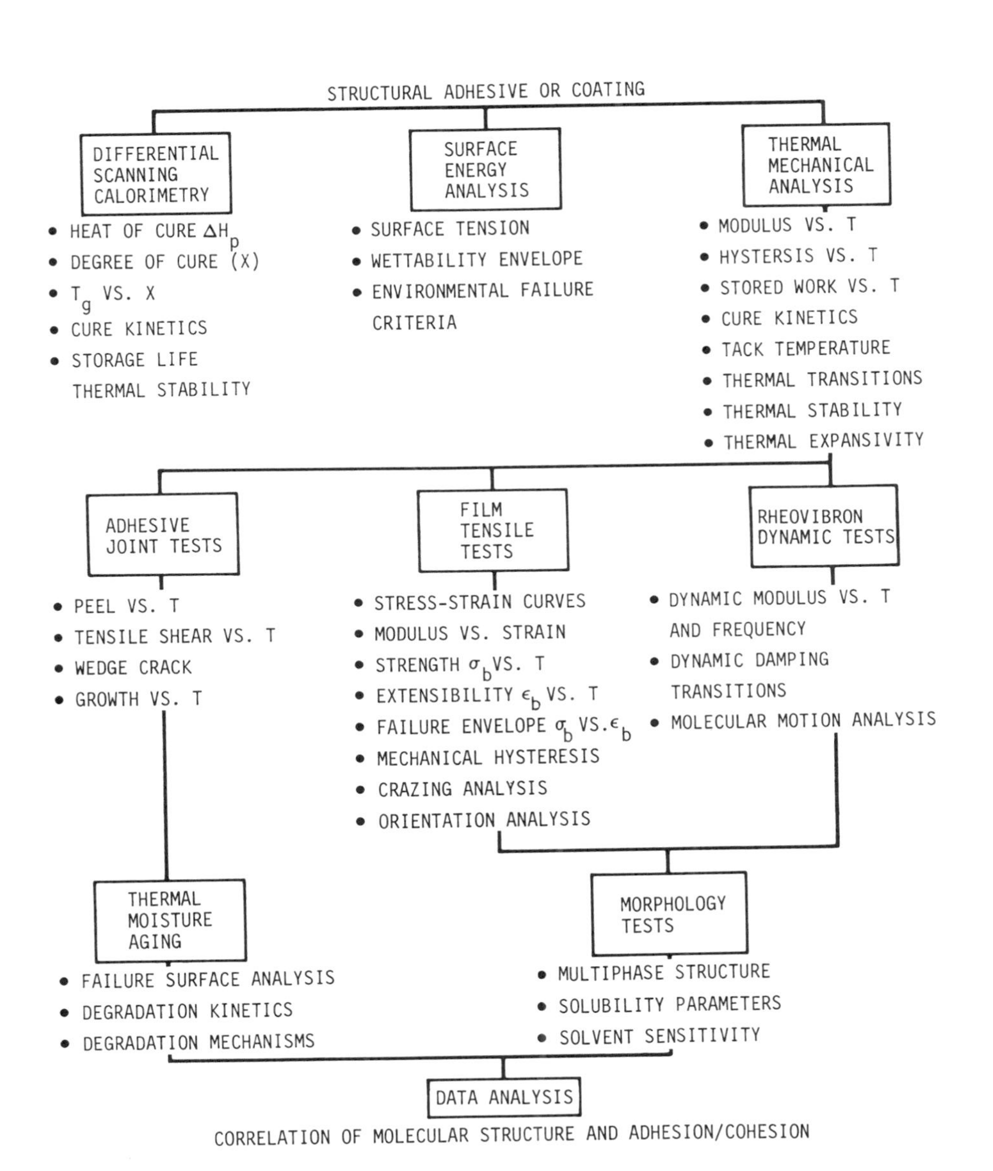

FIG. 1-23 Physical and mechanical analysis flow chart.

A recent workshop on space environment effects on polymeric matrix composites in large scale space structures provides recommendations which also serve as conclusions to this discussion:[43]

1. Prior chemical analysis is necessary for materials identification.

2. Predictive modeling is mandatory.

3. All test designs should be based on the predictive model.

Implementation of the above recommendations promotes organization of present knowledge into an initial mechanistic model with molecular and macroscopic properties correlated. The model can be verified and refined as work progresses. The model elevates the engineering program to the level of conducting science and not simply data gathering.

REFERENCES

1. Exerpts from a poen by Oliver Wendell Holmes, in "The Autocrat of the Breakfast Table," Riverside Press, Cambridge Mass. (1895), pp. 252-258.

2. Proceedings of the Colloquium on Structural Reliability (Editors: J.L. Swedlow, T.A. Cruse, and J.C. Halpin), Carnegie-Mellon University, Pittsburgh, October, 1972.

3. R.A. Heller, "Extreme Value Methods in Reliability Engineering," Ibid., p. 104.

4. E.M. Lenoe and F.I. Barratta, "Recent Studies Toward the Development of Procedures Design of Brittle Materials," Ibid., pp. 433-462.

5. C.F. Tiffany, "The Design and Development of Fracture Resistant Structures," Ibid., pp. 210-216.

6. K.L. Jerima and J.C. Halpin, "Statistical Stress Concentration Effects in Composites," Ibid., pp. 308-322.

7. N.R. Mann, R.E. Shafer, and N.D. Singpurwalla, "Methods of Statistical Analysis of Reliability and Life Data," Wiley, New York (1974).

8. W. Weibull, J. Appl. Mech. 18, (1951), p. 293.

9. C.E. Browning, "The Effects of Moisture on the Properties of High Performance Structural Resins in Composites," Soc. of Plastics Industry, SPI Reinforced Plastics/Composites Division, 28th Ann. Conf. 1973: Section 15A, pp. 1-15

10. J.F. Carpenter and T.T. Bartels, "Characterization and Control of Composite Prepregs and Adhesives," Proc. 7th Nat. SAMPE Conf., Vol. 7, SAMPE, Azusa, CA (1975), pp. 43-52.

11. C.A. May, T.E. Helminiak, and H.A. Newey, "Chemical Characterization Plan for Advanced Composite Prepegs," Proc. 8th Nat. SAMPE Conf., Vol. 8, SAMPE, Azusa, CA (1976), pp. 274-294.

12. D.H. Kaelble and P.J. Dynes, "Preventive Nondestructive Evaluation (PNDE) of Graphite Epoxy Composites," Ceramic Eng. and Sci. Proc. 1(7-8A), (1980), pp. 458-480.

13. D.H. Kaelble, P.J. Dynes, L.W. Crane, and L. Maus, "Composite Reliability," ASTM, STP 580, Amer. Soc. of Testing and Materials, 1975, pp. 247-262.

14. D.H. Kaelble, Polymer Eng. and Science, 17 (1977), pp. 91-95.

15. F.N. Kelley and M.L. Williams, Rubber Chem. and Tech. 42, (1969), p. 1175.

16. D.H. Kaelble, Treatise on Adhesion and Adhesives, Vol. 4, (Editor, R.L. Patrick) Dekker, New York, 1975, pp. 194-208.

17. R.A. Hayes, J. Appl. Poly. Sci. 5, (1961), p. 318.

18. D.H. Kaelble, Physical Chemistry of Adhesion, Wiley-Interscience, New York (1971), Chapter 8, 12.

19. N.W. Johnston, J. Macromol. Sci. Rev. Macromol. Chem. C14 (2), 1976, p. 215.

20. R.F. Boyer, J. Poly. Sci: Symposium No. 50, (1975), p. 189.

21. C.E. Browning, G.E. Husman, and J.M. Whitney, "Moisture Effects in Composites," ASTM Spec. Tech. Pub. STP 617, Philadelphia, 1977.

22. T.G. Fox and P.J. Flory, J. Appl. Phys. 21, 581 (1950).

23. J.D. Ferry, Viscoelastic Properties of Polymers, 2nd. Ed., Wiley, New York, 1970, Chap. 11.

24. J.C. Halpin and H.W. Polley, J. Comp. Materials 1, 64 (1967).

25. J.L. Gardon, J. Appl. Poly. Sci. 7, 625 (1963).

26. R.F. Landel and R.F. Fedors, "Fracture Processes in Polymeric Solids," Editor: B. Rosen, Chap. 3B, Wiley, New York (1964).

27. D.H. Kaelble, Physical Chemistry of Adhesion, Wiley, New York (1971), p. 369.

28. J.C. Bolger, in Treatise on Adhesion and Adhesives (Editor, R.L. Patrick), 3, Chap. 1, Dekker, NY (1973).

29. PABST Roadmap Coordination Meeting, Long Beach, CA, October 8-9, 1976.

30. J.N. Sultan, R.C. Laible and F.J. McGarry, J. Appl. Poly. Sci. 6, 127 (1972).

31. N.J. Sultan and F.J. McGarry, J. Poly. Eng. and Sci. 13, 29 (1973).

32. E.H. Rowe, A.R. Siebert and R.S. Drake, Mod. Plastics 47, 110 (1970).

33. N.K. Kalfoglou and H.L. Williams, J. App. Poly. Sci. 17, 1377 (1973).

34. D.H. Kaelble, in Adhesion Science and Technology, 9A (Editor: L.H. Lee) Plenum Press, NY, 119 (1976).

35. W.D. Bascom and R.L. Cottington, J. Adhesion 7, 333 (1976).

36. D.H. Kaelble and P.J. Dynes, Mat. Eval. 35(4), 103 (1977).

37. J.C. Bolger and A.S. Michaels, in Interface Conversion for Polymer Coatings (Editor, P. Weiss and G.D. Cheever, American Elsevier Pub. Co., NY) (1969) Chapter 1.

38. D.W. Levi, W.C. Tanner, R.C. Ross, R.F. Wegman, and M.J. Bodnar, J. Appl. Poly. Sci. 20, 1475 (1976).

39. T. Smith and D.H. Kaelble, "Mechanisms of Adhesion Failure Between Polymers and Metal Substrates," Technical Report AFML-TR-74-73, Air Force Materials Laboratory, WPAFB, June 1974.

40. D.H. Kaelble and P.J. Dynes, J. Coll. and Interface Sci. 52, 562 (1975).

41. D.H. Kaelble, J. Appl. Poly. Sci. 18, 1869 (1974).

42. D.H. Kaelble, Polymer Eng. and Sci. 17 (7), 474 (1977).

43. Proceedings of the Workshop on Space Environmental Effects on Polymeric Matrix Composites, Langley Research Center, Hampton, VA., April 5-7, 1978.

2
Characterization Techniques for Composite Reliability

2-1. The Scope of Characterization

Advanced composite materials technology has undergone a fundamental transition in the last decade and is now implemented in a wide range of large scale primary structures ranging from composite helicopter rotor blades to composite cargo bay doors for Space Shuttle orbitor. Part of this technology development for composite reliability is a highly organized advancement in the methods and management of characterization methodologies. These characterization methodologies can be listed in the approximate order of their implementation as follows:

1. Chemical Quality Assurance Testing
2. Processability Testing
3. Cure Monitoring and Management
4. Nondestructive Evaluation (NDE)
5. Performance and Proof Testing
6. Durability Analysis and Service Life Prediction

This overview will discuss the detailed characterization methods in the context of the management concept for implementing the specific tests.

Chemical analysis, nondestructive evaluation (NDE) and environmental fatigue testing of composites generates three classes of information on composite reliability which needs to be integrated in a reliability analysis. Several new management methodologies for accomplishing this result are reviewed and discussed. The rapid evolution of computer aided design and manufacturing (CAD/CAM) places new emphasis on automated monitoring and feed-back control during both the manufacture and service usage of composite materials. The idealized feedback control signal from an NDE monitoring system is a structural margin of safety indicator. Computer models for composite durability and environmental fatigue presently contain margin of safety predictions. A major challenge for characterization methodologies is the development of practical structural margin of safety mnonitoring systems which operate in the structure during manufacture and service.

Characterization begins with materials selection and continues through manufacture and use of a composite material. The logic flow chart of Fig. 2-1 shows a typical predictive design methodology which begins on the left with system definition and ends on the right with an accept or reject decision for the manufactured part. The logic flow of Fig. 2-1 shows that materials selection occurs as an early step in component design. Requirements for materials improvement are also shown in Fig. 2-1 to be closely linked to component design, life prediction and materials selection. Physical property data on commercial polymers is extensively tabulated but is generally limited to performance and proof test data developed under ASTM or DIN standard methods. For example, a single publisher systematically compares over 16,000 polymeric materials including adhesives, plastics, foams, films, sheets and laminates, and composite prepregs.[1,2] This discussion is specifically addressed to fiber reinforced composite characterization and covers the full range of special topics from initial quality acceptance of prepreg constitutents to durability analysis and service life prediction of the reinforced composite structure. A recent encyclopedic review provides a comprehensive presentation of commercially available fiber reinforced prepregs and cured laminates with extensive compilations and ranking of performance and proof test properties of composites.[2]

The scope of this discussion is defined by the detailed listing of characterization methods presented in Table 2-1. The classification and methods listing of Table 2-1 includes thirteen test methods for chemical quality assurance, processability, and cure monitoring. An additional 13 tests describe nondestructive evaluation (NDE) and a more extensive list of 47 ASTM-DIN equivalent tests define standard performance and proof testing. Section 7 of Table 2-1 lists several of the current research programs for durability analysis and service life prediction.

The concepts of characterization are reviewed in several general references.[3-7] Billmeyer[3] presents the general principles of polymer chemistry and introduces the essential definitions of polymer physical chemistry measurement which are essential to this discussion. The discussion of Kaelble[4] develops and combines the subject of chemistry, adhesion and polymer rheology into models of composite response which are essential to this discussion. The proceedings of several recent ACS (American Chemical Society) symposiums review the state of progress in resins for aerospace[5] and physical characterization methods.[6] The specific effects of service environments on composite materials has been recently reviewed in an AGARD conference proceedings and this report adequately assesses the current status of proof testing and service life prediction.[7]

2-2. Polymer Physical States and Transitions

It is useful to define five potential physical states for components of polymer composites which are:

vapor (v)	-	involving volatile or condensible components, generally of low molecular weight.
liquid (l)	-	involving the flow state where interchain entanglements or crosslinks do not inhibit macromolecular motion.
rubber (r)	-	involving free motion of polymer segments with superposed restriction of macromolecular motion by interchain entanglements and crosslinks.
glass (g)	-	involving the formation of an amorphous solid state with restricted rotational motion of short chain segments.
crystalline	-	involving the first order transition to a crystalline solid state with restricted rotational and vibrational motion of short chain segments.

In discussing thermal and rheological transitions from one to another of these five states there is often a confusion as to both the type of transition and time direction of transition from initial to final state. For this

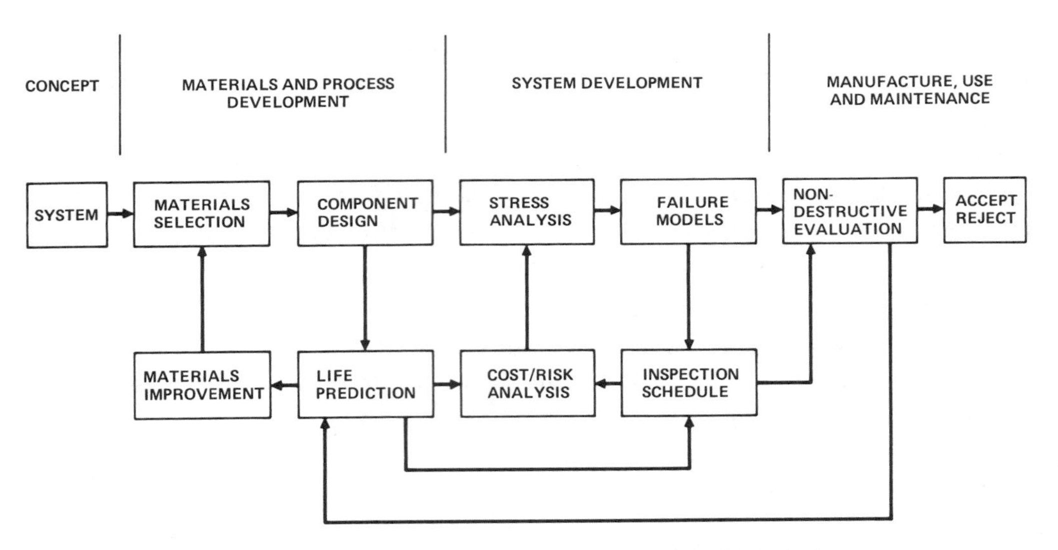

FIG. 2-1 Logic flow chart for predictive design methodology.

discussion we will adopt a nomenclature which details both the type and direction of transition. For example the following transition temperatures are:

T_{lv} = from liquid (l) to vapor (v)
T_{vl} = from vapor (v) to liquid (l)
T_{gr} = from glass (g) to rubber (r)
T_{rg} = from rubber (r) to glass (g)
T_{rl} = from rubber (r) to liquid (l)
T_{lr} = from liquid (l) to rubber (r)

During curing it is often shown that the path of cure involves changes from liquid to rubber to glass to rubber with the following sequence of transition temperatures T_{lr}, T_{rg}, T_{gr}. The detailed time-temperature-transformation state diagrams discussed by Gillham[8] illustrates these multiple transitions and the potential value of the above definitions.

A detailed listing of characterization methods for composites is presented in Table 2-1. In later sections each of these methods will be discussed in more detail. The standard units and conversion factors associated with these methods are listed in Table 2-2. A detailed listing of the characterized properties is summarized in Table 2-3. A brief pictorial review of polymer characterization is shown in Figs. 2-2 through 2-6 which may clarify the interrelations between these numerous test methods.

The upper view of Fig. 2-2 shows the chemical degree of polymerization (left ordinate) versus degree of cure (abscissa). The polymer is soluble up to

TABLE 2-1 Detailed Listing of Characterization Methods

1. Chemical Quality Assurance
    1. HPLC (high performance liquid chromatography)
    2. GC/MS (gas chromatography/mass spectroscopy)
    3. FTIR (Fourier transform infrared spectroscopy)
    4. NMR (nuclear magnetic resonance spectroscopy)
    5. Elemental Analysis
    6. Surface Analysis

2. Processability Testing
    1. DSC (differential scanning calorimetry)
    2. TMA (thermal mechanical analysis)
    3. DMA (dynamic mechanical analysis)
    4. TGA (thermal gravimetric analysis)
    5. SEA (surface energy analysis)

3. Cure Monitoring and Management
    1. Temperature/Pressure/Vacuum
    2. AC Dielectrometry
    3. DC Conductivity
    4. Acoustic Emission

4. Non-destructive Evaluation
    1. US (ultrasonic) immersion C-scan reflector plate
    2. US immersion C-scan through transmission
    3. US contact through transmission
    4. US contact pulse-echo
    5. Fokker bond tester
    6. 210 sonic bond tester
    7. Sondicator
    8. Harmonic bond tester
    9. Neutron radiography
    10. Low KV x-ray
    11. Coin tap test
    12. Acoustic emission
    13. Thermography

5. Surface NDE
    1. Ellipsometry
    2. Surface Potential Difference (SPD)
    3. Photoelectron Emission (PEE)
    4. Surface Remission Photometry (SRP)

6. Performance and Proof Testing

    The following presents a listing of the properties of plastics reported in this book, the ASTM test numbers and the equivalent DIN test:

(continued)

TABLE 2-1 (Continued)

ASTM-DIN Test Equivalents

	Units of Measure			Test	
	English	Metric	SI	ASTM	DIN
Processing					
1. Processing Methods	°F	°C			
2. Comp'n Molding Temp	°F	°C			
3. Inject Stock Melt Temp	°F	°C			
4. Extrusion Temp	°F	°C			
5. Bulk Factor				D1895	D[53466]
6. Linear Mold Shrinkage	in./in.			D955	D[53464]
7. Melt Flow		g/10 min		D1238	D[53735]
8. Melting Point	°Fl	°C			
9. Density	lb/ft^3	g/cm^3	Mg/m^3	D792	D[53479]
10. Specific Volume	$in.^3$/lb	cm^3/g	m^3/Mg	D792	D[53479]
Mechanical Properties					
11. Tensile Str. yield	10^3lb/$in.^2$	10^2kg/cm^2	MPa	D638	
12. Tensile Str. Break	10^3lb/$in.^2$	10^2kg/cm^2	MPa	D638	D[53455]
13. Tensile Str. low temp	10^2lb/$in.^2$	10^2kg/cm^2	MPa	D638	D[53455]
14. Tensile Str. high temp	10^3lb/$in.^2$	10^2kg/cm^2	MPa	D638	D[53455]
15. Elongation %, yield				D638	D[53455]
16. Elongation %, break				D638	D[53455]
17. Tensile Modulus	10^5lb/$in.^2$	10^4kg/cm^2	GPa	D638	D[53457]
18. Flexural Str. yield	10^3lb/$in.^2$	10^2kg/cm^2	MPa	D790	D[53452]
19. Flexural Modulus	10^5lb/$in.^2$	10^4kg/cm^2	GPa	D790	D[53457]
20. Stiffness in Flex.	10^5lb/$in.^2$	10^4kg/cm^2	GPa	D747	
21. Compressive Str.	10^3lb/$in.^2$	10^2kg/cm^2	MPa	D695	D[53454]
22. Izod. notched R.T.	ft lb/in.	kg cm/cm	kJ/m	D256	
23. Izod. low temp	ft lb/in.	kg cm/cm	kJ/m	D256	
24. Hardness	(test)				
Thermal Properties					
25. Thermal Conductivity	BTU in./hr ft^2 °F	10^{-4} cal/sec cm^2 °C/cm	W/Km	C177	D[52612]
26. Specific Heat	BTU in.hr ft^2	cal/g°C	kJ/kg K	C351	
27. Linear Therm. Expan	10^6 in./in.°F	10^{-5} mm/mm	K	D696	D[52328]
28. Vicat Soft Point	°F	°C		D1525	D[53460]
29. Brittle Temp	°F	°C		D746	
30. Continuous Svc Temp	°F	°C			
31. Defl Temp 264 lb/$in.^2$	18.5 kg/cm^2 °F	1.81 MPa °C		D648	D[53461]
32. Defl Temp 66 lb/$in.^2$	4.6 kg/cm^2 °F	0.45 MPa °C		D648	D[53461]
33. U.L. Temp Index		°C/mm			
Electrical Properties					
34. Volume Resistivity		Ohm cm		D257	D[53482]
35. Surface Resistivity		Ohm		D257	D[53482]
36. Insulation Resistance		Ohm		D257	D[53482]
37. Dielectric Strength	V/10^{-3} in.	kV/mm	MV/m	D149	D[53481]
38. Dielectric Constant	50-100 Hz			D150	D[53483]
39. Dielectric Constant	10^2 Hz			D150	D[53483]
40. Dielectric Constant	10^4 Hz			D150	D[53483]
41. Dissipation Factor	50-100 Hz			D150	D[53483]
42. Dissipation Factor	10^3 Hz			D150	D[53483]
43. Dissipation Factor	10^4 Hz			D150	D[53483]
Optical Properties					
44. Refractive Index, Sodium D				D542	D[53491]
45. Clarity					
Environmental Properties					
46. Water Absorp. %, 24 hr				D570	D[53473]
47. Equil. Water Content %				D570	D[53473]

(continued)

TABLE 2-1 (Continued)

7. Durability Analysis and Service Life Prediction
(Some Current Programs)
   1. U.S. Army Composite Materials Research Program (AMMRC).
   2. AFML, "Processing Science of Epoxy Resin Composites," Contract No. F33615-80-C-5021.
   3. AFML/ARPA, "Quantitative NDE," Contract No. F33615-74-C-5180.
   4. AFML, "Integrated Methodology for Adhesive Bonded Joint Life Predictions," Contract No. F-33615-79-C-5088.

---

the gel point. The breadth of the molecular weight distribution is measured by the ratio of weight to number average molecular weight $\overline{X}_w/\overline{X}_n$ which is termed the heterogeneity index of molecular weights. At the gel point the weight average molecular weight, which describes the larger molecules of the polymer, approaches infinity. Further increase of the degree of cure beyond this gel point causes a rapid rise in the insoluble fraction termed gel formed by the crosslinking of these large molecules. At complete cure the gel fraction should constitute the bulk of the polymer with negligible unreacted low molecular polymer. The chemical quality assurance tests in upper Table 2-2 are exploited to verify the chemical changes graphed in upper Fig. 2-2.

The curves in lower Fig. 2-2 outline the characteristic changes in rheological states of liquid flow, rubber, and glass which shift the flow temperature T_m and glass temperature T_g with degree of cure. The lower limit of the liquid flow state extends down to the monomeric glass temperature T_{go}. The uncured resin does not possess a rubbery state at zero degree of cure. With increasing degree of cure the lower curves of Fig. 2-2 show the appearance of a soluble rubbery state which separates the flow state from the soluble glass state. The transition between flow and rubbery state, termed T_m, rises to the limits of thermal stability as the degree of cure approaches the gel point. In cure processing, the elimination of bubbles, entrapped air, and unwetted interface by manipulation of pressure and vacuum must all be accomplished in the flow state and prior to gelation. The gelled polymer has an infinite viscosity and will not flow. At a degree of cure beyond the gel point only the gelled rubber and gelled glass states exist.

Processability tests which measure both thermal and rheological transitions are shown in upper Table 2-2. These tests use small samples of polymer in fully instrumented experiments to define the appearance of characteristic changes in physical properties. The right margin of lower Fig. 2-2 identifies five scan temperatures, T_1 to T_5, for isothermal monitoring of the degree of cure.

TABLE 2-2 Standard Units and Conversion Factors

	To Convert		To Convert		
Metric Units	Multiply By	Si Units	Multiply By	English Units	Property
g/cm^3	1.0	Mg/m^2	0.016	lb/ft^3	Density
kgf/cm^2	10.194	MN/m^2 or MPA	0.0069	lb/in^2	Tensile Strength
kgf/cm^2	10.194	MN/m^2 or MPA	0.0069	lb/in^2	Tensile Modulus
kgf/cm^2	10.194	MN/m^2 or MPa	0.0069	lb/in^2	Flexural Strength
kgf/cm^2	10.194	MN/m^2 or MPa	0.0069	lb/in^2	Flexural Strength
kgf/cm^2	10.194	MN/m^2 or MPa	0.0069	lb/in^2	Compressive Strength
kgf cm/cm	10.194	kJ/m	0.0534	ft lb/in	Izod
kgf cm/cm	101.936	kJ/m^2	0.021	ft lb/in^2	Charpy impact
cal/sec cm C	23.88	W/K m	0.144	BTU in/hr ft^2 F	Thermal Conductivity
cal/g C	0.239	kJ/kgK	4.187	BTU/lb F	Specific Heat
cm/cm C	1.0	m/m K	1.8	in/in F	Linear Expansion
kV/mm	1.0	MV/m	0.0394	$V/10^{-3}$ in	Dielectric Strength

Temperature Conversion:

°F → °C = (°F - 32) ÷ 1.8 °C → °F = (°C x 1.8) + 32

Special names and symbols for a few typical SI units are listed below:

Quantity	Name	Symbol
frequency	Hertz	Hz
power	Watt	W
electrical/resistance	Ohm	Ω
electrical/potential	Volt	V
force	Newton	N
pressure	Pascal	Pa
energy, work	Joule	J
length	meter	m
mass	kilogram	kg
time	second	s
thermodynamic/temperature	Kelvin	K

ASTM TEST DESCRIPTIONS AND UNITS

To Convert			To Convert	
Si Units	Multiply By	English Units	Multiply By	Metric Units
Mg/m^2	62.5	lb/ft^3	0.016	g/cm^3
MN/m^2 or MPa	144.93	lb/in^2	0.0703	kgf/cm^2
MN/m^2 or MPa	144.93	lb/in^2	0.0703	kgf/cm^2
MN/m^2 or MPa	144.93	lb/in^2	0.0703	kgf/cm^2
MN/m^2 or MPa	144.93	lb/in^2	0.0703	kgf/cm^2
MN/m^2 or MPa	144.93	lb/in^2	0.0703	kgf/cm^2
kJ/m	18.73	ft lb/in	5.44	kgf cm/cm
kJ/m^2	47.62	ft lb/in^2	2.141	kgf cm/cm^2
W/Km	6.944	BTU in/hr ft^2 F	3.45×10^{-4}	cal/sec cm C
kJ/kgK	0.239	BTU/lb F	1.0	cal/g C
m/m K	0.555	in/in F	1.8	cm/cm C
MV/m	25.381	$V/10^{-3}$ in	0.0394	kV/mm

$$°C \rightarrow °K = °C + 273.15$$

The following table lists SI units prefixes for decimal multiplication and submultiples:

Factor	Prefix	Symbol
10^{18}	exa	E
10^{15}	peta	P
10^{12}	tera	T
10^{9}	giga	G
10^{6}	mega	M
10^{3}	kilo	k
10^{2}	hecto	h
10^{1}	deka	da
10^{1}	deci	d
10^{-2}	centi	c
10^{-3}	milli	m
10^{-6}	micro	μ
10^{-9}	nano	n
10^{-12}	pico	p
10^{-15}	femto	f
10^{-18}	atto	a

TABLE 2-3 Detailed Listing of Characterized Properties

1. Chemical Quality Assurance
   1. Chemical composition
   2. Degree of cure
   3. Molecular weight distribution
   4. Number average molecular weight
   5. Weight average molecular weight
   6. Entanglement molecular weight

2. Processability
   1. Gel point
   2. Gel faction
   3. Crosslink molecular weight
   4. Glass temperature
   5. Melt (flow) temperature
   6. Dynamic storage modulus
   7. Dynamic loss modulus

3. Cure Monitoring
   1. Temperature/pressure/vacuum
   2. Dynamic dielectric constant
   3. Dielectric loss factor
   4. DC conductivity

4. Nondestructive Evaluation
   1. Internal stress distributions
   2. Damage zone size
   3. Crack growth rate

5. Performance and Proof Testing
   1. Stress and environment dependent T_g
   2. Stress and environment dependent T_m
   3. Isothermal stress-strain-time-response
   4. Strength distribution
   5. Extensibility distribution
   6. Fracture energy distribution

6. Combined Bonding and Failure Testing
   1. Surface energy
   2. Surface chemistry
   3. Surface morphology
   4. Surface roughness

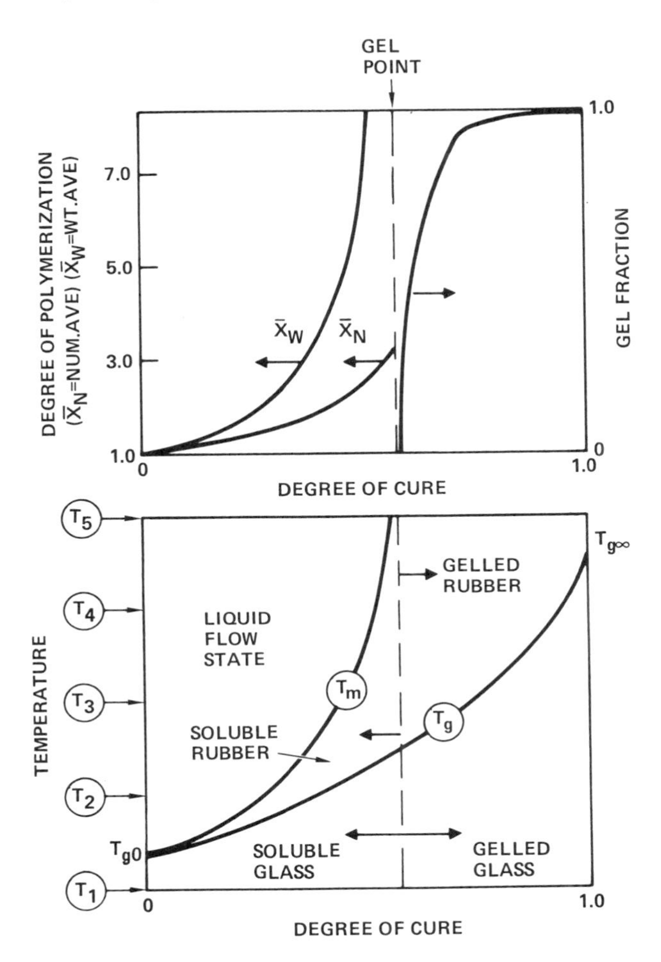

FIG. 2-2 (Upper): Change in molecular weight distribution and sol-gel state with degree of cure (idealized). (Lower): The effect of degree upon glass transition temperature T_g and melt temperature T_m for liquid flow (idealized).

The curves of lower Fig. 2-2 show that the glass transition T_g rises with degree of cure to a final value $T_{g\infty}$ for the fully cured polymer of infinite molecular weight. This shift in T_g with degree of cure is the fundamental change in properties which produces the structural stiffness and strength for the cured polymer.

The idealized isothermal dynamic mechanical monitoring of the degree of cure for the five temperatures T_1 - T_5 is shown in Fig. 2-3. The lowest temperature T_1 is below T_{go} and therefore monitors the glass state throughout cure. The upper curve of Fig. 2-3 plots the nearly constant value of the glass state storage modulus G' and shows cure state does not influence this property. The remaining curves of upper Fig. 2-3 show the characteristic rise in log G' to the

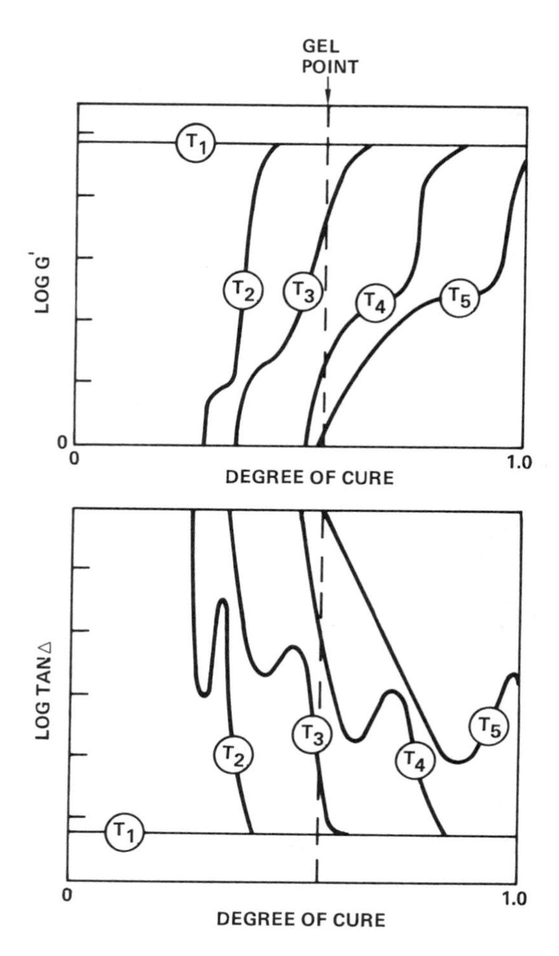

FIG. 2-3 Idealized isothermal dynamic mechanical monitoring of degree of cure in terms of shear storage modulus G' (upper view) and loss tangent tan δ (lower view).

glass state value as the increasing degree of cure raises first T_m and then T_g through the constant monitoring temperatures T_2 - T_5. In the flow state where, for example T_2 is greater than T_m, the expected value of G', which measures the elastic energy of dynamic deformation, should be zero.

The lower curves of Fig. 2-3 plot the idealized trends in the loss tangent $\tan \delta = G''/G'$ where G'' is the loss modulus and a measure of the flow energy dissipated during dynamic deformation. At the low temperature T_1 the polymer is glassy at all states of cure and tan δ is unaffected by state of cure. At the high temperatures T_2 - T_5 where cure starts in the liquid flow state the ideal value of tan δ should be infinity since G' = 0 and then decrease to the glass state value with increased state of cure as shown in lower Fig. 2-3. The first knee on the log G' and log tan δ curves of Fig. 2-3 is specifically associated with the liquid to rubber transition T_{lr} rising through the isothermal test temperature. The second knee on the curves of Fig. 2-3 is associated with the rubber to glass T_{rg} rising through the isothermal monitor temperatures T_2 - T_5. Since T_5 is slightly above $T_{g\infty}$, as shown in Fig. 2-2, the full transition to glass state dynamic response is not obtained at full cure.

In category 3 of Table 2-2 the common cure monitor tests such as AC impedance (dielectrometry) and DC resistance measure rheological properties which relate in a fairly direct way to the curves of Fig. 2-3. The fundamental problem in cure monitoring and cure process management is the reconstruction of the more fundamental chemical and physical information graphed in Fig. 2-2 from cure monitoring data. Optimum process control of temperature, pressure, and vacuum conditions for cure can be visualized as being more direct from data shown in Fig. 2-2 than Fig. 2-3.

Consider now that cure is complete as shown by the properties at the right ordinate of Figs. 2-2 and 2-3. The upper curve of Fig. 2-4 shows the idealized thermal scan of tensile modulus E from well below the glass temperature $T_g = T_{g\infty}$ of the fully cured resin. The types of molecular motion typically available below and above T_g are shown in upper Fig. 2-4. The flow temperature T_m shown in Fig. 2-4 is taken to be associated with high temperature chemical decomposition with network scission to produce a new high temperature flow state. The alternative thermal decomposition process would involve thermal decomposition with additional cross linking which would tend to raise T_g and produce a brittle solid without available network segment motion.

For the cured resin with network segment motion available above T_g the lower curves of Fig. 2-4 show three domains of tensile stress versus temperature response at constant loading time. The lower diagram of Fig. 2-4 shows that both T_g and T_m are reduced by applied tensile stress. A number of theories concerning stress or pressure effects on rheological response indicate the physical state zones defined in lower Fig. 2-4. The region to the left of the T_g curve is the brittle solid state and when stress rises past the solid failure stress σ_b a brittle fracture is predicted. The intersection of the T_g and σ_b curves defines the brittle temperature T_b above which plastic yielding can occur.

Tensile stressing above T_b provides a tensile yield stress when the stress penetrates the T_g curve of lower Fig. 2-4 and craze fracture when stress exceeds σ_b. At a higher temperature in lower Fig. 2-4 such as T_4 the stress can rise through the T_m curve and produces a flow failure mechanism. Idealized stress versus strain curves for the four temperatures T_1 - T_4 plus T_g in lower

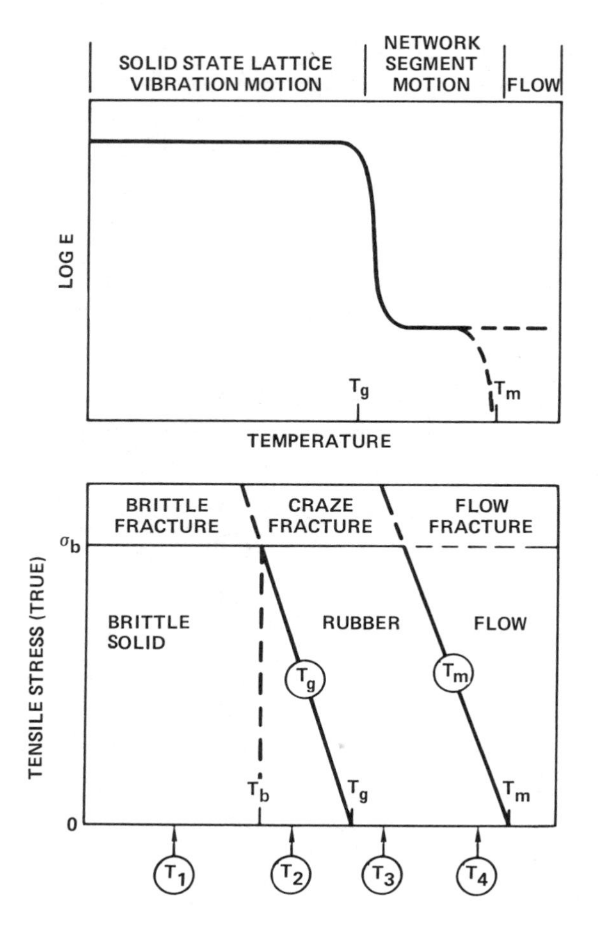

FIG. 2-4 Thermal scanning of fully cured matrix for tensile modulus (upper view) and stress-temperature response (lower view) at constant time of loading (idealized).

Fig. 2-4 can be drawn to illustrate these points. The fundamental point illustrated in lower Fig. 2-4 is that stress and temperature interact to produce characteristic transitions from brittle to rubber to flow states. These stress-temperature transitions in mechanical response fundamentally affect mechanical performance and structural integrity.

The upper view of Fig. 2-5 sketches the expected form of the tensile stress versus strain curves. The stress is represented as the true stress of the deformed cross-section to retain the connection with the curve of lower Fig. 2-4. The area beneath the related curves of nominal stress (of the undeformed area) versus strain define the unnotched fracture energy per unit volume W_b shown by the temperature curve of lower Fig. 2-5. The lower curve of Fig. 2-5 is typical of many structural polymers which display maximum toughness

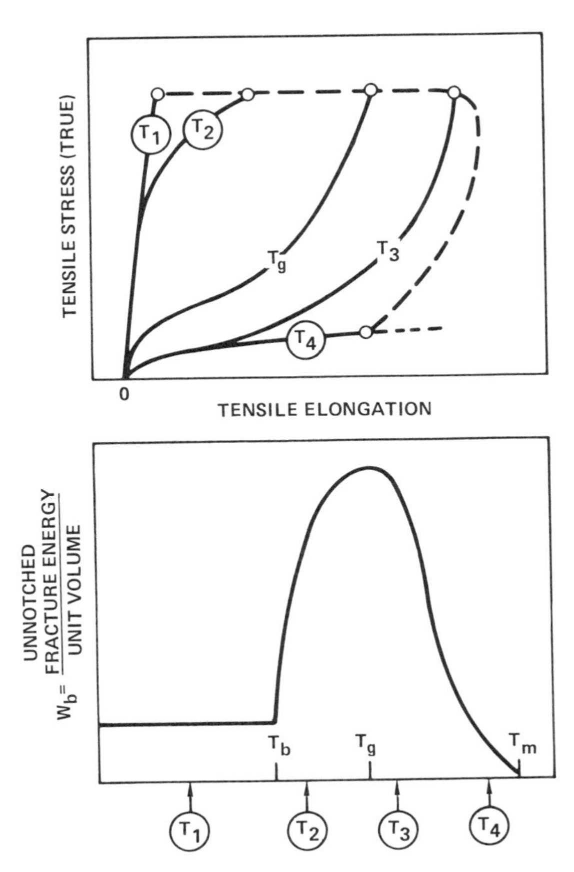

FIG. 2-5 Characteristic tensile stress-strain and fracture response (upper view) and temperature profile of unnotched tensile fracture energy (lower view).

near T_g. The temperature span of high fracture toughness is one of the distinct features of engineering thermosets such as epoxy resins and thermoplastics such as the polycarbonates.

The stress strain curves of upper Fig. 2-5 show the typical shapes shown by engineering plastics. At T_g or higher temperature the upward curvature reflects chain orientation effects which in some polymers are quite pronounced. At the highest temperature, T_4, failure is visualized as caused by a tensile cold drawing which is characteristic of filamentary orientation of cavities at crack tips in fracture mechanics specimens.

The utilization of the thermomechanical responses shown in Fig. 2-4 and Fig. 2-5 in fracture mechanics and stress analysis models is illustrated in the curves of Fig. 2-6. The upper curve of Fig. 2-6 illustrates an elastic-plastic analog curve which is fit to an experimentally measured polymer stress versus strain response. The analog (dashed) curve has the same curve area, since A_1 =

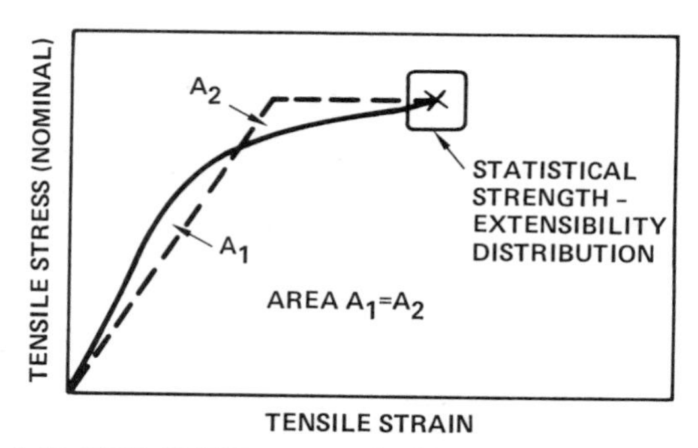

I. ELASTIC-PLASTIC ANALOG STRESS-STRAIN CURVE

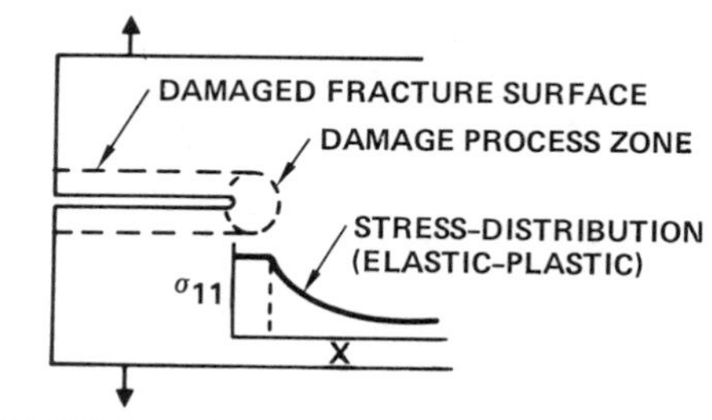

II. FRACTURE MECHANICS (DUGDALE MODEL)

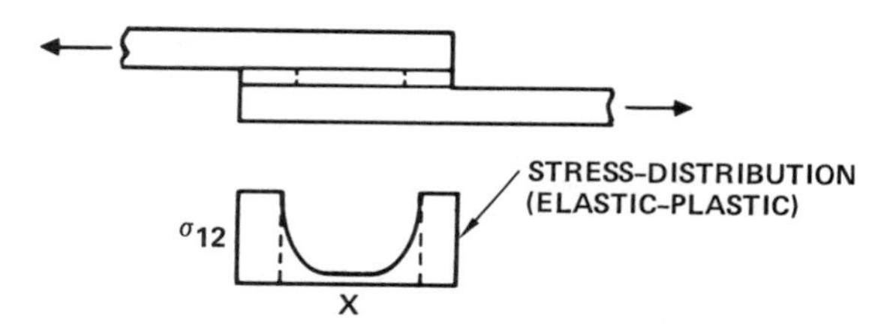

III. STRESS ANALYSIS (HART-SMITH MODEL)

FIG. 2-6 Conversion of measured stress-strain to elastic-plastic analog (I) and introduction into fracture mechanics (II) and stress analysis (III) predictive models.

A_2 and the same strength and extensibility end point as measured by repetitive testing and statistical analysis. For fracture mechanics analysis for crack growth this elastic-plastic analog is introduced into the Dugdale(8) model shown in the middle view of Fig. 2-6. For stress analysis this elastic-plastic analog is introduced into typical Hart-Smith models(9) for adhesive joints designed as shown in lower Fig. 2-6.

The two lower views of Fig. 2-6 thus incorporate all of the data of Table 2-2 into either fracture mechanics or stress analysis models. Durability analysis and service life prediction studies outlined in Part 7 of Table 2-2 utilize mechanics models of this type. The more sophisticated fracture mechanics and stress analysis models currently under development utilize the curvilinear stress strain response pictured in the upper curve of Fig. 2-6 to obtain greater computational precision.

2-3. Chemical Quality Assurance

One essential objective in chemical quality assurance testing is verifying the lot-to-lot reproducibility of material constituents in composite prepregs. Each prepreg chemistry may require a complete development of a chemical quality assurance program. The flow chart for chemical analysis illustrated in Fig. 1-22 involves a sequence of chemical separations combined with quantitative chemical analysis of each of the separated components. The flow chart of Fig. 1-22 was developed by Carpenter and Bartels[11] for analysis of selected types of 350°F (177°C) service ceiling epoxy matrix composite prepregs and adhesives. Chemical characterization programs for composite prepregs are reported by May, Helminiak and Newey[12] which implement a similar approach of successive molecular separation followed by quantitative chemical analysis. The separation methodologies implemented in the analysis of epoxies depends largely upon solvolytic separation by HPLC combined with infrared spectroscopy and elemental analysis. Quite a different approach has been employed by Alston[13] in chemically analyzing fully cured PMR-15 polyimide-graphite composites by use of thermo-oxidative degradation combined with Fourier transform IR and NMR analysis of degradation products.

Several recent reports describe and demonstrate interdisciplinary approaches for composite characterization which incorporate chemical characterization, adhesion criteria, and composite durability analysis in a unified characterization scheme.[14-16] These early attempts to integrate chemical quality assurance into a general composites durability evaluation follows the strategy suggested by Fig. 2-1. A recent special issue of the journal "Polymer Composites" presents a collection of papers which summarize the central position of chemical characterization in the U.S. Army composite materials research program.[17]

What follows are brief summaries and descriptive references to the chemical quality assurance test methods listed in Table 2-1. Gas and liquid chromatography form a class of molecular separation methodologies and these are classified into ten categories in Table 2-4 according to the description of McNair.[19] Surface characterization methods for reinforcing fiber coatings also fall into nine or more categories in Table 2-5 and these are classified according to ability to detect specific surface properties.[19a]

TABLE 2-4 Classification of Chromatographic Methods

I. Gas Chromatography (GC)

   Gas liquid (GLC)
   Gas solid (GSC)

II. High Performance Liquid Chromatography (HPLC)

   A. Planar Chromatography
      Thin layer (TLC)
      Paper (PC)

   B. Column Chromatography
      Exclusion (EC)
         Gel Permeation (GPC)
         Gel filtration (GFC)
      Liquid-solid or adsorption (LSC)
      Liquid-liquid or partition (LLC)
      Bonded phase (BPC)
      Ion exchange (IEC)

From: H.M. McNair, American Laboratory, May 1980, pp. 33-44.

TABLE 2-5 Decision Matrix of Surface Characterization Methods for Reinforcing Fiber Coatings (35 to 70 nm thickness)

4 = Excellent
3 = Acceptable
2 = Marginal
1 = Unacceptable
0 = No Information

	Coating Durability	Molecular Orientations	Surface Concentration of Components	Surface Coatage Uniformity	Fiber Curvatures	Adhesion Strength	Thickness Uniformity	Average Coating Thickness	Row Ave.
Surface Energy Analysis	3	4	3	4	4	2	1	1	2.75
Scanning Elect. Mic. + EDAX	4	1	1	4	4	1	4	1	2.5
Electron Spect. for Chem. Anal.	4	4	4	1	1	1	1	1	2.13
ASTM Adhesion Test	4	1	1	1	1	4	1	1	1.75
Fourier Transform IR	2	2	3	1	1	1	1	1	1.50
Optical Microscopy	1	1	1	1	1	1	1	1	1.0
Secondary Ion Mass Spec.	1	1	1	1	1	1	1	1	1.0
Laser Microprobe Mass Analyser	1	1	1	1	1	1	1	1	1.0
Raman Microspectroscopy	1	1	1	1	1	1	1	1	1.0
Col Ave.	2.33	1.78	1.78	1.67	1.67	1.44	1.33	1.00	

The advantages and limitations for each of the six test methods for chemical quality assurance are listed below with detailed references provided for each method. Surface characterization methods are further subdivided as outlined in Table 2-5.

1. HPLC (High Performance Liquid Chromatography)[18,19,20]

Advantages: Quick separations in seconds to minutes. Very high resolution of 100 or more components. Provide easily performed quantitative analysis with errors less than 1%. Very small sample size with measurements of 10^{-9} to 10^{-12} g reported. Analyses a wide variety of sample types - molecular weights from 18 to 6 million g/mole can be measured. Measurements largely automated.

Limitations: Requires high-resolution column and high-pressure pump. Instruments are expensive. Extensive experience of 6 - 12 months required for proficient operation. HPLC is a poor identifier with other instruments such as mass spectrometer required to identify each peak. There is no universal and sensitive detector. Refractive index detectors are universal but have limited sensitivity. Ultraviolet detectors are sensitive but selective. This method is limited to soluble materials.

2. GC/MS (Gas Chromatography/Mass Spectroscopy)[21,22]

Advantages: Pyrolysis combined with gas chromatography and mass spectroscopy provides a means for direct chemical separation and analysis of insoluble and crosslinked polymers. Controlled pyrolytic breakdown of macromolecular structures furnishes a profile of gas chromatographic peaks for mass spectrum analysis. Recently introduced Fourier transform MS (Nicolet FTMS-1000) increases the speed and resolution of the pyrolytic analysis methodology.

Limitations: Method is quite expensive with a price tag of $300,000 for a computerized FTMS system.

3. FTIR (Fourier Transform Infrared Spectroscopy)[22-25]

Advantages: The high resolution (generally less than 1.0 cm^{-1}) combined with rapid (2-20s) scanning of the IR spectrum makes this a versatile method for chemical analysis. The computer assisted addition and subtraction of reference spectra permit direct isolation of molecular volatile loss, and specific chemical reactions. Reflectance IR can be combined with transmission measurements to provide a variety of sampling methodologies. Kinetics of curing reactions can be measured by FTIR. Combined with molecular separation techniques such as HPLC the versatility of FTIR is greatly extended.

Limitations: There are few limitations to this method and its versatility is a continuing challenge to the analyst.

4. Nuclear Magnetic Resonance[26-27]

Advantages: Proton, ^{19}F and ^{13}C NMR provide important methods for defining the short range stereochemical structure of monomers and polymers. Over the last decade ^{13}C NMR has assumed increasing importance in defining tacticity, comonomer sequence, clas-trans isomerism, branching and cross-linking. High resolution ^{13}C NMR in the solid state is now achieved using proton dipolar decoupling, magic angle spinning and 'H - ^{13}C cross polarization. Where applicable ^{19}F NMR shows high sensitivity to tacticity.

Limitation: Many of the newer ^{13}C techniques are still under development and therefore not amenable to routine use in chemical characterization.

5. Elemental Analysis[22,28]

Advantages: The atomic ratios of C, N, O, Si, P, S, H and the halogens F, Cl, Br, I provide essential information in identifying the organic chemistry of composite materials. Routine elemental analysis is now largely carried out quantitatively in automated commercial instruments (such as the Perkin-Elmer Model 240-B). Atomic adsorption spectroscopy supplements conventional elemental analysis to permit highly sensitive (parts per million sensitivity) and quantitative analysis for over 35 metallic elements. Atomic adsorption (AA) spectroscopy is particularly useful in detecting organometallic catalysts which modify the processability of thermosetting resins.

Limitations: Elemental analysis provides no direct information on molecular structure other than the empirical formula for molar ratio of the elements and must be supplemented by other chemical structure analysis methods.

6. Surface Chemistry Analysis[19a,29,30]

Wettability and Surface Energy Analysis (SEA) via Scanning Wilhelmy Plate Method

Advantages: Measures fiber circumference, estimates surface roughness and component surface concentrations, sensitive to coating removal, indirectly measures surface energies and environmental durability, digital recording permits statistical analysis of wetting forces and predicted adhesion.

Limitations: Sensitive to swelling, liquids analysis complicated by contact angle hysteresis, may emphasize low energy elements of surface chemistry in advancing contact angle, gives no direct chemical information.

Scanning Electron Microscopy (SEM) + Energy Dispersive X-ray (EDAX)

Advantages: Combined high resolution (~ 100Å) and great depth of focus, resolves unevenly distributed coatings, combined low voltage (~ 1.0 kev), low magnification and TV scan can bypass metal coatings.

Limitations: Requires high vacuum and consequent volatiles loss, may cause charging and require metal coatings, EDAX signal from thin (~ 70 nm) coatings too weak for analysis.

Electron Spectroscopy for Chemical Analysis (ESCA)[34,35]

Advantages: Small sampling depth (~ 50 nm) ideal for thin coating analysis, most used elements can be detected and elements (C, N, O) show changes in binding energy with oxidation state, x-ray does not damage the surface, coating thickness can be evaluated by varying take-off angle.

Limitations: Requires high vacuum and consequent volatiles loss, depth profiling restricted by damage produced by ion beams, geometry of fibers and fiber bundles creates problems.

ASTM Adhesion Tests[36]

Advantages: Gives direct measure of apparent bond strength, gives direct measure of durability in terms of strength degradation, can be used to generate reliability statistics for mechanical design.

Limitations: Measures a system response as opposed to an interface property, no direct chemical information, results depend on joint design and test method.

Fourier Transform Infrared Spectroscopy (FTIR) & Internal Reflection and Attenuated Total Reflectance (ATR)[37-39]

Advantages: Little sample preparation, provides direct chemical identification of thicker coatings, digital recording of spectrum provides basis for extended data processing, rapid multiscans increase signal to noise, difference spectra analysis isolates chemistry of thin (~ 35 nm) coatings with reflective strong-absorbance bands.

Limitations: Penetration depth of IR radiation is much larger than typical coating thickness (~ 35 nm), absorption bands at longer wavelength are enhanced by greater depth penetration.

Optical Microscopy

Advantages: Little sample preparation, records color and birefringence, can scan large areas.

Limitations: Low resolution (~ 1 μm), limited depth of focus at high magnification, does not resolve curved surfaces, nor show evenly distributed coatings, no chemical information.

Secondary Ion Mass Spectroscopy (SIMS)[40]

Advantages: Small sampling depth (~ 10Å), potentially broad range of secondary ions can be analyzed.

Limitations: Requires high vacuum and volatiles loss, the high energy inert gas ions which bombard the fiber surface cause degradation and pyrolysis thus rendering this method nonoperative on polymeric fibers.

Laser Microprobe Mass Analysis (LAMMA)[41]

Advantages: Provides analysis of a small fiber area (1 μm diameter), time-of-flight mass analyzer defines secondary ions, limited depth of analysis (~ 0.1 μm), possible to map surface, small sample volume (10^{-19} gm).

Limitations: Requires high vacuum with volatiles loss, the laser energy output produces disintegration of the fiber without isolating coating constituents.

Microprobe Molecular Optics Laser Examiner (MOLE) or Raman Microscopy[42,43]

Advantages: Records Raman spectrum from a small area (~ 1 μm^2), does not require high vacuum so can analyze volatiles, in principle can locate and analyze for coating surface distribution.

Limitations: Requires strong Raman scatterers in coating and weak scatterers in fiber, in thin coatings the dominant sampling volume will come from the fiber.

2-4. Processability Testing

Laboratory tests for composite processability are designed to determine how a sample performs during a simulated manufacturing cure cycle. In general, these tests use small quantities of material (from 0.01 to 1.0 gm), are fully instrumented, and operated by programmed scanning of temperature at constant scan rate. The four thermal analysis methods listed under processability testing in Table 2-1, part 2, are highly complementary and are normally marketed as components of a thermal analysis system (such as Mettler TA 2000, DuPont 900, and Perkin-Elmer DSC-2). Recent studies and comparison on the several commercial systems are provided by Wunderlich and coworkers.[44] The function of processability testing is to define the kinetics of curing, the limits of thermal stability, and the optimum cure cycle which leads to high performance and durability. Very often, procedures of chemical analysis, such as outlined in Fig. 1-22, are implemented to verify thermal analysis data and to define the chemical mechanism of curing. Composite matrix polymers are commonly classified in terms of their temperature range of cure processing or the service ceiling temperature for environmental stability. The earlier discussion of Figs. 2-2 through 2-6 has already introduced the multiple changes induced in thermal and rheological response as a consequence of curing. Considering the complexity of curing processes it is not surprising that logic flow diagrams, such as discussed by Kaelble,[16] and shown in Fig. 1-23 have been developed to better integrate processability testing with evaluation of composite performance and durability.

The upper portion of Fig. 1-23 describes two forms of thermal analysis, DSC (differential scanning calorimetry) and TMA (thermal mechanical analysis) as central components of processability testing. These combined tests characterize the degree of cure and the effect of cure on the melt temperature T_m and glass temperature T_g as shown by the lower view of Fig. 2-2. By clever design of the commercially available accessories for thermal analysis systems many of the ASTM performance and proof tests listed in category 6 of Table 2-1 can be simulated using small samples (0.01 to 0.10 g) in a highly instrumented thermal analysis measurement. The function of processability testing is to find the optimum processing "window" which is a combination of processing times, temperatures, and pressures which both consolidate, form by flow, and chemically cure the composite laminate. These processability studies should always be accompanied by relevant chemical analysis.

The processing of laminates and fiber reinforced composites inevitably involves interface bonding during the process cycle. The utilization of wettability tests and surface energy analysis (SEA) is a subject well developed in the field of adhesive bonding but still largely overlooked in fibrous composite process evaluation. Rather extensive interface integrity studies sponsored by the Army Composites Research Program[17] and the Air Force Materials Laboratory[14] strongly support the implementation of SEA in conjunction with thermal analysis. SEA involves the testing of solid surface wettability by contact angle measurements with a selected set of test liquids. The analysis which accompanies SEA testing gives predictions of bonding ability and bond durability of the composite interface.

Brief descriptions of the commonly used processability test methods listed in Table 2-1 are presented in terms of advantages and limitations as follows:

1. Differential Thermal Analysis (DTA)[6]

Advantages: Measures the temperature difference (ΔT) between sample and reference under programmed thermal scan. High ($\Delta T \leq 0.001C$) sensitivity, wide range of high temperature and pressure, and small sample size, and measurement simplicity are advantages.

Limitations: Generally applied in qualitative analysis, calibrations and data processing are required to determine heats of reactions or specific heat properties of sample. DSC accessories are available to simplify direct evaluation of thermal properties.

2. Differential Scanning Calorimetry (DSC)[6,45]

Advantages: Directly measures the rate of heat release (dH/dt) or absorption of a sample relative to a reference specimen. Measurements provide quantitative measure of heat of reaction ΔH (to 3%) and heat capacity (to 0.5%) properties under isothermal or constant thermal scan rates (up to 50°C/min), directly measures thermal state of cure x, cure kinetics, and cure effects on glass transition T_g.

Limitation: Less sensitive and more limited temperature and pressure ranges than DTA.

3. Thermal Mechanical Analysis (TMA)[6]

Advantages: Measures a sample thermal expansion or penetration softening under isothermal or constant rate thermal scan conditions and adjustable mechanical load. Operates as a modular accessory to DTA and TMA. High displacement sensitivity (0.025 μm) and dilatometer accessories permit small scale simulation of ASTM thermal and mechanical tests. Often used in conjunction with DTA, DSC, and TGA.

Limitations: In general, the physical limitations of accessory modes of measurement for penetrometry, dilation and tensile creep limit the quantitative precision of the "nominal" thermal response measurement.

4. Thermal Gravimetric Analysis (TGA)[6,45]

Advantages: Measures sample weight changes due to volatile loss and thermal decomposition under isothermal or constant thermal scan rate. Detects chemical reactions which change sample weight which cannot be sensed by other methods. High precision (0.1%) and sensitivity (0.01 mg) with small sample size (≈ 10 mg). Often used in conjunction with DTA, DSC, or TMA and can be augmented by chemical analysis such as GC/MS.

Limitations: Weight change measurement needs additional modes of characterization to identify the process chemistry or physics.

5. Surface Energy Analysis (SEA)[31-33,46-49]

Advantages: Tests for bonding ability by liquid-solid contact angle measurement. Method is quantitative and simple to apply. A thermogravimetric analysis (TGA) can be applied to provide automatic measure of advancing and receding contact angle on fibers and sheets at controlled temperature or programmed temperature change. Adhesive bonding and interface durability parameters can be calculated. Method extends and quantifies ASTM tests for surface quality assurance.

Limitations: Method is limited to continuous surfaces, and high microroughness introduces contact angle hysteresis. Interface solubility and chemical reactions limit the surface energy analysis. Small samples required for TGA method and small drops and microscope for large surface analysis. See Table 2-5 for comparative sensitivities.

6. Dynamic Mechanical Analysis (DMA)[5]

Advantages: Isolates the storage (elastic) and loss (viscous) components of rheological response at varied frequency, temperature, and state of cure. Methods applicable to unsupported polymer or polymer supported by braid. Highly sensitive to both the flow T_m and glass T_g transitions at all states of cure. A primary laboratory tool with DSC for defining optimum cure process conditions.

Limitations: No single DMA method is universally applicable so multiple DMA methods are generally required which cover the varied ranges of liquid to solid response. Quantitative interpretation requires concurrent chemical analysis.

2-5. Cure Monitoring and Management

This section is concerned with feasible methods for direct in situ monitoring of cure during press or autoclave molding processes typical of composite production. The related subject of cure management is the utilization of the cure monitor in a closed loop control of the production cure process through the program time (t) versus cure temperature (T), external pressure (P), and internal degassing vacuum (V) imposed on the curing part. Information for "real time" management of this (t-T-P-V) program for production process control can derive from three sources which are:

1. Prior chemical analysis
2. Prior processability testing
3. Current cure monitor data from within the curing part.

The U.S. Army Composites Research Program provides extensive documentation of the combined utilization of the above three approaches to cure management[51-60] of epoxy-glass and epoxy-graphite composites. In these Army studies of matrix resins, prepregs, and composites the minimum chemical analysis includes HPLC (high performance liquid chromatography) and FTIR (Fourier transform infrared spectroscopy) used both individually and jointly as quality assurance tests to verify chemical reproducibility. The minimum processability testing is by DSC (differential scanning calorimetry) to verify the reproducibility of the curing kinetics. The characteristics of the epoxy resins which can be determined by these prior tests are degree of cure, aging of prepreg, effect of cure conditions on the oxidative stability of the resin, products of degradation attack, and effect of processing on the resin chemistry.[51] This information can be exploited to optimize the cure cycle to accommodate variability in resin chemical composition and prepreg aging.

1. <u>AC Dielectrometry</u>[61-66]

<u>Advantages</u>: Close relations exist between AC dielectrometry and DMA response. Commercial instrumentation is available with adequate range and sensitivity of measurement. New microcircuitry is being developed for in situ measurement. Frequency scanning can be exploited in dielectric spectrum analysis and in feed-back control logic for process management.

<u>Limitations</u>: Changes in spacing between electrodes can modify the dielectric output. Increasing the number of prepreg plys between electrodes lowers measurement reliability. In some cases electrodes must be placed in bleeder layers outside the curing part.

2. <u>DC Resistance</u>[51,64]

<u>Advantages</u>: Very simple electrical circuit. Resistance variations appear to correlate with storage component of dynamic modulus.

<u>Limitations</u>: No physical model correlating DC resistance and rheology is available.

3. Acoustic Emission[51]

Advantages: Acoustic emission sensors provide direct information on microcracking processes which occur in the cured composite during improper cool down. This signal can be exploited to control cool down and minimize internal microdamage and internal thermoelastic stresses.

Limitations: This method is new and needs further development. Physical interpretation of the monitor output also needs further study.

In principle, the sensors for all three of the above cure monitors could remain in the cured part for utilization in nondestructive evaluation (NDE) in service. Current reports indicate that AC dielectric monitoring of large autoclave molded parts is feasible, and within limits, can be applied in process control.[66,67] It is evident that cure monitoring and management involves the intelligent incorporation of prior chemical analysis and processability test data. Computerized monitoring and automated cure management are currently under intensive development.

2-6. Nondestructive Evaluation (NDE)

Nondestructive evaluation has recently emerged as a new branch of material science which concentrates in the area of nondestructive testing and analysis to determine residual reliability and durability of manufactured parts. By adopting new modes of computer controlled three dimensional scanning, the ultrasonic response parts with complex surface curvatures can be automatically mapped for flaws.[68] Ultrasonic inspection is often restricted by the necessity of providing liquid acoustic couplants between the US (ultrasonic) transducer and the test specimen. For metal containing substrates the use of electromagnetic induction to launch and detect ultrasonic waves across an air gap permits high speed inspection using electromagnetic acoustic transducers (EMAT's).[69]

Normally no nondestructive testing method presently employed can detect poor adhesion resulting in low bond strength. Poor bond quality produced on phosphoric anodized aluminum by oily contamination or surface damage of the oxide by rubbing with Kraft paper prior to bonding was not detected by standard ultrasonic inspection.[70,71] The recent development of an automated ellipsometer for rapid optical scanning of treated metal surfaces has shown that both of the above types of surface defects can be detected and corrected prior to bonding as part of process monitoring.[72] It is thus evident that process monitoring combines with post production (NDE) to provide a viable reliability and durability test methodology for adhesive bonded structures. Surface NDE will be reviewed as a separate subject in the next section.

The studies of Hagemeier, Fassbender, and Clark[70,71] form one of the most detailed assessments of conventional NDE methods for large area inspection of critical bond-line flaws in laminated structures. This study forms part of the Primary Adhesive Bonded Structure Technology Program (PABST) in which an adhesively bonded and highly loaded primary aircraft fuselage was designed, fabricated and proof tested for reliability and durability. Some specific objectives of the NDE portion of this program were as follows:

1. Define the common flaw types and locations.
2. Determine flaw behavior and growth rates under fatigue loading and hydrothermal exposure.
3. Assess the ability of NDE to detect flaws and monitor flaw growth rates.

The interaction of these three objectives provides the new and interesting depth to the capability assessment of the NDE method under Item 3 in the above objectives. A brief description of the advantages and limitations of these standard NDE methods is presented as follows.

1. Ultrasonic NDE

Advantages: This method uses pulsed ultra sound at 2.25 to 10 MHz. Both contact and immersion techniques of inspection are employed using four specific test geometrics which are:

a. immersion C-scan reflector plate
b. immersion C-scan through transmission
c. contact through transmission
d. contact pulse echo.

These methods may be automated to produce plan view recordings (C-scans) using methods (a) and (b).

Limitations: The ultrasonic method suffers from destructive wave interference. The problem with destructive wave interference is that there is an appearance that elements of a laminate are unbonded when they are not.

2. Fokker Bondtester

Advantages: This tester operates on the principle of resonance impedance and is widely used in the aerospace industry. The instrument is calibrated to respond to a shift in frequency and amplitude between an unflawed specimen and a flawed or unbonded standard. For unbonds less than the diameter of the transducer the instrument response will vary between unflawed and unbonded standard responses. In addition to detecting unbonds or voids this method can detect porosity. Empirical relationships have been shown between Fokker NDE quality readings and shear bond strength and porosity. The Fokker bondtester measures both amplitude and frequency shift to more fully characterize the flaw.

Limitations: This instrument operates in the low kilohertz range with consequent longer wavelength and lower inherent spatial resolution. The test specimen must be manually scanned and flawed areas manually marked.

3. NDT - 210 Bondtester

Advantages: The principle of operation is similar to the Fokker bondtester.

Limitations: This instrument has only an amplitude meter readout and is therefore more restricted than the Fokker bondtester in interpretation of flaw type.

4. Sondicator (detailed description not provided)

5. Shurtronics Harmonic Bondtester

Advantages: Induces an intrasonic wave in an electrically conductive (metallic) substrate through electromagnetic induction (14 - 15 KHz). A change in structural ultrasonic response is detected by a wide-band (28 - 30 KHz) receiving microphone. The instrument is calibrated to read just above zero for good structure and full scale for debonds 1/2 inch in diameter. This method does not require a liquid couplant.

Limitations: This method is limited to metallic structures and has limited sensitivity to first ply debonds.

6. Neutron Radiography

Advantages: Particularly useful when bonding components are not x-ray opaque. Can be used to detect voids and porosity. The hydrogen atoms in water and organic plastics are neutron opaque and therefore can be imaged in terms of intrusion and defects on neutron radiograms.

Limitations: Requires radiographic facility, generally used to supplement x-ray radiography.

7. X-ray Radiography

Advantages: Low voltage (25 to 65 KV) x-ray provides maximum contrast. This method is effective for complex geometries difficult to inspect ultrasonically. The method can be used to detect water intrusion. For honeycomb assemblies radiography is often a primary inspection method. Water intrusion into honeycomb is easily detected.

Limitations: This method is enhanced by use of x-ray opaque materials as adhesive and matrix components.

8. Coin Tap Test

Advantages: Tap testing with a coin or small aluminum rod is useful in locating large voids and disbonds of 1.5 in. (3.7 mm) diameter or larger. The method is applicable for metal-metal or thin skin-honeycomb assemblies.

Limitations: The method is limited to the outer ply disbonds. The method is subjective and may yield variation in test results.

9. Acoustic Emission (Dunegan-301)

Advantages: The method uses a broadband (165 KHz) detector, 50 dB preamplifier, x-y recorder and a hot air gun to generate thermal stress in the part. The method detects wet interface corrosion delamination.

Limitation: To detect wet interface corrosion the detector transducer must be placed over the corroded area. Sensitivity is very dependent on both detector location and location of heat source.

10. Thermography

Advantages: Remote IR camera or liquid crystal coatings construct thermographic maps of a structural part undergoing current mechanical cycling or prior surface heating. The temperature map locates hot spots or cool spots which indicate locations of specific mechanical energy dissipation or thermal diffusivity differences indicative of stress concentrations or structural defects.

Limitations: The method is qualitative and requires physical interpretation by other measurements. The method is not applicable to metal skin laminates due to high thermal diffusivity.

The results of the study of Hagenaier and Fassbender(70) on the ability of NDE methods to detect defects in two types of bonded solid laminates is summarized in the decision matrix of Table 2-6. Each intersection of row and

TABLE 2-6 Decision Matrix Between Nondestructive Evaluation (NDE) and Built-In Defects in Laminate Panels

Direction of Decreasing Correlation:
0 = Defect Not Detected;
1 = Partial Detection;
2 = Detected

Built-In Defects In Laminate Panels	Nondestructive Test (NDT) Method												
	(a) Immersion C-Scan Reflector Plate	(b) 210 Sonic Bond Tester	(c) Immersion C-Scan Through Transmission	(d) Fokker Bondtester	(e) Contact Through Transmission	(f) Contact Pulse-Echo	(g) Immersion C-Scan Pulse-Echo	(h) Sonicator	(i) Harmonic Bond Tester	(j) Neutron Radiography	(k) Coin Tap Test	(l) Low KV X-ray	Row Ave.
(1) Void	2	2	2	2	2	2	2	2	2	2	1	0	1.75
(2) Void (C-14 repair)	2	2	2	2	2	2	2	2	2	2	1	0	1.75
(3) Void (9309 repair)	2	2	2	2	2	2	2	2	2	2	1	0	1.75
(4) Corroded Bond	2	2	2	2	2	2	1	1	1	2	1	2	1.67
(5) Lack of Bond (skin to adhesive)	2	2	2	2	2	2	2	2	2	0	1	0	1.58
(6) Porous Adhesive	2	2	2	0	2	2	2	0	2	2	0	2	1.50
(7) Manufacturer's Separator Sheet	2	1	2	2	1	0	2	2	0	1	0	0	1.08
(8) Burned Adhesive	2	2	2	2	2	2	1	0	0	0	0	0	1.08
(9) Thick Adhesive (1, 2, 3 ply)	2	2	1	2	0	0	0	0	0	0	0	0	0.58
Col. Ave.	2.00	1.89	1.89	1.78	1.67	1.56	1.56	1.22	1.22	1.22	0.56	0.44	

column in Table 2-6 is given a score which is 2 = defect detected, 1 = defect partially detected, and 0 = defect undetected. Nine types of defects which describe the rows of Table 2-6 are ranked relative to their row averaged scores decreasing from top to bottom. Twelve types of NDE methods which form the columns of Table 2-6 are ranked by decreasing column average score from left to right across the table. The best combination for high reliability flaw detection occurs in the upper left region of Table 2-6. The lower reliability sector of NDE methodology is identified in the lower right region of Table 2-6.

Both Table 2-6 and Table 2-7 show that voids are the most reliably detected type of defect. Thickness variations, and solid inclusions (separation sheet) are the least detectable types of defect in both solid and hollow core laminates. Dramatic new information is provided in intercomparing the scores of test methods in the solid laminate (Table 2-6) and skin stressed laminate (Table 2-7). For the solid laminate the ultrasonic test methods receive highest scores and display highest NDE reliability. For the skin stressed laminate (Table 2-7) neutron radiography and the coin tap test receive the highest detection rating. These studies clearly show that the relibility of an NDE method depends on both the type of structure (solid or hollow core) and the type of defect (void or inclusion) being studied. The decision matrix format provides a convenient means of identifying high reliability NDE methodologies with regard to test method, flaw type, and type of composite structure.

2-7. Surface NDE

Standard NDE methods as reviewed in the previous section, are not capable of defining poor interface quality which may lower the durability of a laminated composite structure. A new and rapidly developing area of surface NDE has recently emerged to fill this important gap in standard NDE methodology. Surface NDE methods are in general modifications of the tools of surface characterization to permit automation, rapid surface property mapping, and computerized data storage and processing. The objective of surface NDE is to perform a final inspection of surfaces to be bonded and to make accept-reject decisions on whether the surface will form a reliable-durable bonded joint or whether rejection and recycling through surface treatment is required.

The proceedings of a recent symposium on surface contamination edited by Mittal[73] provides an overview and detailed summaries of progress in this important emerging field of surface NDE. Very specific discussions of surface NDE are developed in this review.[74] In general, surface NDE falls into direct methods which directly identify the nature of the surface contaminant and indirect methods which identify contaminants through a surface property change.

TABLE 2-7 Decision Matrix Between Nondestructive Evaluation (NDE) and Defects in Honeycomb Structures

Direction of Decreasing Correlation:
0 = Defect Not Detected;
1 = Partially Detected;
2 = Detected

Built-In Defects In Honeywell Structure	Nondestructive Test (NDT) Method												
	(a) Neutron Radiography	(b) Coin Tap Test	(c) Contact Through Transmission	(d) Immersion C-Scan Pulse-Echo	(e) Immersion C-Scan Through Transmission	(f) Fokker Bondtester	(g) Low KV X-Ray	(h) Harmonic Bond Tester	(i) 210 Sonic	(j) Sondicator	(k) Contact Pulse-Echo	(l) Contact Shear Wave	Row Ave.
(1) Void (Foam to Closure)	2	2	2	2	2	2	2	2	2	2	2	0	1.83
(2) Void (Adhesive to Skin)	2	2	2	2	2	2	2	2	2	2	2	0	1.83
(3) Inadequate Tie-In of Foam to Core	2	2	2	0	2	2	2	2	2	2	0	0	1.50
(4) Void (Adhesive to Core)	2	2	2	2	2	0	2	2	0	0	0	0	1.17
(5) Separator Sheet (Skin to Adhesive)	2	2	2	2	2	0	0	1	0	0	1	0	1.00
(6) Water Intrusion	2	2	2	0	2	0	2	0	0	0	0	0	0.83
(7) Crushed Core (After Bonding)	1	2	1	1	1	1	2	1	0	0	0	0	0.83
(8) Inadequate Foam Depth At Closure	2	1	0	0	0	2	2	0	1	2	0	0	0.83
(9) Separator Sheet (Adhesive to Core)	2	2	0	2	1	0	0	0	0	0	2	0	0.75
(10) Chem-Mill Step Void	2	0	0	0	0	0	0	0	0	0	0	0	0.17
Col. Ave.	1.90	1.60	1.30	1.30	1.20	1.10	1.00	1.00	0.9	0.8	0.7	0	

Direct methods for surface NDE have been reviewed under surface chemical analysis and process monitoring. Indirect methods measure a surface property change which correlates with a specific class of surface degradation which lowers bond reliability and durability. Of the several indirect surface NDE measurements the following four methods have complementary advantages and limitations.

1. Ellipsometry[73,77]

Advantages: The method is noncontacting and nondestructive. A beam of polarized monochromatic light is reflected from the surface. The phase shift of the reflected polarized light and reflection coefficients are analyzed to measure surface roughness, contaminant film thickness (from 0.0 nm to 500 nm), and optical properties of the contaminant film. This method is automated and developed for rapid computer controlled surface mapping.

Limitations: Sensitivity is limited by the difference in refractive index of film and substrate. Maintaining the proper angle of incidence may require Z-Y-Z indexing for curved surfaces.

2. Surface Potential Difference (SPD)[73-79]

Advantages: This method is noncontacting and nondestructive. SPD is the difference between the work function of the test surface and a reference electrode and is extremely sensitive to the outer dipole layer of surface contamination. Commercial NDE instruments (Fokker contamination tester and Monroe Electronics ISO Probe) are available and computerized surface mapping has been developed.

Limitations: This method requires other measurements to make a physical interpretation of data. Electrode contamination and capacitance gap misalignment can affect the measurement.

3. Photoelectron Emission (PEE)[73-76]

Advantages: The method is sensitive to both substrate and surface film photoemission properties. It is extremely sensitive to thickness effects of electron attenuating contaminants. Methods for automation and surface mapping have been developed.

Limitations: Method requires an intense UV, (250 nm) light source. High sensitivity requires differences in photoemission properties of substrate and contaminant.

4. Surface Remission Photometry (SRP)

Advantages: Remission photometry permits surface spectral analysis at 200 - 800 nm wavelengths. The influence of surface roughness is small. The test surface is compared to a reference to minimize nonlinear spectral sensitivity of the photometer.

Limitations: Requires use of a light integrating sphere and twin beam optics. This method remains to be automated and computerized for rapid surface NDE mapping.

The combination of surface NDE and surface chemical analysis (see Table 2-5) promises to provide a valid approach to reliability and durability analysis of structural adhesive bonding. At the moment surface NDE is an emerging technology which needs further development and integration of measurement and analysis methodologies to provide quantitative reliability and durability predictions.[23]

2-8. Performance and Proof Testing

Performance and proof testing of composite reliability by standard ASTM methods involves a group of 47 test methods as summarized in Part 6 of Table 2-1. These ASTM test methods fall into six categories of response which are:[2]

1. Processing
2. Mechanical Properties
3. Thermal Properties
4. Electrical Properties
5. Optical Properties
6. Environmental Properties

A current and comprehensive set of brief descriptions of the advantages and limitations of these test methods are available.[2,80] Rank ordered summaries of commercial prepreg and composite laminate properties as measured by these test methods are organized in tables for convenient reference and use in materials selection.[2] The great importance of these ASTM performance and proof tests are very largely related to the fact that they are commonly accepted and utilized thus providing a common fund of characterization data. Experience shows that ASTM tests used alone are an expensive and generally inadequate means of testing for composite reliability and durability. On the other hand, extensive quantitative characterization without ASTM testing provides a data base without a general technology reference. The appropriate solution is, of course, to design a test program based upon appropriate selection of the six test categories of Table 2-2 which fulfills the requirements of the generalized predictive design methodology shown in Fig. 2-1. Based on the earlier discussion of physical states and transitions the minimum aim of performance and proof testing

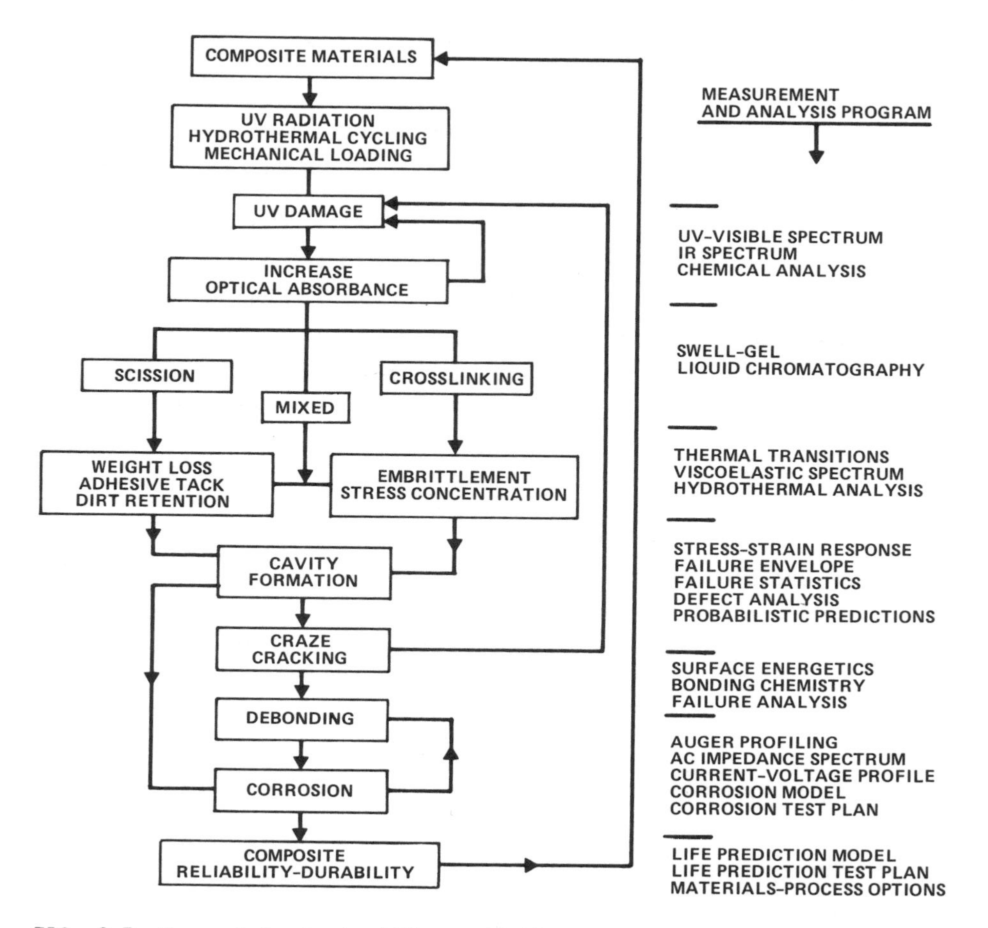

FIG. 2-7 General laminate life prediction program.

is to verify the predictions generated from chemical analysis and nondestructive evaluation. The detailed listing of significant characterized properties proposed in Table 2-3 also summarizes the critical types of properties required from performance and proof testing. A full understanding of the stress and environment dependent glass temperature T_g and flow temperature T_m is essential in terms of performance and proof test conditions. The connection of T_g and T_m characterization by ASTM methods and micro characterization techniques form an important logic link between scientific characterization and end-use testing. The utilization of standard ASTM tests to establish the isothermal stress-strain-time response is essential to fully define the mechanical responses graphically outlined in Figs. 2-4 through 2-6 in end-use related test geometries. Finally, the statistical distributions of strength, extensibility, and fracture energy are required for defining statistical margins of safety in stress analysis and strength analysis of predict reliabilty and durability. Section 7 of

Table 2-1 indicates four research programs which are continuing to advance the state-of-the-art for durability analysis and service life prediction. More detailed development of chemically based models for computer aided design and manufacture (CAD/CAM) of advanced composites is part of the continuing research program being conducted as an extension of this overview. In place of a redundant review of ASTM performance proof test methods the reader is referred to the detailed and continuously up-dated references on this important subject particularly developed for composite materials with commercial source and compiled test data.[1,2]

The descending flow chart of Fig. 2-7 is an outline of composite material interactions under UV radiation, hydrothermal cycling, and mechanical loading which leads to property changes, damage mechanisms, and results in diminished reliability and durability. The types of chemical, physical, and mechanical tests which accompany the several stages of degradation are listed on the right side of Fig. 2-7. While still in an early development stage a general laminate life prediction program which combines measurement and analysis appears to form an important avenue for future research. An important subject presented as a central issue in lower Fig. 2-7 is chemical corrosion degradation within the composite system. While this subject is only recently receiving detailed attention in composite life studies it appears to warrant increased attention in life prediction modelling.

References

1. "Commercial Names and Sources for Plastics and Adhesives," Cordura Publications, San Diego (1980).

2. "Films, Sheets and Laminates, Desk Top Data Bank," Cordura Publications, San Diego (1980).

3. F.W. Billmeyer, Textbook of Polymer Science, Interscience, New York, 1962.

4. D.H. Kaelble, Physical Chemistry of Adhesion, Wiley-Interscience, New York (1971).

5. C.A. May (Editor), Resins for Aerospace, ACS Symposium Series 132, 1980.

6. Symposium on Physical Characterization of Macromolecules, ACS Organic Coatings and Plastics Chemistry Preprints, 44, (April 1981), pp. 491-798.

7. "Effect of Service Environment on Composite Materials," AGARD Conference Proceedings No. 288, 50th Meeting of the AGARD (NATO Advisory Group for Aerospace Res. and Dev.) Structures and Materials Panel, Athens, Greece, April 1980 (Report B-005777 Defense Technical Information Center, Alexandria, VA).

8. J.K. Gillham, in Ref. 6, pp. 185-191.

9. D.S. Dugdale, J. Mech. Phys. Solids, 8, 100 (1960); see also Ref. 6, pp. 309-317.

10. L.J. Hart-Smith, "Adhesive-Bonded Single-Lap Joints," NASA Langley Report CR-112236, January 1973.

11. J.F. Carpenter and T.T. Bartels, "Characterization and Control of Composite Prepregs and Adhesives," Proc. 7th National SAMPE Conf., Vol. 7, (1975), pp. 43-52.

12. C.A. May, T.E. Helminiak, and H.A. Newey, "Chemical Characterization Plan for Advanced Composite Prepregs," Proc. 8th National SAMPE Conf., Vol. 8, (1976), pp. 274-294.

13. W.B. Alston, "Characterization of PMR-15 Polyimide Resin Composition in Thermo-oxidative Exposed Graphite Fiber Composites," AVRDCOM Tech. Report. 80-C-10, Lewis Research Center (1980).

14. S. Eddy, M. Lucarelli, W. Jones, and L. Picklesimer, "An Evaluation of an Acetylene Terminated Sulfone Oligimer," Organic Coatings and Plastics Chemistry Preprints, American Chem. Soc., 42 (1), (March 1980), p. 502-508.

15. D.H. Kaelble, in Ref. 5, pp. 395-417.

16. D.H. Kaelble and P.J. Dynes, "Preventative Nondestructive ERvaluation (PNDE) of Graphite Epoxy Composites," Ceramic Eng. and Sci. Proc., 1 (1980), pp. 458-472.

17. N. Schneider (Editor), Polymer Composites, 1(2) (1980), pp. 65-109.

18. L.R. Snyder and J.J. Kirklund, Introduction to Modern Liquid Chromatography, Wiley - Interscience, New York (1974).

19. H.M. McNair, American Laboratory, May 1980, pp. 33-34.

19a. G. Gillberg and D. Kemp, "Surface Characterization of Polyester Fibers," J. Appl. Poly. Sci., 26, (1981), pp. 2023-2051.

20. Ref. 6, pp. 611-622.

21. Ref. 6, pp. 581-603.

22. "Instrumentation 81," Chem. and Eng. News, (March 23, 1981), pp. 43-74.

23. Ref. 6, pp. 623-639.

24. J.L. Koenig, "Application of Fourier Transform Infrared Spectroscopy to Chemical Systems," Applied Spectroscopy, 29 (4), (1975) pp. 293-308.

25. P. Griffiths, Fourier Transform Infrared Spectroscopy, Wiley, New York (1975).

26. Ref. 6, pp. 553-581, 705-706.

27. J. Schaefer and E.O. Stejskal in "Topics in Carbon-13d NMR Spectroscopy," Vol. 3, (Editor: G.C. Levy), Wiley, New York, (1979), pp. 284-324.

28. H. Kahn and W. Slavin, "Atomic Adsorption Analysis," Int. Science and Technology, November 1962.

29. "New Methods Probe Surface Phenomena," Chem. and Eng. News, (Sept. 22, 1980), pp. 27-30.

30. Ref. 6, 192-4, 540-552, 641-651.

31. D.H. Kaelble, Physical Chemistry of Adhesion, Wiley, New York, 1971, Chap. 5.

32. G.E. Hammer and L.T. Drysal, "Graphite Fiber Surface Analysis by X-ray Photoelectron Spectroscopy and Polar/Dispersive Free Energy Analysis," Appl. of Surface Sci., 4, (1980) pp. 340-355.

34. W.M. Riggs and M.J. Parker, in "Methods of Surface Analysis," Editor: C.W. Czanderna, Elsevier, New York, (1975) Chap. 4.

35. M.M. Millard in "Characterization of Metal and Polymer Surfaces," Editor: L-H. Lee, Academic Press, New York, 1977, p. 86.

36. ASTM Test Methods.

37. H.A. Willis and V.J. Zichy, in "Polymer Surfaces," Editors: D.T. Clark and W.J. Feast, Wiley, New York, (1978), Chap. 15.

38. G.W. Urbanzyk, J. Poly Sci., Polymer Symposium 58, 311 (1977).

39. J.P. Siblia in "Surface Characteristics of Fibers and Textiles," Editor: M.J. Schick, Dekker, New York, (1975), Part 1, Chap. 8.

40. G.K. Wehner, in "Methods of Surface Analysis," Editor: C.W. Czanderna, Elsevier, 1975, Chap. 1.

41. I. Luderwald and H. Urrutia, in "Analytical Pyrolysis," Editors: C.E. Roland and C.A. Cramer, Elsevier, (1977), p. 139.

42. M. Delhaye and M. Leclereq, Industrial Res., 19, (1977) p. 69.

43. P. Dhamelincourt, F. Wallert, M. LeClereq and D.O. Landon, "Laser Raman Molecular Microprobe," Anal. Chem., 51, 1978, p. 414A.

44. A. Mehta, R.C. Bopp, D. Gaur, and B. Wunderlich, J. Thermal Anal., 13, (1978), p. 197.

45. Ref. 6, pp. 652-661, 670-679, 396-401.

46. L.T. Dryzal, J.A. Mescher and D.L. Hall, "The Surface Composition and Energetics of Type HM Graphite Fibers," Report No. AFWAL-TR-80-4030, Air Force Wright Aeronautical Laboratories, (March 1980).

47. L.T. Dryzal, M.J. Rich, J.D. Camping, and W.J. Park, "Interfacial Shear Strength and Failure Mechanisms in Graphite Fiber Composites," Report No. AFWAL-TR-81-4003, Air Force Wright Aeronautical Laboratories, (January 1980).

48. J.F. Mandell, J.H. Chen, F.J. McGarry, "A Microdebonding Test for In Situ Assessment of Fiber/Matrix Bond Strength in Composite Materials," Int. J. Adhesion, 1, (1980), pp. 40-44.

49. D.H. Kaelble, "Interface Degradation Processes and Durability," Poly Eng. and Sci., 17, (1977), pp. 474-478.

50. Ref. 6, pp. 185-191, 402-408, 492-539.

51. G.R. Thomas, B.M. Halpin, J.F. Sprouse, G.L. Hagnauer, and R.E. Sacher, "Characterization of Epoxy Resins Prepregs and Composites Using HPLC FTS-IR and DSC," Proc. 24th National SAMPE Symposium, (1979), pp. 458-505.

52. B.M. Halpin, J.F. Sprouse, and G.L. Hagnauer, "Characterization of Epoxy Resins, Prepregs, and Composites Using HPLC and FTS-IR," Proc. 33rd Annual Tech. Conf., Reinforced Plastics/Composites Institute, SPI Inc. (1978).

53. G.L. Hagnauer and I. Setton, "Compositional Analysis of Epoxy Resin Formulations,: J. of Liquid Chromatography, 1(1), (1978), pp. 55-73.

54. G.L. Hagnauer and D.A. Dunn, "Dicyanamide Analysis and Solubility in Epoxy Resins," J. Appl. Poly. Science (1981), In Press.

55. G.L. Hagnauer and D.A. Dunn, "Quality Assurance of an Epoxy Prepreg Using HPLC,"Proc. 12th National SAMPE Technical Conf. (October, 1980), pp. 648-655.

56. G.L. Hagnauer and D.A. Dunn, "HPLC-A New Reliability Tool for Epoxy Resin Prepreg Analysis,": Ind. and Eng. Chem. Product R&D, (Feb. 1981).

57. J.F. Sprouse, B.M. Halpin, Jr., and R.E. Sacher, "Cure Analysis of Epoxy Composites Using Fourier Transform Infrared Spectroscopy," AMMRC-TR-78-45, Army Materials and Mechanics Reserach Center (November 1978).

58. G.L. Hagnauer, J.M. Murray and B.M. Bowse, "HPLC Monitoring of Graphite-Epoxy Prepreg Aging," AMMRC TR-79-33, Army Materials and Mechanics Research Center (May, 1979).

59. G.L. Hagnauer, "HPLC and GPC Analysis of Epon 828 Epoxy Resins," AMMRC TR-79-59, Army Materials and Mechanics Research Center, (November 1979).

60. G.L. Hagnauer, J.F. Sprouse, R.E. Sacher, I. Setton and M. Wood, "Evaluation of New Techniques for the Quality Control of Epoxy Resin Formulations," AMMRC TR-78-8, Army Materials and Mechanics Research Center, (January, 1978).

61. N.G. McCrum, B.E. Read, G. Williams, "Anelastic and Dielectric Effects in Polymer Solids," Wiley, New York (1967).

62. P. Hedwig, "Dielectric Spectroscopy of Polymers," Halstead-Wiley, New York, (1977).

63. Ref. 6, pp. 402-408, 492-539.

64. G.W. Lawless, "Dielectric and Phaseometric Monitoring of Adhesives," Adhesives Age (April, 1981) pp. 26-29.

65. J. Chottiner, Z.N. Sanjana, M.R. Kodani, K.W. Lengel, and G.B. Rosenblatt, "Monitoring Cure of Large Autoclave Molded Parts by Dielectric Analysis," Proc. 26th Nat. SAMPE Symposium, 26, (April 1981), pp. 65-76.

66. N.F. Sheppard, S.L. Garverick, D.R. Ray, and S.D. Senturia, "Microdielectrometry: A New Method for In Situ Cure Monitoring," Proc. 26th National SAMPE Symposium, 26, (April, 1981), pp. 65-76.

67. R. Hinricks, "Interactive Computer Process System for Composite Autoclave Fabrication," Critical Review: Techniques for the Characterization of Composite Materials," Mass. Inst. Tech. (June 1981).

68. J.F. Martin, "An Automated Ultrasonic Testbed: Application to NDE in Graphite-Epoxy Materials," Proc. 26th National SAMPE Symposium, 26, (April 1981), pp. 12-23.

69. G.A. Alers, "Applications of Electromagnetic Acoustic Transducers," Ibid., pp. 34-44.

70. D. Hagemeier and R. Fassbender, "Nondestructive Testing of Adhesive Bonded Structure," SAMPE Quarterly, 9, (July 1978), pp. 36-58.

71. H.T. Clark, "Definition and Nondestructive Detection of Critical Adhesive Bond-Line Flaws," Air Force Flight Dynamics Laboratory, Report No. AFFDL-TR-78-108 (July, 1978).

72. T. Smith, "NDE Method for Characterizing Anodized Al Surfaces," Air Force Materials Laboratories Report No. AFML-TR-78-146 (January, 1979).

73. K.L. Mittal, "Surface Contamination-Genesis Detection and Control," Vols. 1 and 2, Plenum Press, New York (1979).

74. Ref. 73, Vol. 1, pp. 3-45, Vol. 2, pp. 697-895.

75. T. Smith, "Surface Tools for Automated Non-Destructive Inspection of Contamination," Surface Technology, 9, (1979), pp. 1-29.

76. T. Smith and R.L. Crane, "NDE and Effect of Contamination and Process Errors on Bond Strength and Durability," Proc. 25th National SAMPE Symposium, 25, (May 1980), pp. 25-38.

77. Ref. 74, pp. 697-712, 749-768.

78. Ref. 74, pp. 697-712, 723-748.

79. Ref. 74, pp. 716-721.

80. ASTM Test Methods, American Society of Testing and Materials, Philadelphia, 1980.

3

Relations Between Atomic and Molecular Properties

3-1. Elementary Properties

Computer aided methods are presented for estimating the energies of ionic-covalent chemical bonding, properties of the molecular crystalline state, and acid-base properties, surface interaction, and molecular dipoles. The primary objective of this atomic to molecular properties model is to provide direct numeric estimation of the chemical, environmental, and corrosion stability of metals and their oxides and protective encapsulants. The analysis of ionic-covalent bonding includes chemical bond energy, percent ionic character, and bond lengths. The molecular analysis includes calculation of heats of formation, molecular weight, and molar and specific volumes. The analysis of acid-base interactions includes surface isoelectric point (in pH units), Coulomb adsorption energies, and surface dipole moments. This analysis includes new estimates for the bond energies of metallic elements. The adhesion and hydrophobic or hydrophillic properties of metal oxides are related to their acid-base character and consequent orientation effects on chemisorbed films.

Extensive tabulations of the chemical and physical properties of metallic and polymeric materials are limited to the major classes of commercially important materials. Very often the chemist or materials specialist wishes to obtain a qualitative estimate of corrosion or environmental durability which is developed directly from the chemical structure of the material. In this discussion, the homonuclear bond energies of metals are estimated from their heats of vaporization and combined with the Pauling values of single bond energy for nonmetallic elements. The table of elemental properties is stored in computer memory and automatically retrieved by the computer program for estimating the chemical and physical response of chemical compounds.

The elemental properties for the 67 elements included in this study are listed in Table 3-1. The first column of Table 3-1 lists a code number which is employed as a retrieval number in the computer program. The second and third columns of Table 3-1 identify the atomic number and chemical symbol of the element. The right seven columns of Table 3-1 tabulate the elemental properties utilized in the computation. From left to right these properties are defined as follows:

W = atomic weight
D = single (2 electron) bond energy
X = electronegativity
R = atomic covalent radius
V = most stable valence
MV = maximum valence
S = MV/R = the ionic potential relating to acid-base properties.

All of the above properties of the elements are taken from a recent version of the periodic table[1] with the exception of the single bond energy D. The single bond energies for the nonmetallic elements in Table 1 are based on values reported by Pauling.[2] The apparent bond energies for the metals are obtained by a modification of a convention applied by Vijh[3] in the following relation:

$$D(M\text{-}M) = \frac{2\Delta H_S}{C.N.} \qquad (3\text{-}1)$$

where D(M-M) is the metal-metal bond energy, ΔH_S is the heat of sublimation (sum of the heats of melting and vaporization), and C.N. is the maximum lattice coordination number of the metal. The factor 2/C.N. in Eq. (1) corrects for the fact that each bond would otherwise be counted twice in the simple summing of all nearest neighbor interactions. Inspection of the bond energies listed by Vijh[3] indicates what appear to be values with too low a magnitude for metal bond energies.

An alternative form of Eq. (3-1) can be written to incorporate the cohesive energy density δ^2 and molar volume V_m of metals as follows:

$$D(M\text{-}M) = \frac{2\,\delta^2 V_m}{C.N.} \qquad (3\text{-}2)$$

where extensive listings of both δ^2 and V_m for metals are compiled by Hildebrand.[4] A comparative listing of single bond energies for ten elements is shown in Table 3-2 where the first column summarizes single bond energies of Pauling.[2] The second column provides calculated values from Eq. (3-2) for an assumed maximum coordination number C.N. = 12. The notable feature of Table 3-2 is that the Pauling values of bond energy are higher than the estimates of Eq. (3-1) or Eq. (3-2) by a nearly constant multiplier factor which appears in the right column of Table 3-2. This result implies that a chemical coordination number of C.N. = 3 to 4 is more appropriate in calculating chemical bond energies by Eq. (2).

This result was applied to provide the calculated values of metal bond energies in Table 3-1. Where Pauling values of bond energy were not available the cohesive energy density δ^2 and molar volume V_m data of Hildebrand were applied in Eq. (2) with a chemical coordination number C = 3 for elements of Groups IA-VIA and C = 4 for elements of Groups VIIA-VIIIA and IB-VB of the revised periodic table.[1] These provisional values of metal single bond energy are parenthesized in Table 3-1 to distinguish them from the Pauling reference values.

3-2. Ionic-Covalent Bonding

Having a complete listing of homonuclear bond properties as presented in Table 3-1 permits application of the Pauling relations for calculation of heteronuclear single bond energy D(A-B) and percentage ionic character (%I). The ionic component of single bond energy I(A-B) is defined by the following standard relation:[2]

$$\Delta_{AB}\ (\text{kJ/mol}) = 96.5\ (X_A - X_B)^2 \qquad (3\text{-}3)$$

The total single bond energy D_{AB} between atoms A and B is defined as the sum of the covalent and ionic bond energies by the following standard relation:[2]

$$D_{AB}\ (\text{kJ/mol}) = 0.5(D_A + D_B) + \Delta_{AB} \qquad (3\text{-}4)$$

The percent ionic character I(%) of the single bond energy is calculated from the ratio of Eq. (3-1) and Eq. (3-2) as proposed by Kaelble:[5]

$$I(\%) = \frac{100\ \Delta_{AB}}{D_{AB}} \qquad (3\text{-}5)$$

TABLE 3-1 Properties of the Elements

Code No.	Z	SY	W G/Mole	D 10^5 J/Mole	X	R 10^{-10} m	V	MV	S
1	1	H	1.008	4.35	2.20	0.32	1	1	3.13
2	3	LI	6.941	1.11	0.98	1.23	1	1	0.81
3	3	BE	9.012	(2.28)	1.57	0.90	2	2	2.22
4	5	B	10.81	(2.53)	2.04	0.82	3	3	3.66
5	6	C	12.01	3.48	2.55	0.77	4	4	5.19
6	7	N	14.01	1.61	3.04	0.75	3	5	6.67
7	8	O	16.00	1.39	3.44	0.73	2	2	2.74
8	9	F	19.00	1.53	3.98	0.72	1	1	1.39
9	11	NA	22.99	0.753	0.93	1.54	1	1	0.65
10	12	MG	24.31	(0.971)	1.31	1.36	2	2	1.47
11	13	AL	26.98	(2.06)	1.61	1.18	3	3	2.54
12	14	SI	28.09	1.77	1.90	1.11	4	4	3.60
13	15	P	30.97	2.15	2.19	1.06	5	5	4.72
14	16	S	32.06	2.13	2.58	1.02	6	6	5.88
15	17	CL	35.45	2.43	3.16	0.99	1	7	7.07
16	19	K	39.09	0.552	0.82	2.03	1	1	0.49
17	20	CA	40.08	(1.15)	1.00	1.74	2	2	1.15
18	21	SC	44.96	(2.58)	1.36	1.44	3	3	2.08
19	22	TI	47.90	(2.64)	1.54	1.32	4	4	3.03
20	23	V	50.94	(3.36)	1.63	1.22	5	5	4.10
21	24	CR	52.00	(2.38)	1.66	1.18	3	6	5.08
22	25	MN	54.94	(1.43)	1.55	1.17	2	7	5.98
23	26	FE	55.85	(2.03)	1.83	1.17	3	3	2.56
24	27	CO	58.93	(2.20)	1.88	1.16	2	3	2.59
25	28	NI	58.70	(2.12)	1.91	1.15	2	3	2.61
26	29	CU	63.55	(1.72)	1.90	1.17	2	2	1.71
27	30	ZN	65.38	(0.653)	1.65	1.25	2	2	1.60
28	31	GA	69.72	(1.36)	1.81	1.26	3	3	2.38
29	32	GE	72.59	1.57	2.01	1.22	4	4	3.28
30	33	AS	74.92	1.34	2.18	1.20	3	5	4.17
31	34	SE	78.96	1.84	2.55	1.16	4	6	5.17
32	35	BR	79.90	1.93	2.96	1.14	1	7	6.14
33	37	RB	85.47	0.519	0.82	2.16	1	1	0.46
34	38	SR	87.62	(1.05)	0.95	1.91	2	2	1.05
35	39	Y	88.91	(2.74)	1.22	1.62	3	3	1.85
36	40	ZR	91.22	(3.45)	1.33	1.45	4	4	2.76
37	41	NB	92.91	(4.85)	1.60	1.34	5	5	3.73

TABLE 3-1 (Continued)

Code No.	Z	SY	W G/Mole	D 10^5 J/Mole	X	R 10^{-10} m	V	MV	S
38	42	MO	95.94	(4.30)	2.16	1.30	6	6	4.62
39	43	TC	98.0	(3.35)	1.90	1.27	7	7	5.51
40	44	RU	101.07	(3.35)	2.20	1.25	3	8	6.40
41	45	RH	102.91	(3.24)	2.28	1.25	3	4	3.20
42	46	PD	106.4	(1.93)	2.20	1.28	2	4	3.13
43	47	AG	107.87	(1.44)	1.93	1.34	1	1	0.75
44	48	CD	112.41	(0.552)	1.69	1.48	2	2	1.35
45	49	IN	114.82	(1.18)	1.78	1.44	3	3	2.08
46	50	SN	118.69	1.43	1.96	1.41	4	4	2.84
47	51	SB	121.75	1.26	2.05	1.40	3	5	3.57
48	52	TE	127.60	1.38	2.10	1.36	4	6	4.41
49	53	I	126.90	1.51	2.66	1.33	1	7	5.26
50	55	CS	132.91	0.448	0.79	2.35	1	1	0.43
51	56	BA	137.33	(1.12)	0.89	1.98	2	2	1.01
52	57	LA	138.91	(2.48)	1.10	1.69	3	3	1.78
53	72	HF	178.49	(4.72)	1.30	1.44	4	4	2.78
54	73	TA	180.95	(5.56)	1.50	1.34	5	5	3.73
55	74	W	183.85	(5.61)	2.36	1.30	6	6	4.62
56	75	RE	186.21	(3.97)	1.90	1.28	7	7	5.47
57	76	OS	190.2	(3.64)	2.20	1.26	4	8	6.35
58	77	IR	192.22	(3.48)	2.20	1.27	4	6	4.72
59	78	PT	195.09	(2.79)	2.28	1.30	4	4	3.08
60	79	AU	196.97	(1.86)	2.54	1.34	3	3	2.24
61	80	HG	200.59	(0.301)	2.00	1.49	2	2	1.34
62	81	TL	204.37	(0.866)	2.04	1.48	1	3	2.03
63	82	PB	207.2	(0.992)	2.33	1.47	2	4	2.72
64	83	BI	209.0	(1.03)	2.02	1.46	3	5	3.42
65	90	TH	232.04	(3.42)	1.30	1.65	4	4	2.42
66	92	U	238.03	(3.56)	1.38	1.42	6	6	4.22
67	94	PU	244.0	(2.29)	1.28	1.21	4	6	4.96
68	7	(N2)/2	14.01	4.73	3.04	0.55	3	5	6.67
69	8	(O2)/2	16.00	2.01	3.44	0.62	2	2	2.74

TABLE 3-2 Comparison of Single Bond Energies

Element	Group	Single Bond Energy (kcal/mol) Ref. 2	Eq. (2) (C.N.=12)	Ratio
Lithium	IA	25.6	6.3	4.06
Sodium	IA	18	4.3	4.19
Potassium	IA	13.2	3.3	4.00
Rubidium	IA	12.4	3.4	3.65
Cesium	IA	10.7	3.0	3.57
				3.89 ± 0.27
Boron	IIIB	25.0	15.1	1.66
Germanium	IVB	37.6	13.2	2.85
Arsenic	VB	32.1	9.5	3.38
Tin	IVB	34.2	11.4	3.00
Antimony	VB	30.2	10.6	2.85
				2.75 ± 0.65

The more standard approach is to define the percent ionic character of the single bond energy purely from electronegativity differences by the following standard relation:[1,2]

$$I(\%) = 100 \{1 - \exp [-0.25 (X_A - X_B)^2]\}$$

The obvious deficiency of the above expression is, of course, that no information on total bond energy is contained in the relation. A comparison of experimental and calculated values of %I for 18 diatomic compounds with experimental values of I(%) = 4 to 92 shows that Eq. (3-5) provides an improved agreement with experiment.[5] Consequently, Eq. (3-5) is incorporated in the computer program of this analysis.

The calculation of single bond length L_{AB} utilizes the standard Shoemaker-Stevenson relation as follows:[2,6]

$$L_{AB}\ (10^{-10}\text{m}) = R_A + R_B - 0.09\ |X_A - X_B| \qquad (3\text{-}6)$$

The absolute difference in electronegativities is shown by Eq. (3-6) to reduce the single bond length relative to purely covalent bonding. Values for both covalent radii R and electronegativity are tabulated in Table 3-1.

Due to the importance of molecular nitrogen N_2 and molecular oxygen O_2 in many chemical reactions, the atomic properties of these molecules are described after the elements as Code No. 68 and 69. For the purpose of elemental computations, the triple bond energy of N_2 of D_{N2} = 946 kJ/mol is divided by 2 for the elemental constituent $(N_2)/2$ shown in Table 3-2 and the elemental radius is taken as one-half the triple bond length $L_{N2} = 1.10 \cdot 10^{-10}$ m. Equivalently, the elemental bond energy and radius of $(O_2/2)$ is taken as one half the double bond energy $D(O_2)$ = 404 kJ/mol and bond length $L(O_2) = 1.24 \cdot 10^{-10}$ m. This convention makes allowance for the loss of resonance energy and increase in bond length on conversion to singly bonded states.

3-3. Molecular Analysis

A computer program has been written to provide a simple chemical and physical analysis of chemical compound structures. The program first inputs a description of the compound in terms of the number of moles N_i of each of the elements, which is equivalent to writing the empirical chemical formula. The program then inputs a complete description of the number of moles N_{AB} of each of the chemical bonds and the A and B elements for each bond, which is equivalent to writing a chemical structure formula. The individual bond properties are calculated by means of Eq. (3-1) through Eq. (3-4). The total bond energy is obtained by summation:

$$U_T = \sum N_{AB} D_{AB} \tag{3-7}$$

The molecular weight is determined by a summation of the elemental constituents with the following expression:

$$W_T = \sum N_i W_i \tag{3-8}$$

A defect free lattice formed by N moles of homogeneous atomic units all with uniform chemical bond length L_{AA}, has a volume V which is defined by the following relation:[5]

$$V = \frac{N_o \, (N_i L_{AA}^3)}{C}$$

where $N_o = 6.023 \cdot 10^{23}$ is Avogadro's number and C is a lattice packing factor which varies with lattice type as shown in Table 3-3. The volume of a lattice with a mixture of chemical bond lengths will be approximated by summing the partial volumes for each fraction of bond types. An appropriate expression for

TABLE 3-3 Lattice Types and Packing Factors

Lattice Type	Coordination Number	Packing Factor (C)
Face centered cubic	12	1.414
Body centered cubic	8	1.299
Simple cubic	6	1.000
Tetrahedral	4	0.650

calculating a compound lattice of mixed elements and chemical bonds is given as follows:

$$V_m = \frac{N_o \sum N_i}{C} \frac{\sum (N_{AB} L_{AB}^3)}{\sum N_{AB}} \tag{3-9}$$

and the specific volume V_s is computed by the ratio of Eqs. (3-8) and (3-9) as follows:

$$V_s = V_m / W_T \tag{3-10}$$

The cross plot of homonuclear bond energy D_{AA} vs elemental electronegativity shown in Fig. 3-1 provides a summarizing illustration of the distribution of the elemental properties from Table 3-1 relative to ionic-covalent bonding.

The data displayed in Fig. 3-1 is referenced to the atomic properties of hydrogen by the dashed horizonal and vertical curves. Elements with electronegativity less than $X < 2.2$ are metallic, and elements with $X > 2.2$ are semimetallic or nonmetallic. Chemical compounds formed between metallic and nonmetallic elements, as indicated by Eq. (3-3), become increasingly ionic with the increased separation of X_A and X_B; this property is easily visualized by the horizontal separation of data points in Fig. 3-1. The covalent bond energies D_{AA} shown on the vertical axis of Fig. 3-1 define many properties of the metallic elements such as melting point, elastic modulus, hardness, and abrasion resistance.[3] As shown in Fig. 3-1, only three refractory metals, tantalum Ta, niobium Nb, and tungsten W, display D_{AA} values greater than hydrogen H. One notes in Fig. 3-1 no particular trends between D_{AA} and electronegativity X indicating that these are essentially independent material selection properties of the elements.

Some of the useful outputs of the computer program are illustrated in the computations shown in Table 3-4 for the energy (or heat) of formation of oxides from their elements. The computations illustrate the following chemical reactions:

$$Be + 1/2\ O_2 \rightarrow BeO \quad + \Delta H_f$$

$$Ti + O_2 \quad \rightarrow TiO_2 \quad + \Delta H_f$$

$$2Al + 1.5\ O_2 \rightarrow Al_2O_3 + \Delta H_f$$

where ΔH_f is the heat of formation (negative for stable product). The properties of the elements as stored in Table 3-1 are presented at the top of each example in Table 3-4. The appropriate disappearance of the homonuclear bonds is shown by the negative moles and the production of the reaction product by positive moles. The total bond energy computed by Eq. (3-7) sums over both negative (bonds broken) and positive (bonds formed) moles of bonds to obtain the net energy $U_T \approx -\Delta H_f$ of formation. As shown in Table 3-4, the computer model also reports the bond energy, % ionic character, and bond length for each component of the above chemical reactions.

Table 3-5 compares calculated and reference values of the heats of formation for ten metal oxides. The right column records the differences. The sum of the differences is seen to be small, indicating no large tendency to either over or under estimate the heat of formation. The standard deviation of $\pm 2.04 \cdot 10^5$ J/mol, which is about 40% of the hydrogen single bond energy appears typical of this simple calculation.

Table 3-6 compares calculated and reference values of the heats of formation for metal chlorides. The differences in calculated and reference energies are similar to the oxides. By accepting these qualitative results, one has available a simple method of computing the chemical bond properties for an enormous range of chemical compounds.

In characterizing the corrosion protection properties of oxides one important parameter is the oxidation dilation factor ϕ for various metals as defined in Table 3-7. The oxidation dilation factor ϕ was introduced by Pilling and Bedworth[8] as an important corrosion resistance index of metals. When $\phi \leq 0.7$ the oxide film is in tension and cracks. Conversely, when $\phi \geq 2.0$, the oxide film is under compression and tends to blister and peel from the substrate metal. The computer program, by Eq. (3-9), estimates molar volume V_m of both metals and oxides, which permits a rapid evaluation of ϕ.

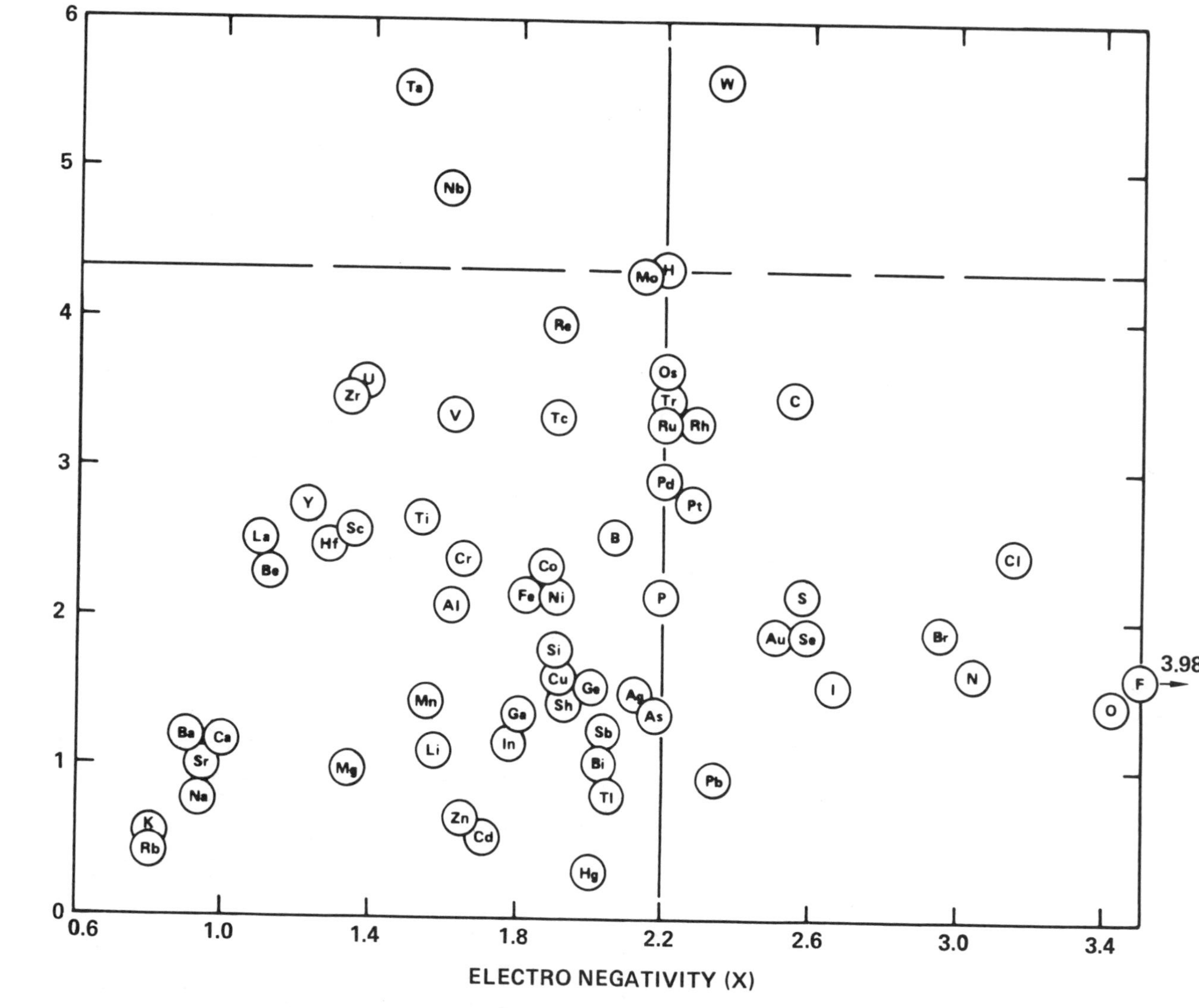

FIG. 3-1 Covalent bond energy D_{AA} and electronegativity X for the elements.

TABLE 3-4 Calculation of Heat of Formation for BeO, TiO_2, and Al_2O_3

```
Z, SY, W, D/1E5, X, R/1E-10, V, PH =
 4 BE 9.012  2.28  1.57  0.9  2  7
 8  0  16  1.39  3.44  0.73  2  2
 To continue press ENTER
 ?
 Chemical Analysis:
 Bonding           Bond          % Ionic      Bond        Moles
Elements          Energy         Energy      Length
A     B           (J/mole)                  (M*1E-10)
BE    BE          228000         0            1.8          -1
(02)/2(02)/2      201000         0            1.24         -1
BE    0           520951         64.776       1.4617        2
Total             612902                                    0
To continue press ENTER
Z, SY, W, D/1E5, X, R/1E-10, V, PH =
 22 TI 47.0  2.64  1.54  1.32  4  7
 8  0  16  1.39  3.44  0.73  2  2
 To continue press ENTER
 ?
 Chemical Analysis:
 Bonding           Bond          % Ionic      Bond        Moles
Elements          Energy         Energy      Length
A     B           (J/mole)                  (M*1E-10)
TI    TI          264000         0            2.64         -2
(02)/2(02)/2      201000         0            1.24         -2
TI    0           549865         63.3547      1.879         4
Total             1.26946E+06                               0
To continue press ENTER
Z, SY, W, D/1E5, X, R/1E-10, V, PH =
 13 AL 26.98  2.06  1.61  1.18  3  7
 8  0  16  1.39  3.44  0.73  2  2
 To continue press ENTER
 ?
 Chemical Analysis:
 Bonding           Bond          % Ionic      Bond        Moles
Elements          Energy         Energy      Length
A     B           (J/mole)                  (M*1E-10)
AL    AL          206000         0            2.36         -3
(02)/2(02)/2      201000         0            1.24         -3
AL    0           495669         65.1986      1.7453        6
Total             1.75301E+06                               0
To continue press ENTER
? -
```

TABLE 3-5 Comparison of Calculated and Experimental Heats of Formation for Oxides

Compound	$-\Delta H_f$(calc.) (10^5 J/mole)	$-\Delta H_f$ (Ref. 7) (10^5 J/mole)	Difference (10^5 J/mole)
Al_2O_3	17.5	16.3	1.20
Fe_2O_3	13.1	11.2	1.90
TiO_2	12.7	9.11	3.59
MgO	8.13	5.20	2.93
SiO_2	7.91	8.56	-0.65
BeO	6.13	6.10	0.03
MoO_2	5.08	5.43	-0.35
WO_2	3.26	5.70	-2.44
Au_2O_3	2.83	-0.80	3.63
SeO_2	1.82	2.29	-0.47
		Sum:	0.94
		Std. Dev:	± 2.04

TABLE 3-6 Comparison of Calculated and Experimental Heats of Formation for Chlorides

Compound	$-\Delta H_f$(calc.)	$-\Delta H_f$ (Ref. 7) (10^5 J/mole)	Difference
$AlCl_3$	6.95	6.95	0.0
$FeCl_3$	5.12	4.05	1.07
$TiCl_4$	10.1	7.50	2.60
MgCl	3.30	6.41	-3.11
$SiCl_4$	6.12	6.10	0.02
$BeCl_2$	4.87	5.11	-0.24
$MoCl_4$	3.86	3.30	0.56
$AuCl_3$	1.11	1.18	-0.07
		Sum:	0.10
		Std. Dev:	± 1.60

TABLE 3-7 Calculation of the Oxidation Dilation Factor ϕ for Metals

Oxide	Calculated (for Z = 12)			Exper. (7,8)	
	V(Me) (CC)	$V(MeO_x)$ (CC)	ϕ	$V(MeO_x)$ (CC)	ϕ
K_2O	57.02	20.55	0.36	40.6	0.45
BaO	26.45	13.00	0.49	26.8	0.67
MgO	8.57	6.34	0.74	11.3	0.81
Al_2O_3	11.19	11.32	1.01	25.7	1.28
TiO_2	7.84	8.48	1.08	18.7	1.78
Fe_2O_3	10.92	11.52	1.05	30.5	2.14
Ta_2O_5	16.40	20.31	1.24	53.9	2.50
Nb_2O_5	16.40	20.60	1.26	59.5	2.68
MoO_3	7.49	11.96	1.60	30.7	3.30
WO_3	7.49	12.30	1.64	32.4	3.35

$$\phi = \frac{\text{molecular volume of metal compound } MeX_x}{\text{atomic volume of equal moles of metal Me}}$$

The right columns of Table 3-7 report experimental values of molar volumes for various oxides and their characteristic ϕ values.[8,9] The left columns of Table 3-7 report the calculated values of molar volume for metals and their oxides and the computed estimate of ϕ. These computations represent a maximum coordination number Z = 12 for both metal and oxide in order to minimize both V_m and ϕ. One notes a consistent and agreeable trend between the calculated and experimental values of ϕ. Oxides tend to display lower coordination numbers than their parent metals. With the four high ϕ oxides in lower Table 3-7, assuming a Z = 6 for the oxide would double the calculated values of ϕ to provide good agreement with experimental data. The computer program calculates values of V_m for Z = 12, 8, 6, and 4.

3-4. Acid-Base Interactions

Many adhesion and corrosion properties of metals are dominated by the acid-base interactions of the metal oxide which covers all metals, except gold, under ambient atmospheric exposure. As pointed out by Parks[10] and Bolger and Michaels,[11] experience shows that different metals and their oxides display widely differing types of interactions with adhesives and corrosive environments based on the isoelectric point (IEP) or zero point of surface charge (ZPC) under aqueous immersion. The IEP of a surface is defined as the pH = - log $[H^+]$ where the immersed oxide surface has zero net surface charge due to equivalent concentrations of positive (cation) and negative (anion) complexes.

In this discussion the objective is to develop specific relations for coulombic charge displacement between atoms A and B. Atom A is defined as more electropositive than atom B. In the ionized state atom A will act as a Lewis acid (A = electron acceptor or cation) and B as a Lewis base (B = electron donor or anion). Through the use of a simple electrostatic model for coulombic interactions Parks derives a two parameter model for IEP with the following relation:[10]

$$\text{IEP} = \text{pH} = C_1 - C_2(V/R_I) \tag{3-11}$$

where V is the metal cation valence, R_I is the "effective" ionic radius of the oxolated cation, and C_1 and C_2 are constants. The ratio, (V/R_I), termed the ionic potential was introduced by Cartledge[12] and, as indicated in a review by Moeller,[13] has found broad applications in evaluations of atomic contributions to acid-base properties. A comprehensive review of IEP values for solid oxides, solid hydroxides, and aqueous hydroxylated oxides by Parks[10] shows the characteristic ranges of acid-base properties summarized in Table 3-8.

The essential form of Eq. (3-11) and the data summary of Table 3-8 follows from the standard Coulomb equation:[10]

$$U_E = - \frac{dF}{dL_{AB}} = \frac{M\ N_o\ V_A V_B \varepsilon^2}{DL_{AB}} \tag{3-12}$$

where U_E is electrostatic potential energy, F is the Coulomb electrostatic

TABLE 3-8 Correlation Between Metal Oxidation State and IEPs

Oxide	IEPS Range (pH Units)	Acid-Base Character
M_2O	pH > 11.5	strong base
MO	8.5 < pH < 12.5	intermediate base
M_2O_3	6.5 < pH < 10.4	weak base
MO_2	0 < pH < 7.5	intermediate acid
M_2O_5, MO_3	pH < 0.5	strong acid

force, and L_{AB} is the distance separation of cation A and anion B. The moles of acid-base bonds per mole of adsorbate is M, N_o is Avagadro's number, V_A and V_B are the respective cation and anion valences, D the dielectric constant of the medium, and ε the charge of the electron.

A combined consideration of the Coulomb equation, Eq. (3-12), and Eq. (3-11) provides a new empirical form for Eq. (3-11) as follows:

$$\text{IEP} = \text{pH} = 16.0 - 3(V/R) \qquad (3\text{-}13)$$

where V is atomic valence and R is covalent radius as listed in Table 3-1. In Eq. (3-13) the limits of IEP are bounded by the following ionization properties of ambient water:[13]

$$H_2O \rightarrow OH^- + H^+ \quad ; \; pK_A = 15.97$$

$$H_3O^+ \rightarrow H_2O + H^+ \quad ; \; pK_A = -1.74$$

which tend to limit the acid-base responses of solvated ions. The atomic properties of the hydrogen cation with V/R = 3.13 provide a value of IEP = 6.61 for a neutral surface as compared with the bulk ambient state of water with pH = 7.0. Application of the covalent radius R and the maximum valence V = MV values of Table 3-1 in Eq. (3-13) will tend to emphasize the specific acidic-base properties of the elements relative to hydrogen. For this extreme case of highest oxidation state, Eq. (3-13) will describe the minimum IEP as follows:

$$\text{IEP} = \text{pH} = 16.0 - 3S \qquad (3\text{-}14)$$

where the atomic values of S = MV/R are listed in the right column of Table 3-1.

A plot of S = MV/R vs electronegativity X for the elements is shown in the data display of Fig. 3-2. Despite a broad scatter of the data, one notes an underlying trend of increasing S and related oxide acidity with increasing electronegativity. The dashed horizontal and vertical curves of Fig. 3-2 again reference the atomic properties of hydrogen and the pH of water as a reference. Of special reference and importance in the adhesion and corrosion properties of the metal oxides are those metals of intermediate electronegativity X = 1.4 to 2.6 which show high or low values of S and which display extremes of acidic (high S) or basic (low S) surface properties.

These extreme values of S for the elements are more clearly shown in Fig. 3-3 when plotted as a function of atomic number. The dashed horizontal

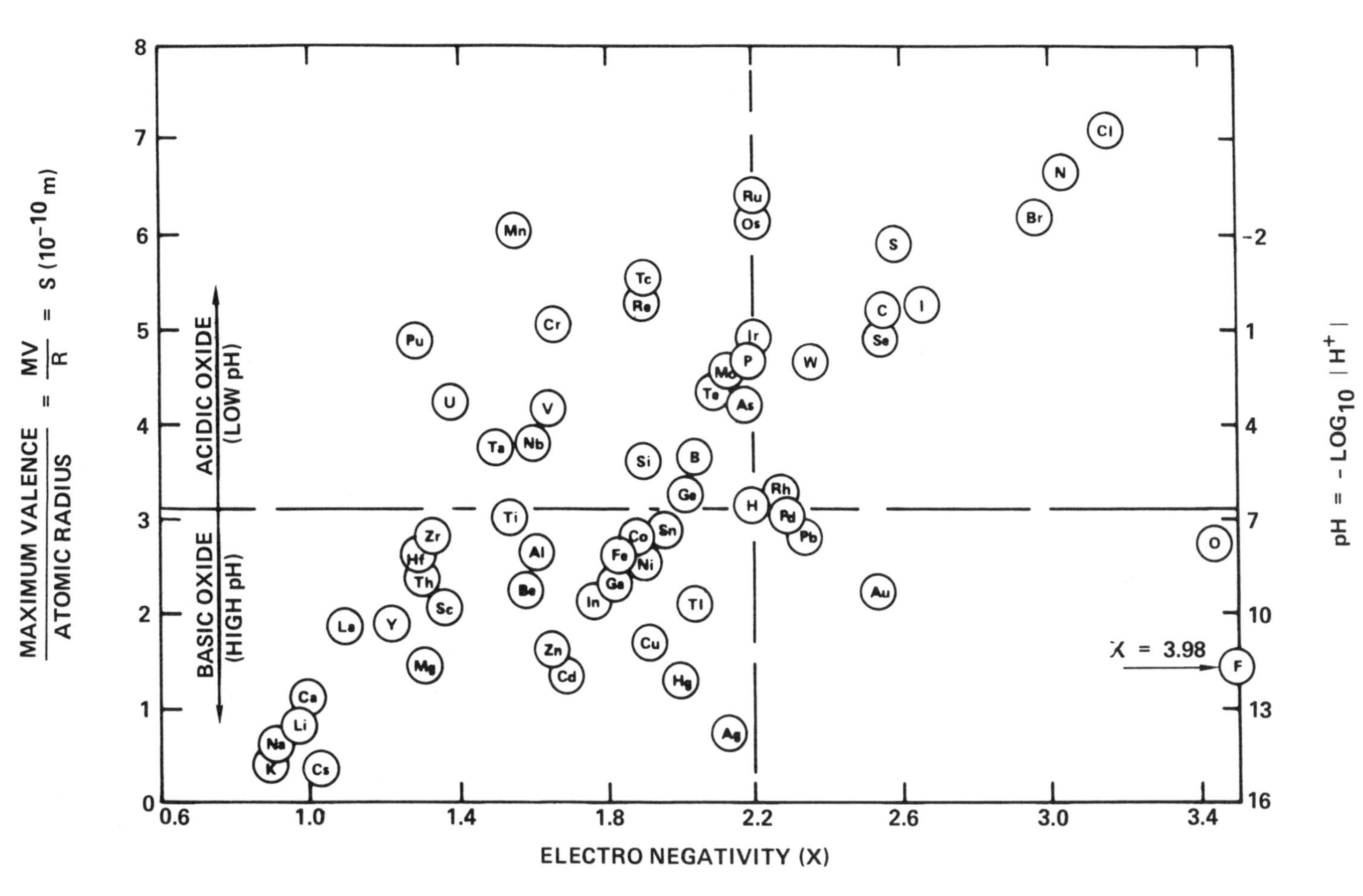

FIG. 3-2 The maximum valence to atomic radius ratio, MV/R vs electronegativity.

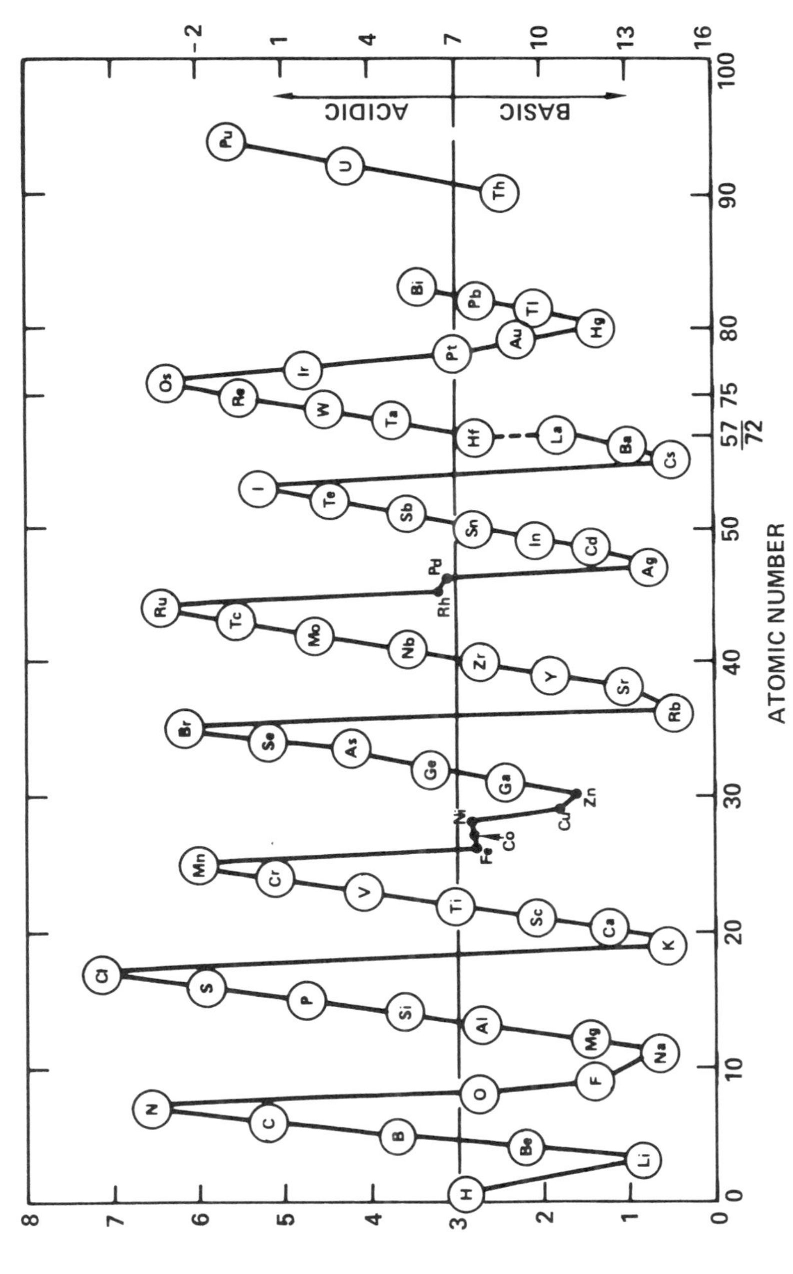

FIG. 3-3 Acidity index pH = 16 - 3S vs atomic number.

curve of Fig. 3-3 represents the neutral properties of hydrogen with S = 3.13 forming the dividing line between elemental acid or base character. One notes in both Fig. 3-2 and Fig. 3-3 that acidic elements with S > 5 and of intermediate electronegativity X = 1.4 to 2.2 are restricted to four, chromium Cr, manganese Mn, technetium Tc, and rhenium Re. The high oxidation state chromates such as zinc chromate are commonly used as corrosion suppressing pigments, and this function may be primarily due to strong acid-base interaction with neutral or basic oxides such as formed on iron, steel, aluminum and titanium. Manganese, which forms a strongly acidic oxide, is commonly used as an alloying constituent in copper, brass, and bronze alloys that are subject to marine applications and salt water corrosion. Manganese is considered an excellent deoxidizing agent for steel manufacture. Technetium is informally reported to have exceptional corrosion properties, with essential immunity to salt water corrosion and an unique resistance to barnacle growth in marine environments.[14]

According to the definitions applied here, all metals in the metallic state V = 0 are strongly basic and strong electron donors. Metals with a low maximum valence MV ⩽ 2 remain strongly basic in their fully oxidized state, as shown in the plots of Fig. 3-2 and Fig. 3-3. Metals of intermediate electronegativity Z = 1.4 to 2.2 and low S ⩽ 2 tend to form basic oxides which are hydrophobic and electrically conductive. The extensive use of copper oxide as a hydrophobic substrate in lithography is well known. The oxides of zinc, copper, and silver, all tend to display hydrophobic character and semi-conductive properties. One notes that lead (Pb) and gold (Au), although of higher electronegativity, tend to fit this latter category of definition, particularly in their lower valence states.

The general qualitative validity of Eq. (3-13) and Eq. (3-14) is shown in the three examples illustrated in Figs. 3-4 through 3-6. Figure 3-4 plots the reference pH values of concentrated acids and bases[7] vs the ratio of MV/R for the cation as listed in Table 3-1. The dashed curve is the predicted relation of Eq. (3-14), which shows a reasonable correlation with experimental data.

Figure 3-5 plots data from Moeller[13] for a wide variety of aqueous dissociation constants as $pK_A \approx pH$ where the principal anion or cation is not necessarily in its maximum valence state. For example, sulphur has a valence V = -2 in hydrogen sulphide (H_2S) and V = +6 in sulphuric acid (H_2SO_4). Also shown are several valence states for iodine where V = +1 in HIO and V = +5 in HIO_3 and V = +7 in H_5IO_6. These data indicate a reasonable correlation with the dashed curve of Eq. (3-13), where the valence V is assumed variable.

Figure 3-6 plots a number of the dissociation states of phosphorous and phosphoric acids, as reported by Moeller,[13] to illustrate the broad range of

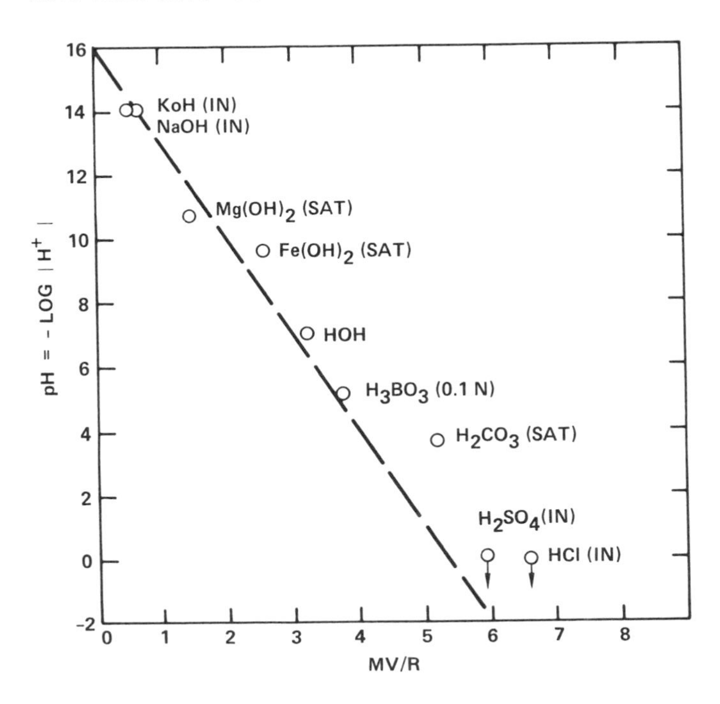

FIG. 3-4 pH vs (MV/R) for acids and bases.

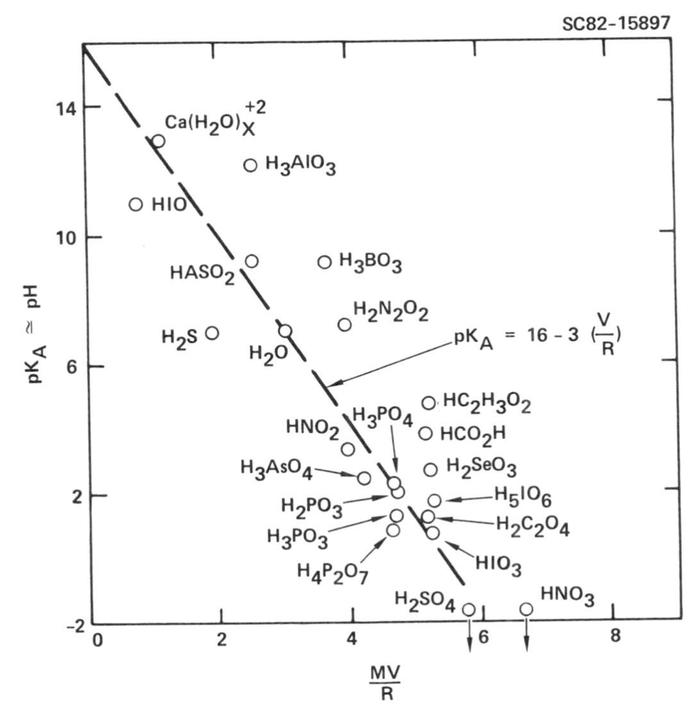

FIG. 3-5 Acid dissociation index pK_A vs (V/R) for miscellaneous acids and bases at varied valence.

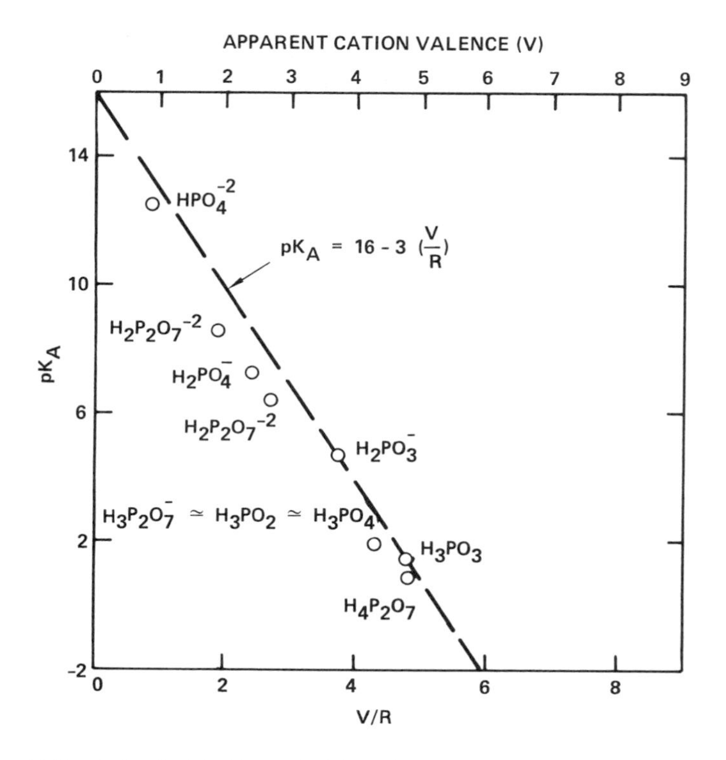

FIG. 3-6 Acid dissociation index pK_A vs apparent cation valence V and V/R.

$pK_A \approx$ pH and apparent cation valence states that are available in some hydrated oxides with multiple valence states. The extensive review listings of IEP data by Parks[10] also indicate that the acid-base properties of metal oxides can display widely varied values of surface pH which evidently depend on metal oxidation states and surface morphology. The data of Fig. 3-6 thus illustrate the important idea that a metal oxide surface can display a spectrum of acid-base characteristics which are bounded by a maximum acid character and minimum pH characteristic of MV/R.

3-5. Surface Adsorption

As an excellent brief review of the corrosion resistance of materials by Hoar[15] points out, the acid-base character of the oxide is one of a number of important variables in corrosion protection. A simple approach to estimating the acid-base character of molecular adsorption to metals and their oxides is available by a special use of the Coulomb energy calculation as defined by Eq. (3-12). By appropriate substitution of physical units we can rewrite Eq. (3-12) in the following special form:

$$U_M \left(\frac{kJ}{mole}\right) = \frac{12.2\ M\ V_A V_B}{R_A + R_B} \tag{3-15}$$

Our objective is to calculate the maximum value of U_M characteristic of the limit of close approach of cation A and anion B. We assume the dielectric constant D of Eq. (3-12) is D = 1.0 and L_{AB} is the sum of the covalent radius, so that $L_{AB} = R_A + R_B$ with R values from Table 3-1. By applying the classical values of valence for hydrogen bonding in water where $V_A = 1$ for hydrogen and $V_B = 2$ for oxygen and $R_A + R_B = 0.32 + 0.73 = 1.05 \cdot 10^{-10}$ m, one finds that the energy U_M for a single hydrogen bond where M = 1 is U_M = 23.2 kJ/mol = 5.5 kcal/mol. This computed U_M value is in reasonable agreement with experimental values for the H-bond energy of bulk water of 5.0 kcal/mole.[5] The O-H- -O bond length of $1.99 \cdot 10^{-10}$ m produced by this simple calculation is, of course, less than the experimental length determined for the hydrogen bond in bulk water,[5] where the O-H- -O hydrogen bond length is $2.76 \cdot 10^{-10}$ m.

Recognizing that Eq. (3-15) calculates an upper estimate for the Coulomb energy, a second relation which estimates the nominal Coulomb energy U_N can be written as follows:

$$U_N \left(\frac{kJ}{mol}\right) = \frac{12.2\ M\ (\sum V_A I_A)(\sum V_B I_B)}{R_A + R_B} \tag{3-16}$$

where $\sum V_A I_A$ and $\sum V_B I_B$ appropriately sum the product of valence and ionic character for each bond within the molecule to reflect the fractional ionic character of molecular structures. From Eq. (3-16) the two O-H bonds of water would produce a sum $\sum V_A I_A = 0.341$ for each H atom and $\sum V_B I_B = 0.682$ for the oxygen atom. The resulting nominal Coulomb energy is U_N = 2.70 kJ/mole for the single hydrogen bond in water. Equation (3-16) thus provides a more conservative estimate of electrostatic bonding which appropriately becomes $U_N = 0$ when either the acid or base are purely covalently bonded.

Estimates of the values of U_M and U_N between various adsorbates and substrates can be readily calculated by Eq. (3-15) and Eq. (3-16). The upper portion of Table 3-9 illustrates model monobase and diacidic orientations of water on five substrate oxides. The lower portion of Table 3-9 tabulates the calculated values of U_M and U_N. We would expect that the actual orientation (acid or base) of water is the orientation which gives the highest respective value of U_M or U_N. The calculations indicate the diacid conformation of water on Ag_2O and CuO which are basic oxides. The values of U_M and U_N become nearly equivalent for the acid and base interaction of water on neutral Fe_2O_3, indicating no strong molecular orientation effect. However, on acidic SiO_2 and CrO_3 the calculation shows that the monobasic interaction with water is favored.

TABLE 3-9 Coulomb Bond Energies Between Water and Various Oxides

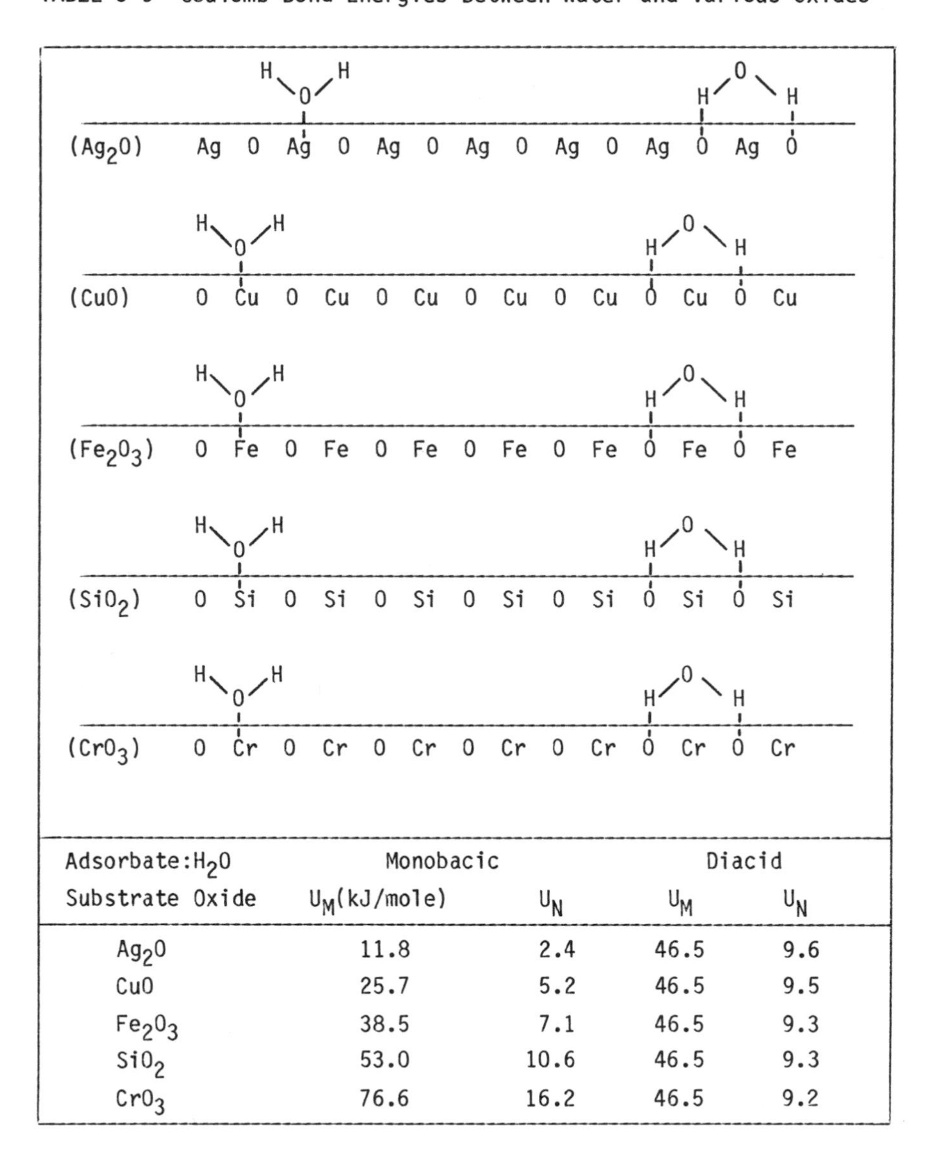

Adsorbate: H_2O	Monobacic		Diacid	
Substrate Oxide	U_M(kJ/mole)	U_N	U_M	U_N
Ag_2O	11.8	2.4	46.5	9.6
CuO	25.7	5.2	46.5	9.5
Fe_2O_3	38.5	7.1	46.5	9.3
SiO_2	53.0	10.6	46.5	9.3
CrO_3	76.6	16.2	46.5	9.2

Different absorbate molecules can be represented by these calculations as shown respectively in Tables 3-10 through 3-12 for ammonia, chrome oxide, and an organic silanetriol. Ammonia represents a cleaning agent often added to water. Chromium trioxide is a common corrosion suppressing pigment constituent. The trihydroxy silane is a postulated intermediate form of trimethoxy silanes commonly employed as adhesion promoters which suppress corrosion and increase adhesive bond strength in the presence of high moisture.[11] By noting only the higher of the U_M values for acid or base orientations of the adsorbates of Tables 3-9 through 3-12, we can plot U_M vs $S = MV/R$ for the five substrate oxides. The curves of Fig. 3-7 show the variation of U_m for each adsorbate. Chromium trioxide adsorbate displays the highest U_M estimate and, by inference, would be expected to displace the other adsorbates from the substrate oxides. The transition from acid to base orientation of the adsorbate is indicated by the change in slope of each curve. Both ammonia and the trihydroxy silane are predicated to displace water from neutral or acidic substrates such as Fe_2O_3, SiO_2, and CrO_3 but not from basic oxides such as Ag_2O and CuO.

The curves of Fig. 3-8 show the trends of adsorption predicted by the appropriate selection of nominal Coulomb U_N values for acid-base interactions as calculated by Eq. (3-16). The curves of Fig. 3-8 show that application of the covalent-ionic criteria for available valence lowers the bond energies and that ammonia NH_3 is more weakly bonded than water H_2O to all five substrates. The trihydroxy silane and chrome oxide are estimated to bond more strongly than water. The inference to be drawn based on the curves of Fig. 3-8 is that both chromium trioxide and trihydroxy silane should displace water from both basic (low V/R) and acidic (high V/R) oxide surfaces which is also the conclusion from numerous studies of adhesion[11] and corrosion.[15]

3-6. Molecular Dipole Properties

A set of simple relations permits the numeric estimation of molecular dipole properties directly from atomic properties. The absolute value of the dipole moment μ_{AB} for a chemical bond is approximately expressed by the following relation:[16]

$$\mu_i = 4.8\, L_{AB} I_{AB} \tag{3-17}$$

where L_{AB} is bond length as defined by Eq. (3-6) and I_{AB} is the fractional ionic character of the chemical bond as defined by Eq. (3-5). As discussed by Hedvig,[16] the effective dipole moment of the rigid molecule $\bar{\mu}$ can be readily

TABLE 3-10 Coulomb Bond Energies Between Ammonia and Various Oxides

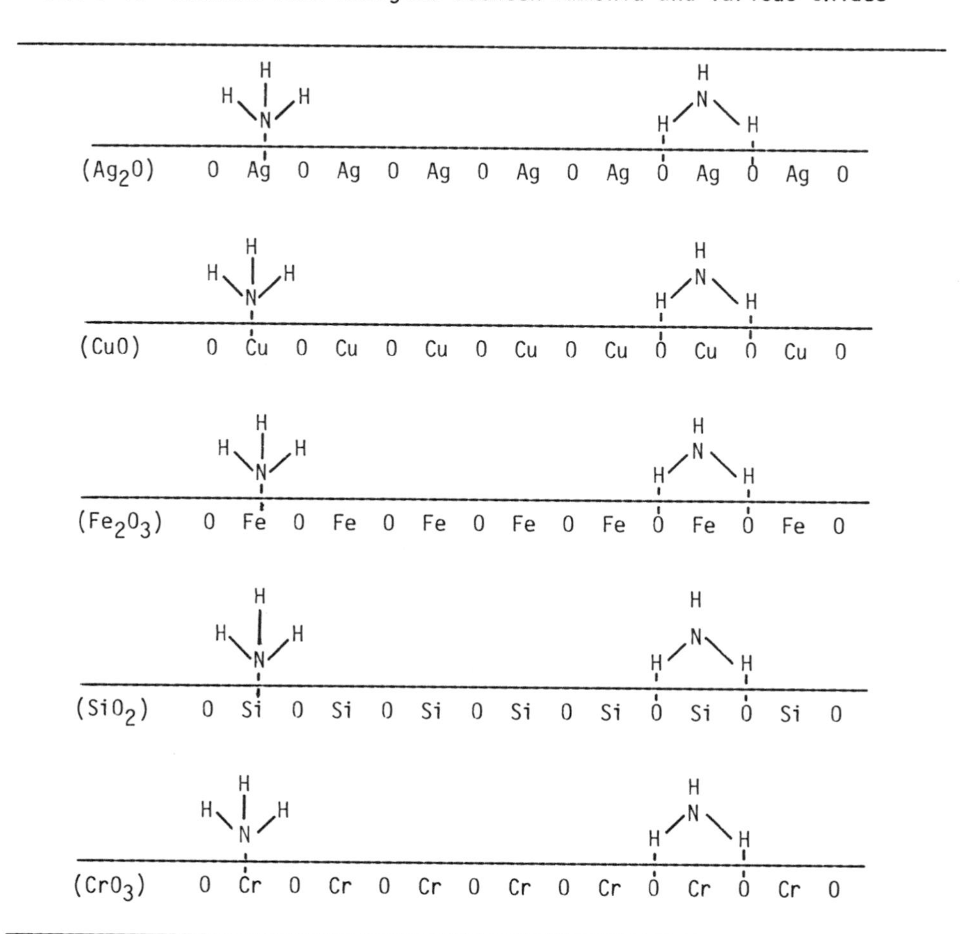

Adsorbate: NH_3 Substrate Oxide	Monobacic U_M(kJ/mole)	U_N	Diacid U_M	U_N
Ag_2O	17.5	2.0	46.5	9.6
CuO	38.1	4.3	46.5	9.5
Fe_2O_3	57.2	6.4	46.5	9.3
SiO_2	78.7	8.8	46.5	9.3
CrO_3	113.8	13.4	46.5	9.2

TABLE 3-11 Coulomb Bond Energies Between Chromium Trioxide and Various Oxides

```
               0                                          0
            O=Cr=O                                     O= || =O
            |    |                                        Cr
            |    |                                        |
(Ag2O)   0  Ag  0  Ag  0  Ag  0  Ag  0  Ag  0  Ag  0  Ag  0

               0                                          0
            O=Cr=O                                     O= || =O
            |    |                                        Cr
            |    |                                        |
(CuO)    0  Cu  0  Cu  0  Cu  0  Cu  0  Cu  0  Cu  0  Cu  0  Cu

               0                                          0
            O=Cr=O                                     O= || =O
            |    |                                        Cr
            |    |                                        |
(Fe2O3)  0  Fe  0  Fe  0  Fe  0  Fe  0  Fe  0  Fe  0  Fe  0  Fe

               0                                          0
            O=Cr=O                                     O= || =O
            |    |                                        Cr
            |    |                                        |
(SiO2)   0  Si  0  Si  0  Si  0  Si  0  Si  0  Fe  0  Fe  0  Fe

               0                                          0
            O=Cr=O                                     O= || =O
            |    |                                        Cr
            |    |                                        |
(CrO3)   0  Cr  0  Cr  0  Cr  0  Cr  0  Cr  0  Cr  0  Cr  0  Cr
```

Adsorbate: $RSi(OH_2)$	Dibasic		Monoacid	
Substrate Oxide	U_M(kJ/mole)	U_N	U_M	U_N
Ag_2O	24.9	8.8	76.6	27.1
CuO	40.9	14.2	76.6	26.7
Fe_2O_3	81.7	28.0	76.6	26.2
SiO_2	112.8	38.6	76.6	26.2
CrO_3	162.7	58.5	76.6	27.6

TABLE 3-12 Coulomb Bond Energies Between $R\text{-}CH_2\text{-}Si(OH)_3$ and Various Oxides

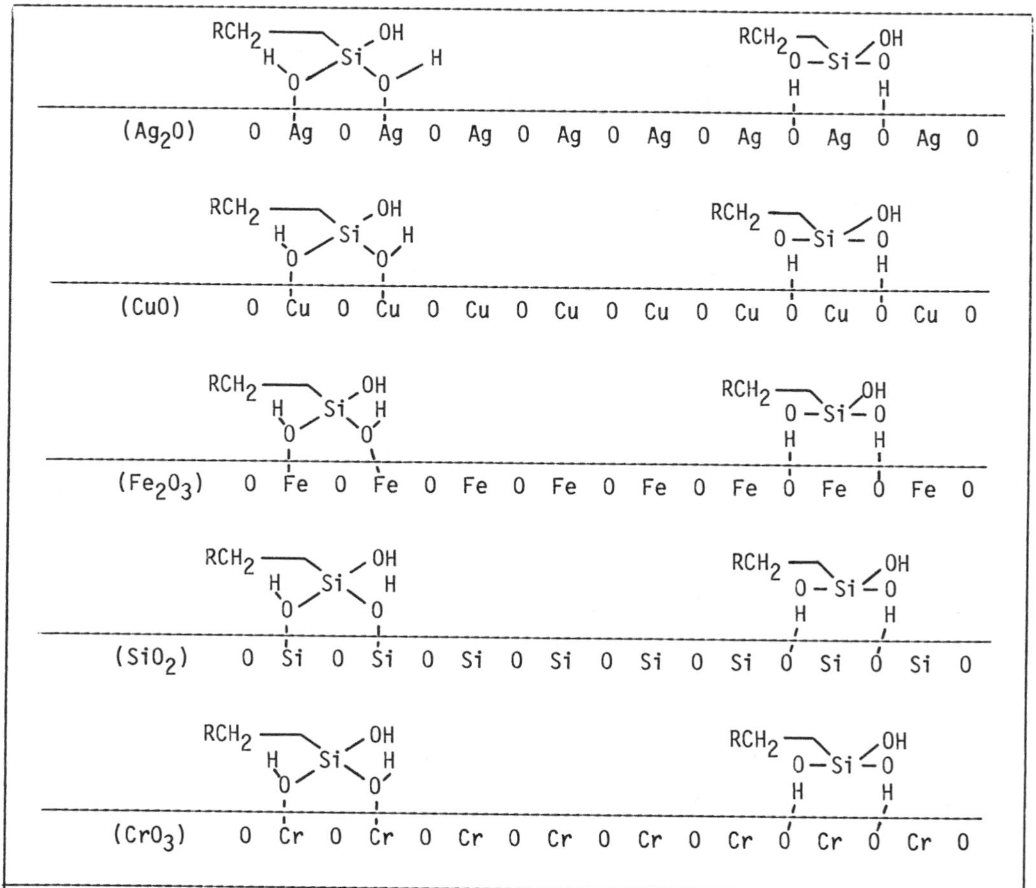

(Ag_2O) O Ag O Ag O Ag O Ag O Ag O Ag O Ag O Ag O

(CuO) O Cu O Cu O Cu O Cu O Cu O Cu O Cu O Cu O

(Fe_2O_3) O Fe O Fe O Fe O Fe O Fe O Fe O Fe O Fe O

(SiO_2) O Si O Si O Si O Si O Si O Si O Si O Si O

(CrO_3) O Cr O Cr O Cr O Cr O Cr O Cr O Cr O Cr O

Adsorbate: $RSi(OH_2)$	Dibasic		Diacid	
Substrate Oxide	U_M(kJ/mole)	U_N	U_M	U_N
Ag_2O	23.6	6.63	45.6	16.4
CuO	51.4	14.6	45.6	16.1
Fe_2O_3	77.0	21.8	45.6	15.9
SiO_2	106.0	30.0	45.6	15.9
CrO_3	153.2	45.6	45.6	16.7

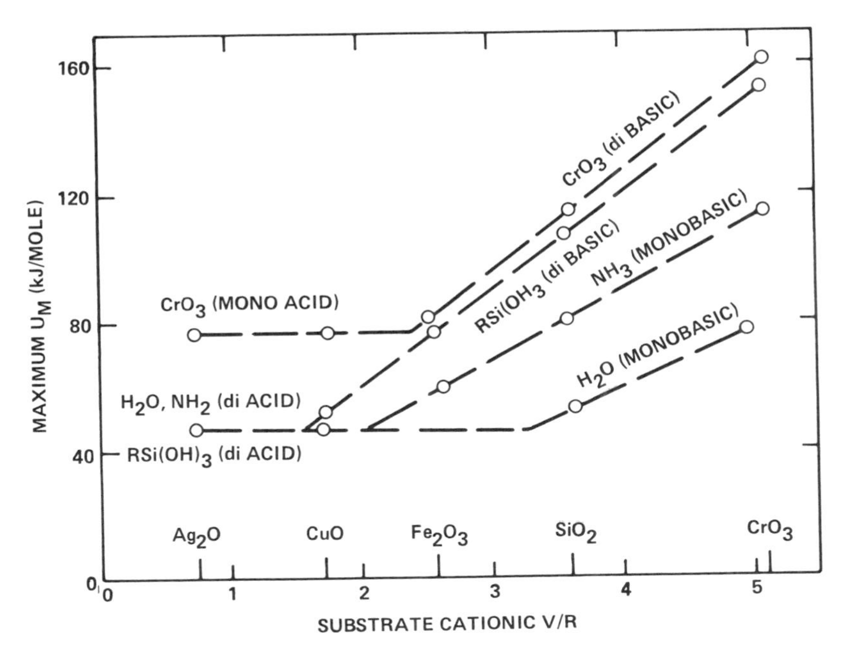

FIG. 3-7 Calculated maximum Coulomb energy U_m between adsorbate and substrate oxides.

calculated from the individual bond moment components μ_{xi}, μ_{yi}, μ_{zi} along cartesian coordinates x, y, z by the following relation:

$$\bar{\mu} = \left(\sum_{i=1}^{n} \mu_{xi}^2 + \sum_{i=1}^{n} \mu_{yi}^2 + \sum_{i=1}^{n} \mu_{zi}^2 \right)^{1/2} \qquad (3\text{-}18)$$

A simple and extraordinarily sensitive measurement for molecular adsorption at surfaces involves the measurement of surface potential difference ΔV. Bewig and Zisman[17] discuss the effects of adsorbed films on ΔV as defined by the classical Helmholtz relation:[17]

$$\Delta V = \frac{4\pi\bar{\mu}_p}{A} \simeq \frac{4\pi}{A} \sum_{i=1}^{n} \mu_{yi} \qquad (3\text{-}19)$$

where $\bar{\mu}_p$ is the effective molecular dipole moment perpendicular to the substrate surface and A is the area per molecule in the close packed adsorption layer. By defining the y-axis as perpendicular to the surface, as shown in Eq. (3-18), it is evident that ΔV can be numerically estimated from the bond properties described in Eq. (3-19). This calculation is not demonstrated, but is an obvious extension of the surface analysis of molecular adsorption shown in Tables 3-9

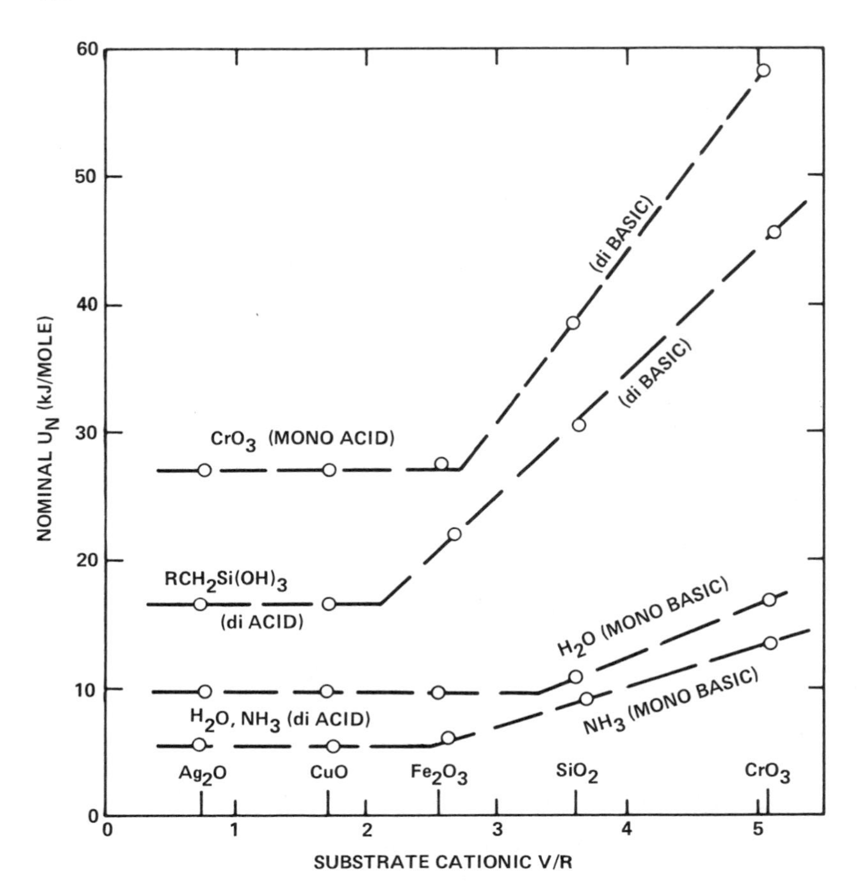

FIG. 3-8 Calculated nominal Coulomb energy U_N between adsorbate and substrate oxides.

through 3-12. The bond dipole moments for all the acid-base bonds are calculated by Eq. (3-17) and included in the data summary of Table 3-13. The charge orientation of each dipole is, of course, defined easily from electronegativities. The additional information required to solve Eq. (3-18) or Eq. (3-19) are the individual bond angles and orientations which are not available directly from Table 3-1.

This discussion has introduced and demonstrated a computer aided model for creating and evaluating molecular compounds, acid-base characterization, surface adsorption energies, and molecular dipole properties which characterize the adsorption layer. The illustrative examples developed by computations from atomic properties show reasonable qualitative agreement with experiment. The real power and versatility of the model discussed here is, of course, the extremely wide range of materials that can be numerically evaluated by this method. This same observation applies to the numeric investigation of surface

TABLE 3-13 Bond Properties for Adsorbates and Substrate Oxides

Bond + - -	D_{AB} (kJ/mole)	%I	L_{AB}	μ (debye)	R_A	R_B
H - O	435	34.1	0.94	1.54	0.32	0.73
H - N	366	18.6	0.99	0.88	0.32	0.75
Cr = O	1051	58.2	1.64	4.58	1.18	0.62
Ag - O	362	60.9	1.93	5.64	1.34	0.73
Cu - O	384	59.5	1.76	5.03	1.17	0.73
Fe - O	421	59.4	1.76	4.56	1.17	0.73
Si - O	387	59.1	1.70	4.82	1.11	0.73
Cr - O	494	61.9	1.75	5.20	1.18	0.73
Si - C	303	13.4	1.82	1.17	1.11	0.77

chemical processes such as oxidation, surface adsorption, and molecular dipole response.

The fundamental question of valency, bond ionicity, and bond length estimation are currently very lively subjects of research and discussion.[18-21] Inspection of Table 3-14 which compares the current (1980) revised estimates[1] of electronegativity with the earlier Pauling[2] estimates shows that the properties of some elements such as tungsten W (Z = 74) have been substantially redefined. This process of revision of atomic properties is, of course, improved by computer models such as discussed here. A recent review by Morgan[22] provides an excellent discussion and recent bibliography of this broad subject as it relates to analysis of chemical bonding in ceramics.

The transcending importance of achieving corrosion resistant interfaces and stable long-term adhesive bonding is defined in a recent review of basic research needs on interfaces in solar energy conversion systems (SECS).[23] Present energy needs of the United States, about 85 quads, would require exposure of 73000 km^2 (28000 sq miles) of solar collectors operating at 20% efficiency. Economics require a stable and non-degrading performance life of about 30 years. The computer assisted model discussed here was developed as part of a research program to develop long term environmental durability and corrosion protection in polymer encapsulated photovoltaic SECS. Related computer based models for selection of polymer chemistry are discussed in separate reports.[24,25] The combination of the three computer aided models are as follows: (1) atomic and molecular properties (discussed here), (2) polymer chemistry and physical properties, and (3) polymer chemistry and mechanical proper-

TABLE 3-14 Comparison of Revised (Ref. 1 = X) and Pauling (Ref. 2 = X_p) Values of Elemental Electronegativity

AT. No.	SY	X	X_p	$X-X_p$	AT. No.	SY	X	X_p	$X-X_p$
1	H	2.20	2.1	0.10	37	RB	0.82	0.8	0.02
					38	SR	0.95	1.0	-0.05
3	LI	0.98	1.0	-0.02	38	Y	1.22	1.2	0.02
4	BE	1.57	1.5	0.07	40	ZR	1.33	1.4	-0.07
5	B	2.04	2.0	0.04	41	NB	1.60	1.6	0.0
6	C	2.55	2.5	0.05	42	MO	2.16	1.8	0.36
7	N	3.04	3.0	0.04	43	TC	1.90	1.9	0.0
8	O	3.44	3.5	-0.06	44	RU	2.20	2.2	0.0
9	F	3.98	4.0	-0.02	45	RH	2.28	2.2	0.08
					46	PD	2.20	2.2	0.0
11	NA	0.93	0.9	0.03	47	AG	1.93	1.9	0.03
12	MG	1.31	1.2	0.11	48	CD	1.69	1.7	-0.01
13	AL	1.61	1.5	0.11	49	IN	1.78	1.7	0.08
14	SI	1.90	1.8	0.10	50	SN	1.96	1.8	0.16
15	P	2.19	2.1	0.19	51	SB	2.05	1.9	0.15
16	S	2.58	2.5	0.08	52	TE	2.10	2.1	0.0
17	CL	3.16	3.0	0.16	53	I	2.66	2.5	0.16
19	K	0.82	0.8	0.02	55	CS	0.79	0.7	0.09
20	CA	1.00	1.0	0.0	56	BA	0.89	0.9	-0.01
21	SC	1.36	1.3	0.06	57	LA	1.10	1.1	0.0
22	TI	1.54	1.5	0.04	72	HF	1.30	1.3	0.0
23	V	1.63	1.6	0.03	73	TA	1.50	1.5	0.0
24	CR	1.66	1.6	0.06	74	W	2.36	1.7	0.66
25	MN	1.55	1.5	0.05	75	RE	1.90	1.9	0.0
26	FE	1.83	1.8	0.03	76	OS	2.20	2.2	0.0
27	CO	1.88	1.8	0.08	77	IR	2.20	2.2	0.0
28	NI	1.91	1.8	0.11	78	PT	2.28	2.2	0.08
29	CU	1.90	1.9	0.0	79	AU	2.54	2.4	0.14
30	ZN	1.65	1.6	0.05	80	HG	2.00	1.9	0.10
31	GA	1.81	1.6	0.21	81	TL	2.04	1.8	0.24
32	GE	2.01	1.8	0.21	82	PB	2.33	1.8	0.53
33	AS	2.18	2.0	0.18	83	BI	2.02	1.9	0.12
34	SE	2.55	2.4	0.15					
35	BR	2.96	2.8	0.16	90	TH	1.30	1.3	0.0
					92	U	1.38	1.7	-0.32
					94	PU	1.28	-	-

ties. The objective of these models is to provide a numeric estimation of bulk and interfacial properties of materials directly from chemical composition and chemical structure. The computer aided models add great speed and convenience to these numeric analysis and property evaluations.

References

1. "Table of the Periodic Properties of the Elements," Sargent-Welsh Scientific Co., Skokie, Ill. (1980).

2. L. Pauling, "The Nature of Chemical Bond," 3rd ed., Cornell University Press, Ithaca, N.Y. (1960)

3. A.K. Vijh, J. of Material Sci. 10, 998 (1975).

4. J.H. Hildebrand and R.L. Scott, "The Solubility of Nonelectrolytes," 3rd ed., Reinhold, New York (1950).

5. D.H. Kaelble, "Physical Chemistry of Adhesion," Wiley-Interscience, New York, Chap. 1, 2, 5 (1971).

6. V. Shoemaker and D.P. Stevenson, J. Am. Chem. Soc. 63, 37 (1941).

7. Handbook of Chemistry and Physics, 54th ed, CRC Press, p. D-61, D-114 (1973).

8. N.B. Pilling and R.E. Bedworth, J. Inst. Met. 29, 529 (1923).

9. O. Kubaschewski and B.E. Hopkins, "Oxidation of Metals and Alloys," Butterworths, London, Chapt 1 (1962).

10. G.A. Parks, Chem. Rev. 65, 177 (1965).

11. J.C. Bolger and A.S. Michaels, Interface Conversion (Editor: P. Weiss), Elsevier, Amsterdam, pp. 3-51 (1968).

12. G.H. Cartledge, J. Amer. Chem. Soc. 52, 3076 (1930).

13. T. Moeller, "Inorganic Chemistry," Wiley, New York, p. 210 (1952).

14. F. Mansfeld, Science Center, personal communication (1981).

15. T.P. Hoar, J. Electrochem. Soc. 117(1), p. 17c (1970).

16. P. Hedwig, "Dielectric Spectroscopy of Polymers," Halstead Press, New York, Chap. 1 (1977).

17. K.W. Bewig and W.A. Zisman, J. of Phys. Chem. 68, p. 1804 (1964).

18. L. Pauling, Amer. Minerologist, 65, p. 321 (1981).

19. R.F. Steward, M.W. Whitehead and G. Donnay, Ibid, 65, p. 324 (1980).

20. J.A. Tossell, Ibid, 65, p. 163 (1980).

21. G.V. Gibbs, "Molecules as Model for Bonding in Silicates," American Mineralogist, 67, (1982), pp. 421-450.

22. P.E.D. Morgan, "Bonding in Nitrgen Ceramics," Nitrogen Ceramics (Editor: F.L. Riley), Nordhoff, Leiden, pp. 23-40 (1977).

23. R.J. Gottschall and A.W. Czanderna, "Basic Research Needs and Opportunities on Interfaces in Solar Materials: An Executive Summary," J. of Metals p. 27 (October 1981).

24. D.H. Kaelble, "Relations Between Polymer Chemistry and Physical Properties," ACS Organic Coatings and Appl. Poly. Sci. Proc. 46 (April, 1982), pp. 241-245.

25. D.H. Kaelble, "Relations Between Polymer Chemistry and Mechanical Properties," Ibid 46, (April, 1982), pp. 246-253.

4

Computer Aided Design and Manufacture

4-1. Manufacturing Science

As already discussed in Chapter 2 and outlined in Fig. 2-1, a generalized strategy already exists for the predictive design of polymers and polymer composites. The detailed elements of this design and their implementation in direct computer aided design and manufacture (CAD/CAM) of polymers is the subject of this chapter. The principles which underlie the CAD/CAM models for polymers presented here were previously developed and presented in molecular models for polymer adhesion and cohesion in a separate textbook by Kaelble.[1] Frequent reference will be made to these molecular models and the interested reader will want to review the derivations which are summarized in Chapter 11 of that text entitled "Mechanical Properties and Cohesion" in order to fully understand the details of the CAD/CAM models presented here.

Methods are presented for computer assisted estimations of polymer specific volume V_p, solubility parameter δ, glass temperature T_g, and molecular weight between entanglements M_e. These properties are computed from 33 functional groups, ranging from methylene to sulfone, which can be combined to describe the main chain and side chain structures of model linear polymers. The additional effects of molecular weight and crosslinking upon glass transition temperature are computationally defined. The functional group properties are stored in computer memory and recalled by an interactive computer program which permits construction of the polymer repeat unit chemistry for both main chain and side chain structures. The computation then describes the above four physical properties for the infinite linear polymer in its amorphous glass state. This monomer-polymer prediction program is self-instructive and useful in qualitative exploration of chemical structure effects on polymer processability and performance.

A model for computer aided cure management is discussed and demonstrated. This model is based upon generic relations between polymer chemical structure and physical properties. The glass transition T_g and the flow T_m temperatures of the curing polymer are defined in terms of curing reaction path (chain extension/crosslinking ratio) and cure state. The monitoring of both T_g and T_m is defined as the basis for controlling the curing kinetics. The reaction path for competing chain extension and crosslinking reactions is shown to be determined by the relation of cure control temperature T relative to T_g and T_m as curing progresses. Cure management is proposed to involve two fundamental objectives which are 1) prior chemical analysis of the curing materials to delineate and optimize the cure path options and 2) the dynamic real time monitoring and control of cure temperature, hydrostatic pressures, and shear stresses. The cure management system combines the above chemical and process monitoring data to optimize the thermomechanical properties of the cured matrix and minimize the voids and debonds at the composite interfaces. Curing of epoxy resins and acetylene terminated sulfone (ATSP) polymers are discussed in terms of computer aided management.

A computer model is presented for direct estimation of time dependent modulus, strength, extensibility, and fracture toughness of polymers based upon a description of chemical structure. The model analyzes the above mechanical properties in simple shear and tension. Polymer-dilutent interactions are defined and properties can be computed for varied temperature and time conditions of mechanical loading. Polymer density, solubility parameter, glass temperature, and interchain entanglement molecular weight are computed from functional group structures of the polymer repeat unit. Assignment of polymer molecular weight and crosslink density permits computation of the time dependent modulus functions for shear and tension. Appropriate integration of variable modulus and strain produces the computed stress vs strain curve and the fracture condition. The energy integral of the stress versus stain curve defines the fracture, energy of the unnotched specimen.

Computer based models are applied to develop relations between the chemical structure of acetylene terminated polysulfone (ATS) resin and the thermomechanical response as a function of cure state. The glass temperature T_g of the ATS resin increases with cure along a theoretical curve characteristic of chain extension and branching from zero to 80% of complete cure. Between 80%

cure and complete cure, the T_g rises abruptly from 150°C to a final value of T_g = 350°C for the fully cured resin. The chemical structure of cured ATS resin is introduced into a computer based model for numeric estimation of time dependent modulus, tensile strength, extensibility and unnotched fracture energy. The computed fracture energy (unnotched) and measured fracture energy (notched) are shown to display similar thermal responses from 25 to 375°C. Computer analysis of model ATS resins structures with high molecular weight and low crosslinking display large increase in fracture energy, which are accompanied by reduced yield stress and strength. ATS resin chemistry which provides optimized modulus, strength and toughness is suggested by the computer analysis.

The combination of the computer models discussed above provides a potential method of interconnecting the technologies of composite manufacture, service maintenance and repair. The clock face diagram of Fig. 4-1 depicts the

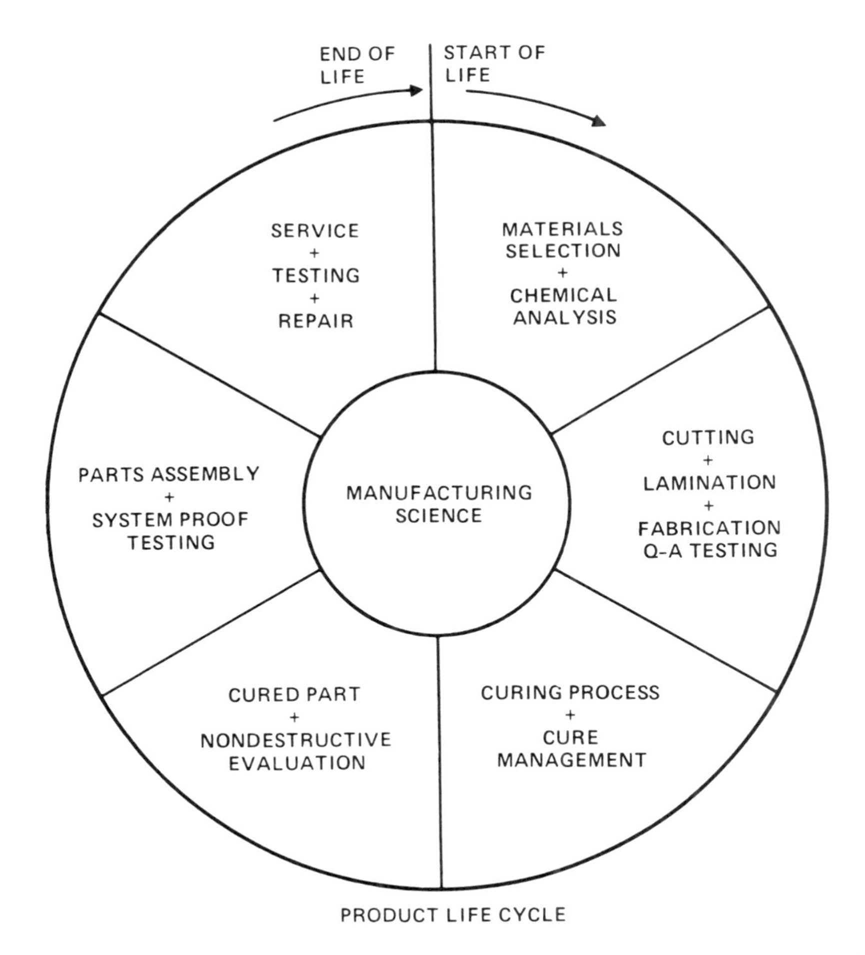

FIG. 4-1 Central role of manufacturing science in the product life cycle.

proposed interrelation between seven branches of a unified composite manufacturing science. The chemorheological cure models integrate the three sectors on the right side of Fig. 4-1 by interrelating all the subjects from materials selection to cure management. The composite cohesive reliability model integrates all subjects on the left side of Fig. 4-1 by specifying all of the mechanical properties of the composite matrix directly from polymer chemical structures as defined by a computer based model.

4.2 Relations Between Polymer Chemistry and Physical Properties

Extensive tabulations of polymer properties very often are limited to the major classes of commercially important polymers. Very often, the chemist or materials specialist wishes to obtain a qualitative estimate of several important polymer properties such as specific volume V_p, solubility parameter δ, and glass temperature T_g, simply by writing the chemical structure of the polymer repeat unit. The characteristic molecule weight between interchain entanglements, M_e, serves as an important index of true high polymer response, and is defined by a new relation which correlates M_e with the apparent cross section area of the polymer chain.

The functional group properties of the 33 functional groups utilized in this computation are listed in Table 4-1. The first column of Table 4-1 lists the group number which is employed as a retrieval number in the computer program. The second column lists the structure group in terms of a simplified upper case code with parenthesis to indicate side group structures and parenthesized numbers to indicate multiple side units. The reference polymer for each structure group shown in the right column of Table 4-1 further identifies the functional group chemistry. The central five columns of Table 4-1 tabulate the group properties utilized in the computation. From left to right these group properties are defined as follows:

- U = the cohesive energy per group.
- H = group contribution to internal degrees of freedom related to rotational motion.
- N = number of main chain atoms per group.
- V = group contribution to molar volume.
- M = molar weight per group.

The substantial proportion of the group values of U and H are based on the published volumes of Hayes[2] and the rest derived by inspection of known polymer properties. The values of N and M are directly deduced from chemical

TABLE 4-1 Functional Group Properties for Polymers

Unit No.	(R = 8.314 J/K*mole) Structure Group	U (J/mole)	H	N	V (m^3/mole)	M (kg/mole)	Polymer Unit
1	-CH2-	4.14E3	8	1	2.22E-5	1.4E-2	ethylene
2	-CH(CH3)-	1.28E4	11	1	4.44E-5	2.8E-2	propylene
3	-C((CH3)2)-	1.19E4	14	1	6.66E-5	4.2E-2	isobutylene
4	-CH(C6H5)-	3.01E4	15	1	1.11E-4	9.0E-2	styrene
5	-P-C6H4-	2.38E4	5	4	8.86E-5	7.6E-2	terephthlate
6	-M-C6H4-	2.58E4	10	3	8.86E-5	7.6E-2	isophthalate
7	-C(CH3)CH-	1.15E4	11	2	5.92E-5	4.0E-2	isoprene
8	-CHCH-	7.49E3	8	2	3.70E-5	2.6E-2	1,4-butadiene
9	-CH(CHCH2)-	1.29E4	11	1	5.90E-5	4.0E-2	1,2-butadiene
10	-CH(C6H11)-	2.56E4	21	1	1.48E-4	9.6E-2	vinyl cyclohexane
11	-CH(C(O)OCH3)-	2.81E4	23	1	7.57E-5	7.2E-2	methacrylate
12	-C(CH3)(C(O)OCH3)-	4.60E4	26	1	9.79E-5	8.6E-2	methylmethacrylate
13	-CH(CH3)O-	1.39E4	17	2	5.54E-5	4.4E-2	propylene oxide
14	-C(O)O-	1.41E4	12	2	3.32E-5	4.4E-2	ethylene adipate
15	-CH(OC(O)CH3)-	3.17E4	23	1	7.57E-5	7.2E-2	vinyl acetate
16	-C(O)-	7.32E3	6	1	2.22E-5	2.8E-2	ketone
17	-CH(C(O)OH)-	3.51E4	20	1	5.64E-5	5.8E-2	acrylic acid
18	-CH(OH)-	2.66E4	14	1	4.90E-5	3.0E-2	vinyl alcohol
19	-CH(OCH(O))-	2.86E4	20	1	5.54E-5	5.8E-2	vinyl formate
20	-O-	6.82E3	6	1	1.06E-5	1.6E-2	ether
21	-NHC(O)-	4.44E4	13	2	3.79E-5	4.3E-2	amide
22	-NHC(O)O-	2.63E4	19	3	4.89E-5	5.9E-2	urethane
23	-CH(CN)-	2.41E4	8	1	4.89E-5	3.9E-2	acrylonitrile
24	-CH(CL)-	1.75E4	8	1	4.07E-5	4.85E-2	vinyl chloride
25	-C(CL)CH-	1.26E4	8	2	5.55E-5	6.05E-2	neoprene
26	-C((CL)2)-	1.13E4	8	1	5.92E-5	8.30E-2	vinylidene chloride
27	-CF2-	4.81E3	8	1	3.48E-5	5.0E-2	tetrafluoroethylene
28	-CH2CF2-	1.48E4	16	2	5.70E-5	6.4E-2	vinylidene fluoride
29	-CF(CF3)-	1.84E4	13	1	6.96E-5	1.0E-1	perfluoropropylene
30	-SI((CH3)2)O-	1.72E4	30	2	8.62E-5	7.4E-2	dimethylsiloxane
31	-N((C(O))2)C6H2((C(O))2)N-	1.10E5	62	7	2.01E-4	2.14E-1	imide
32	-S-	8.26E3	8	1	2.56E-5	3.2E-2	sulfide
33	-S((O)2)	4.54E4	23	1	4.04E-5	6.40E-2	sulfone

structure and the magnitude of V is based upon LeBas values of atomic molar volume as reviewed by Kaelble.[1]

The polymer specific volume for the amorphous glass state is calculated by a simple empirical relation demonstrated by Kaelble, which is given as follows:[1]

$$V_p = 0.69 \left(\frac{\Sigma V}{\Sigma M}\right) \quad (4\text{-}1)$$

where ΣV and ΣM describe the summations of V and M that describe the polymer repeat unit. The constant 0.69 minimizes the deviations between experimental and calculated values of V_p for 13 polymers and generally gives agreement within five percent.

The polymer solubility parameter squared, δ^2, is termed the cohesive energy density and calculated by the following relation:

$$\delta^2 = 1.44 \left(\frac{\Sigma U}{\Sigma V}\right) \quad (4\text{-}2)$$

where ΣU is the summation of the group cohesive energies and ΣV the summation of the group molar volumes. The constant 1.44 reduces the Lebas values of molar volume to represent the polymer structure in the solid state. This computation calculates the value of cohesive energy density and expresses this result as the square of the familiar solubility parameter δ.

The Kelvin scale value of the glass transition temperature is calculated by a slight modification of the equation proposed by Hayes with the following form:[2]

$$T_g = \frac{2}{R} \left(\frac{\Sigma U}{\Sigma H}\right) + C(t) \quad (4\text{-}3)$$

In Eq. (4-3) R is the gas constant and ΣU and ΣH are the respective summations of cohesive energies and degrees of freedom for the functional groups of the polymer repeat unit. The constant C(t) = 25K in Eq. (4-3) is a temperature compensation which reflects the time scale of T_g measurements. The rule of thumb is that T_g is increased by roughly three degrees centigrade by each factor of ten reduction in measurement time.[3] Slow thermal expansion measurement of T_g is associated with C(t) = 25K and is used in the computer computations. More rapidly scanning calorimeter measurements might require C(t) = 20K as a normalizing correction.

The relation for calculating the molecular weight between entanglements M_e, in kg/mole, is given as follows:

$$M_e = 5.4E4 \ \frac{\Sigma M}{\Sigma N}\left[\frac{\Sigma V}{\Sigma N} - 1.48E\text{-}5\right]^{1/2} \tag{4-4}$$

In Eq. (4-4) the ratio ($\Sigma M/\Sigma N$) describes the averaged molecular weight per atom of main chain structure for the polymer repeat unit. The ratio ($\Sigma V/\Sigma N$) within the square root brackets of Eq. (4-4) correlates with the molar volume per main chain. If a functional group from Table 4-1 is introduced as a side chain unit, the computation sets N = 0 for that group, but introduces M and V for that group into Eq. (4-4). The constants 5.4E4 and 1.48E-5 in Eq. (5) are empirical factors which minimize the differences in experimental and calculated values of M_e. Since the chain length per main chain unit is roughly constant, the square root term of Eq. (4-4) represents a measure of the cylindrical cross section area of the averaged polymer repeat unit. The type entanglement described by Eq. (4-4) is associated with overlapping of the statistical chain coils. Interchain bonding by strong polar or hydrogen bonding interactions would tend to reduce M_e in a manner not described by Eq. (4-4).

The data of Table 4-1 and Eqs. (4-1)-(4-4) have been incorporated in an interactive microcomputer program with two sample computations shown in Table 4-2. The upper case of Table 4-2 shows the computational inclusion of eleven

TABLE 4-2 Sample Computations for Methacrylates (Upper Case) and Butadiene-Styrene Copolymers (Lower Case)

Unit No.	Moles	Structure Unit	Polymer Reference
		I. Main Chain Units	
1	1	-CH2-	ethlyene
12	1	-C(CH3)(C(O))CH3)-	methyl methacrylate
		II. Side Chain Units	
1	11	-CH2-	ethylene
Glass Spec. Vol. (M*M*M/kg = 9.89634E-04 (CC/G) = 0.989634			
Glass C.E.D. (J/M*M*M) = 3.7802E+08 (CAL/CC) = 90.3904			
Glass Temp (K) = 214.007 (C) = -59.1928			
Entang. MW (kg/mole) = 88.7177 (g/mole) = 88717.7			
		I. Main Chain Units	
1	0.87	-CH2-	ethylene
8	0.87	-CHCH-	1,4-butadiene
1	0.87	-CH2-	ethylene
1	0.13	-CH2-	ethylene
4	0.13	-CH(C6H5)-	styrene
Glass Spec. Vol. (M*M*M/kg) = 1.00516E-03 (CC/G) = 1.00516			
Glass C.E.D. (J/M*M*M) = 2.96893E + 8 (CAL/CC) = 70.9575			
Glass Temp (K) = 208.462 (C) = -64.7382			
Entang. MW (kg/mole) = 2.58618 (g/mole) = 2586.18			

TABLE 4-3 Comparison of Calculated and Experimental T_g (Ref. 2,4)

Methacrylates		V_p (g/cc)	δ^2 (cal/cc)	T_g (C)	M_e (kg/mole)	T_g (exp) (C)
methyl		0.829	144	107	18.2	105
ethyl		0.861	131	63	23.1	61
propyl		0.886	122	33	28.4	31
butyl		0.907	115	12	34.0	12
hexyl		0.938	105	-17	46.1	-19
octyl		0.989	90	-36	59.3	-38
dodecyl		0.989	90	-59	88.7	-62
Butadiene-Styrene Copolymers						
Mole (B)	Mole (S)					
0	1	0.883	88	110	20.2	100
0.2	0.8	0.902	86	69	12.8	-
0.4	0.6	0.923	82	28	8.2	-
0.61	0.39	0.954	77	-14	5.0	-12
0.64	0.36	0.959	77	-20	4.7	-13
0.72	0.28	0.973	75	-35	3.9	-34
0.77	0.23	0.983	74	-45	3.4	-37
0.87	0.13	1.01	71	-64	2.6	-51,-60
0.95	0.05	1.03	68	-80	2.0	-71,-76
0.99	0.01	1.04	67	-88	1.8	-74
1.00	0	1.04	67	-90	1.7	-79,-87

methylene groups in the side chain of polymethylmethocrylate to create poly-n-dodecylmethacrylate. The lower case of Table 4-2 illustrates the computational procedure for the typical SBR elastomer copolymer of styrene (S) and butadiene (B). Table 4-3 shows the computed property trends for a series of alkylmethacrylate polymers and the copolymers of stryene and 1,4-butadiene. For both polymer series Table 4-3 shows good agreement between computed and experimental T_g values which is typical of such computations. The trends for M_e change with T_g are seen to be opposite for the two polymer series. A comparison of computed and experimental values of M_e is summarized in Table 4-4. The agreement is reasonable considering the broad range experimental estimates of M_e.

An interactive computer model for estimating four polymer physical properties from polymer chemical structure is introduced and demonstrated. This model is useful in computer aided polymer design and in extending or interpolating experimental data on polymer properties. Monomer sequence distribution as discussed by Johnston(5) and semicrystallinity as discussed by Boyer(6) limits the use of this computational model for some polymers.

TABLE 4-4 Calculated and Experimental Values of M_e

Polymer	M_e (kg/mole)	M_e (exp) (kg/mole)
poly-n-octylmethacrylate	59.3	87
poly-n-hexylmethacrylate	46.1	33.9
polymethylmethacrylate	18.2	4.7-10.0
polystyrene	20.2	17.3-18.1
styrene-butadiene copolymer (0.87 mole St, o.13 mole Bd)	2.6	3.0
poly-1,4-polybutadiene	1.7	1.7-2.9

4-3. Computer Aided Cure Management

The analysis and management of curing (simultaneous polymerization and crosslinking of thermosets resins) remains as one of the major engineering development areas in polymer composite materials development. Ennes and Gillham[7] have recently introduced the TTT (time-temperature-transformation diagrams) to aid in the interpretation of the complex interaction of changing chemical structure and rheological (glass-rubber-flow) state during cure as recorded by torsion braid analysis (TBA) of curing. Macosko, Miller and Valles[8] have recently developed and demonstrated models for computing the average molecular properties of polymer structures during cure. The computer models discussed here join the chemical structure models and the rheological models so that the complete chemical-thermal-rheological property spectrum of the curing polymer can be displayed at any point in the cure cycle. The primary function of the computer aided cure model is initially to illustrate the effects of varied cure paths (defined by crosslink/chain extension ration c/r), and the extent of reaction, F, as measured by fractional heat of polymerization.[9]

The first important consequence of curing is the progressive increase in the glass transition T_g and flow temperature T_m with increase in the extent of cure. The effect of cure path upon T_g has recently been shown to follow a theoretical prediction as follows:[9]

$$T_g = T_{g\infty} \left(\frac{r(z - 2)}{r(z - 2 - \frac{2c}{r}) + 2z} \right) \tag{4-5}$$

where r is the number average chain extension, c/r is the crosslink/chain extension ratio, z is the lattice coordination number and $T_{g\infty}$ is the glass transition of the infinite linear polymer. As polymerization progresses an entanglement network will form to produce a rubbery state response prior to gelation. The "effective" fraction of chain segments x_e entrapped by entanglements is given by the following relation:[10]

$$x_e = 1 - 2(M_e/M_w) \geq 0 \qquad (4\text{-}6)$$

where M_e is the molecular weight between entanglement, and M_w is the weight average molecular weight of the curing polymer.

The ratio of the maximum flow relaxation time τ_m to the glass transition-relaxation time τ_g, at constant temperature, is given by the following relation:[10]

$$\frac{\tau_m}{\tau_g} = \left(\frac{M_w}{M_o}\right)^2 \left(\frac{M_w}{2M_e}\right)^{2.4\, x_e} \qquad (4\text{-}7)$$

where M_w is the weight average molecular weight of the curing polymer and M_o is the molecular weight per chain atom. Knowing M_o, M_e, M_n, and M_w provides for direct calculation for the time range of rubbery response up to the point of gelation. The flow temperature T_m can be calculated relative to the glass temperature at constant measurement time, by the following modified form of the familiar William, Landel, Ferry (WLF) relation:[10]

$$T_m = T_g + \frac{51.6 \log(\tau_m/\tau_g)}{17.4 - [\log(\tau_m/\tau_g)]} \qquad (4\text{-}8)$$

as incipient gelation occurs it follows that M_w rises toward infinity and from Eq. (4-7) and Eq. (4-8) one sees that T_m increases toward infinity also.

The uncured polymer is defined as having an initial state of chain extension r_o which is associated with the glass temperature T_{go} of the uncured material. Equation (4-5) can be rearranged to define r_o as follows:[9]

$$r_o = \frac{2z}{z - 2} \; \frac{T_{go}}{T_{g\infty} - T_{go}} \qquad (4\text{-}9)$$

This initial average value of r_o correlates with the following standard description of the reaction system:[11]

$$N_o = N_1 + N_2 \tag{4-10}$$

$$M_o = (N_1M_1 + N_2M_2)/N_0 \tag{4-11}$$

$$s = N_1f_1/N_2f_2 \tag{4-11}$$

$$f = (N_1f_1 + N_2f_2)/N_0 \tag{4-12}$$

$$F_1 = F_2/s \tag{4-13}$$

where N_1, N_2 and M_1, M_2 and f_1, f_2 describe the respective moles, molecular weights, and coreaction functionality of coreacting molecules type 1 and 2. The coreacting system is described by the initial moles N_0, average molecular weight M_0, molecular ratio of reactive groups s, and average functionality f. The polymer structure is defined by both the reacted fractions F_1 and F_2 and the branching coefficient α as follows:[11]

$$\alpha = ASF_1 + (1 - A)sF_1^2 \tag{4-14}$$

where A is defined as the effective fraction of molecules with $f > 2$. The number average, r_n, and weight average, r_w, degrees of chain extension are defined as follows:[11]

$$r_n = \frac{1}{1 - (f\alpha/2)} \tag{4-15}$$

$$r_w = \frac{1 + \alpha}{1 - (f - 1)\alpha} \tag{4-16}$$

and the point of incipient gelation is defined where r_w approaches infinite values when:[11]

$$\alpha \cong \alpha_c = \frac{1}{f - 1} \tag{4-17}$$

The relation between branching coefficient α and sol fraction W_s is given by a relation of Gordon:[12,13]

$$\alpha = \frac{1 - W_s^{(1/f)}}{1 - W_s^{(1-(1/f))}} \tag{4-18}$$

where the gel fraction $W_g = 1 - W_s$ and W_s is the weight fraction of soluble material. The effective molecular weight between crosslinks is proposed by the following provisional relation:

$$M_c = \frac{2M_o}{W_g} \frac{\alpha_c}{\alpha - \alpha_c} \geq 0 \qquad (4\text{-}19)$$

The above relations are approximations to more accurate and specific descriptions of polymer network structure. These relations are useful in illustrative computer modelling of curing processes. The computer model inputs values of F_1 as the independent variable when $\alpha < \alpha_c$. When $\alpha > \alpha_c$ the computer model inputs values of W_s into Eq. (4-18) and solves for α and F_1 as dependent variables.

The relations presented above have been consolidated in a computer model for cure management. This computer model defines a diagram of transition temperatures T_g and T_m vs the extent of chemical reaction F. For a cure reaction in which the diffusional mobility of reacting species is controlled by the free volume state, the rate of reaction will depend upon the temperature difference $T - T_g$. When the glass transition rises during cure so that $T - T_g \leq 0$ the system is cured into its glass state and the reaction is slowed by diffusional contraints. The standard rate equation for chemical reaction can be written as follows:

$$\frac{dC_i}{dt} = A_T C_o^n (1 - F_i)^n \exp(-E/RT) \qquad (4\text{-}20)$$

where A_T is a temperature dependent frequency factor, C_o is the initial concentration of reactants, n is the kinetic order of reaction, E is the chemical (Arrhenius) activation energy, R the gas constant, and T is absolute temperature. The rate of cure (dC_i/dt) is related to the current fraction reacted F_i at time t. Normally A_T is assumed constant with a value $A_T \simeq 10^{13}$ to 10^{15} jumps/sec characteristic of local intramolecular motion. For reaction kinetics controlled by intermolecular segment motion, an appropriate relation for A_T is provided by a modified form of the familar WLF (Williams, Landel, Ferry) relation:[3]

$$\ln A_T = \ln A_{T_g} + \frac{40.7\ (T - T_g)}{51.6 + |T - T_g|} \qquad (4\text{-}21)$$

where a jump frequency at $T = T_g$ of $A_{T_g} \simeq 10^{-3}$ jumps/sec defines the T_g represented by the definitions of Eq.(4-21). Combination of Eq. (4-20) and Eq. (4-21) defines a kinetics of curing which can become dominated by a rise in T_g which minimizes $(T - T_g)$ in Eq. (4-21) and consequently lowers A_T in Eq. (4-20). The inclusion of the absolute value of $T - T_g$ in Eq. (4-21) permits application of this relation both above and below T_g as suggested by the review discussion of Hedvig.[14]

All of the above relations can enter a single cure management model which defines the optimum balances of cure rheology, network structure, and reaction kinetics. A phase diagram of rheological state vs cure state developed from the above model as previously shown in Fig. 2-2. The diagram of Fig. 2-2 presents a graphical correlation between the effects of curing upon the molecular weight distributions (upper view) and the rheological states (lower view) of a thermosetting resin. The polymer is soluble up to the gel point as defined by the dashed vertical line in Fig. 2-2 and by Eq. (4-14) and Eq. (4-17) in the above model. The breadth of the molecular weight distribution is measured by the ratio of weight to number average molecular weight X_W/X_N as shown in the upper left curves of Fig. 2-2. At the gel point the weight average molecular weight, which describes the larger molecules of the reacting polymer approaches infinity. Further increases in the degree of cure beyond this gel point causes a rapid rise in the insoluble fraction termed gel formed by the crosslinking of these large molecules as shown in the right curve of upper Fig. 2-2 and Eq. (4-18) of the computational model. At complete cure the gel fraction should constitute the bulk of the polymer with negligible unreacted low molecular weight polymer.

The curves of lower Fig. 2-2 outline the characteristic changes in rheological states of liquid flow, rubber, and glass states. For the uncured polymer the liquid flow state extends down to the monomeric glass transition T_{go} as shown in the lower left view of Fig. 2-2. The uncured resin does not possess a rubbery state at zero degree of cure. With increasing degree of cure the lower curves of Fig. 2-2 show the characteristic changes in flow temperature T_m and glass temperature T_g predicted by the computational model. The separation of T_m from T_g with increasing cure defines the appearance of the soluble rubbery state which separates the flow state from the soluble glass state. The transition from flow to rubbery state, defined by T_m, rises to the limits of thermal stability as the degree of cure approaches the gel point. The elimination of bubbles, entrapped air and unwetted interface by manipulation of pressure and vacuum must all be accomplished in the flow state and prior to gelation. The gelled polymer has an infinite viscosity and will not flow. At a degree of cure beyond the gel point only the gelled rubber and gelled glass states exist.

The Society of Manufacturing Engineers (SME) recently held a three day conference relating to computer aided manufacturing of large-scale polymer composite structures.[15] These proceedings can be schematically represented by the illustrative computer aided manufacture-service-repair cycle shown in Fig. 4-2. The cycle illustrated in Fig. 4-2 starts, in the upper left with materials acquisition and ends in the lower right with the end of life and scrap. Each of the decision points numbered from 1 to 5 in the diamond boxes of Fig. 4-2 describe the option of accept or rejection based upon prior monitoring of chemical or physical properties.

The connection between Fig. 2-2 and Fig. 4-2 can be based upon the temperature scales shown in lower Fig. 2-2. In general, uncured adhesive or composite prepreg will be stored at a reduced temperature T_1 which is below T_{go} shown in lower Fig. 2-2 in order to quench the curing reaction. Cutting and lamination will occur at temperature T_2 which for the uncured material is in the flow state above T_m. The next step of manufacture is the cure cycle where the degree of cure is advanced to raise the polymer glass transition to $T_{g\infty}$. The cure will be temporarily raised to T_5 (see Fig. 2-2) in order to exceed $T_{g\infty}$ and complete the cure in the crosslinked rubbery state.

Normally this $T_{g\infty}$ represents the thermal ceiling for structural properties and the cured composite continues its service life at temperatures below $T_{g\infty}$. It is not unusual for a cured part emerging from the cure cycle shown in Fig. 4-2 to have a cost of over a million dollars.[15] In Fig. 4-2, the nondestructive evaluation (NDE) of the cured part only identifies the unacceptable parts and thus prevents their incorporation in the assembled structure. The preventative actions to produce good parts must appear in the first three monitor-management points associated with chemical analysis, fabrication quality assurance, and in the cure cycle.

The mathematical models described above in reference to both Fig. 4-1 and Fig. 4-2 combine to provide computational aids in these early preventative stages of manufacturing where process management can minimize or eliminate defects which influence performance and service live. Applications of these computer models to experimental curing studies of acetylene terminated polysulfones (ATS) and epoxy resin composites[16] will be discussed in terms of computer aided cure management.

4-4. Relations Between Polymer Chemistry and Mechanical Properties

Generation of mathematical models to predict the mechanical response of polymers directly from chemical structure represents an important objective of polymer physics.[17,18] In this discussion methods are presented for computer

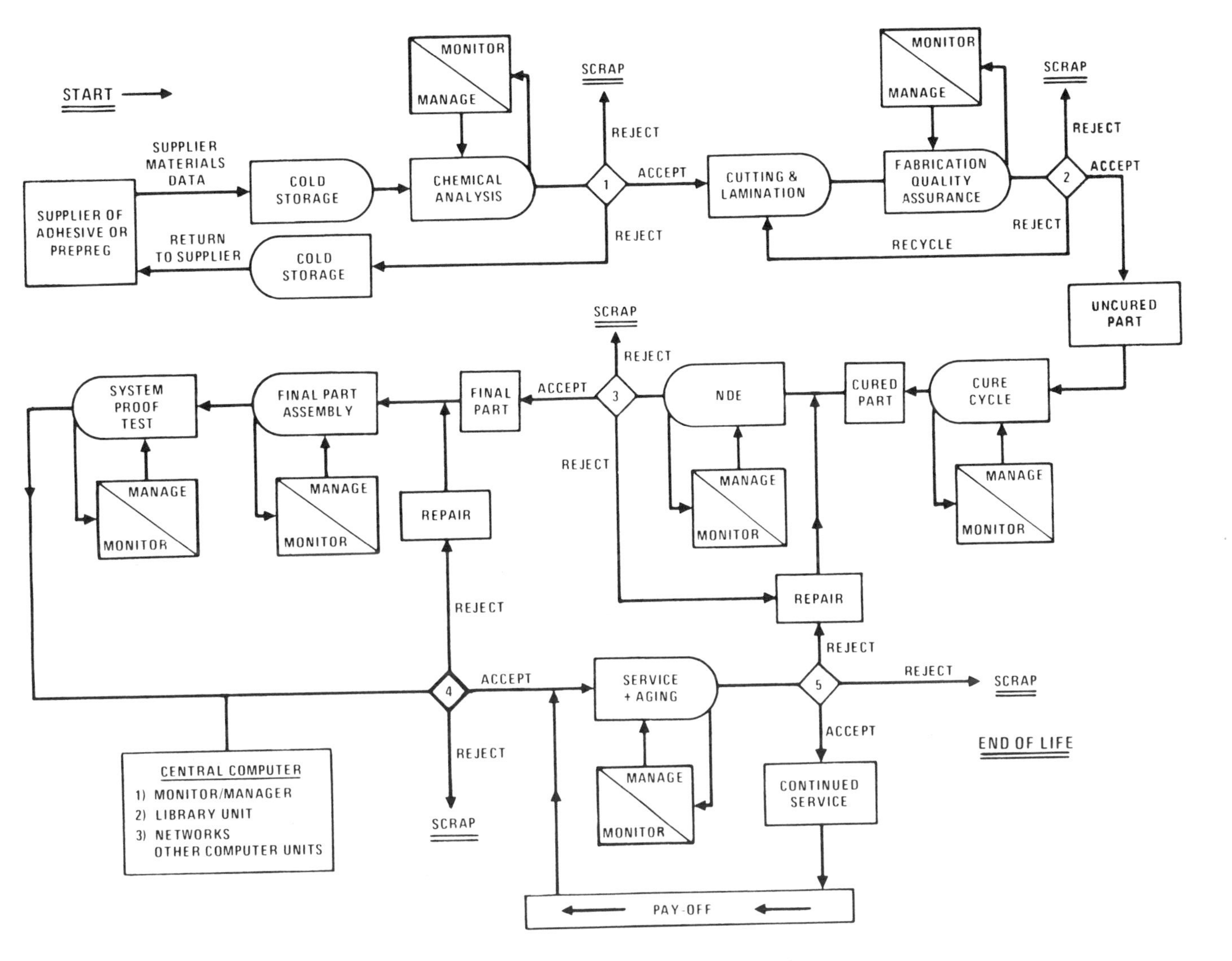

FIG. 4-2 Illustrative computer aided manufacture-service-repair cycle.

assisted estimation of the time dependent modulus, strength and extensibility and the unnotched fracture toughness of polymers based upon chemical composition and macromolecular structure. A molecular model for mechanical properties and cohesion of polymers proposed by Kaelble[10] is implemented in these computer estimations. This computer model thus assembles quite diverse aspects of liquid-solid state theory and polymer rheology to describe mechanical properties and cohesion. The objective of this computer model is to provide a rapid and meaningful translation of chemical structure information into direct estimations of mechanical performance useful in plastics engineering.

The computer model requires as a minimum description of chemical structure the following six chemical structure properties:[10]

V_p = polymer specific volume below T_g
δ_p^2 = polymer solubility parameter squared (cohesive energy density) below T_g
T_g = glass transition temperature
M_e = interchain entanglement molecular weight (num. ave.)
M_n = polymer moleculor weight (num. ave.)
M_c = molecular weight between chemical crosslinks (number ave.)

The upper four properties V_p, δ_p^2, T_g, and Me can often be calculated for the linear high polymer from five functional group properties which are:[19]

U = the cohesive energy per group
H = group contribution to rotational degrees of freedom
N = number of main chain atoms per group
V = group contribution to molar volume
M = molecular weight per group

A combination of polymer (P) and soluble diluent (D) can be described in the above terms plus the addition of the following two parameters which are:[10]

$\phi_P = 1 - \phi_D$ = volume fraction polymer
Φ = polymer - solvent interaction parameter (a pure number usually between 0.5 and 1.0)

With this fairly simple description of the polymer or polymer-diluent combination the computer model proceeds to calculate the "master curve" of time dependent shear modulus such as shown in the upper curve of Fig. 4-3. A second computation generates the stress versus temperature functions for mechanical

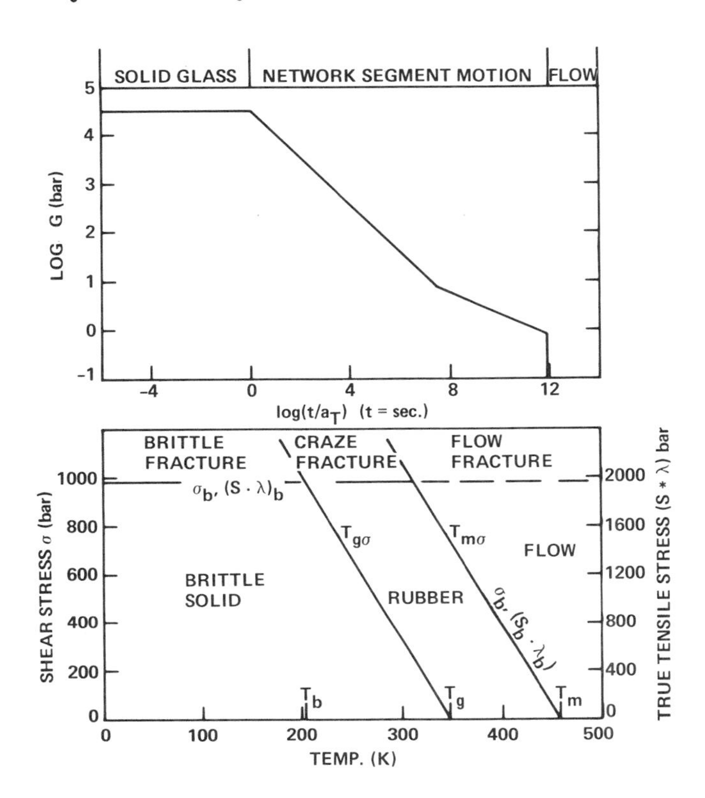

FIG. 4-3 Calculated shear modulus G vs reduced relaxation time t/a_T (upper curve) and stress-temperature functions for yield and fracture (lower curves).

yielding, fracture and flow as shown in the lower curves of Fig. 4-3. These two sets of curves in Fig. 4-3 represent the first translation of the polymer chemistry into polymer physical state diagrams which are otherwise obtained by extensive rheological and mechanical testing.[10,17,18]

The mechanical response model proposed by Kaelble[10] treats stress effects upon free volume. The predicted effect of mechanical loading in shear or tension is to increase free volume, and as a consequence increase molecular motion. The predicted transitions from solid to rubber to flow state thus appear in a stress versus strain curve as the yield stress followed by the flow or fracture stress. A simple variation on the familiar WLF equation of Williams, Landel, and Ferry is employed as follows:[10]

$$\log\left(\frac{1}{a_T}\right) = \log\left(\frac{\tau_g}{\tau_{g\sigma}}\right) = \frac{17.4(T - T_{g\sigma})}{51.6 + |T - T_{g\sigma}|} \qquad (4\text{-}22)$$

where a_T is the dimensionless time shift factor, τ_g the glass transition relaxation time at zero stress ($\sigma = 0$), and $\tau_{g\sigma}$ the glass transition relaxation time at stress σ, T = test temperature and $T_{g\sigma}$ the glass temperature at stress σ.

As shown in the lower curves of Fig. 4-3 the magnitude of $T_{g\sigma}$ is decreased by increased stress thereby increasing the value of $\log(1/a_T)$ as defined by Eq. (4-22). In upper Fig. 4-3, with the test time t either constant or increasing with stress σ, the decrease in $\log(1/a_T)$ will increase the magnitude of $\log(t/a_T)$ displayed on the absicca of the diagram. The prediction is that increased stress σ will increase $\log(t/a_t)$ and thereby decrease the shear modulus G as shown by the master function of upper Fig. 4-3. This general concept that stress and temperature affect the rheological response through Eq. (4-22) is exploited in the numeric analysis of a computed stress versus strain and fracture response.

The relation between shear stress and strain is defined by the following integral equation:[10]

$$\sigma = f(\lambda) \int_0^{\gamma} G(t/a_T)\, d\gamma \tag{4-23}$$

where $f(\lambda)$ is an appropriate function of extension ratio λ which accounts for chain orientation. The reduced time shear module $G(t/a_T)$ is the tangent slope of the shear stress σ versus strain γ curve at external test time t and concurrent internal polymer reduced time a_T as defined by Eq. (4-22). The relation between nominal tensile stress S and strain ε is defined by a similar integral equation:[10]

$$S = f_1(\lambda) \int_0^{\varepsilon} E(t/a_T)\, de \tag{4-24}$$

where $f_1(\lambda)$ is the appropriate function of extension ratio for tensile deformation and chain orientation. The reduced time tensile modulus $E(t/a_T) \approx 3\,G(t/a_T)$ forms the tangent slope of tensile stress-strain curve at time t and reduced time a_T.

The stress-strain curves defined by Eq. (4-23) and Eq. (4-24) are generated by numeric integration where the external test time t and polymer response time $(1/a_T)$ are combined to define the incremental slope $E(t/a_T) \approx 3$ $G(t/a_T)$. The respective fracture energy per unit volume in shear W_s and tension W_T are defined by the stress-strain integrals as follows:[10]

$$W_s = \int_{\gamma=0}^{\gamma_b} \sigma\, d\gamma \tag{4-25}$$

$$W_T = \int_{\varepsilon=0}^{\varepsilon_b} S \, d\varepsilon \qquad (4\text{-}26)$$

which represent the areas beneath the respective stress-strain curves. The magnitudes of W_S and W_T represent the fracture energy per unit volume of unnotched polymer deformed under uniform stress and strain.

Structural reliability R_W is defined here as the statistical probability of load supporting ability (nonfailure) under a defined set of general environmental conditions which include mechanical stress and strain, temperature, and loading time. A multidimensional failure envelope can be defined by the extreme value statistics introduced by Weibull into the widely used models of structural reliability.[20,21] Three principal failure responses can be incorporated into the following Weibull type definition for structural reliability as follows:[22]

$$\begin{aligned} R_W &= \exp\left(-\left(\frac{\sigma}{\sigma_\sigma}\right)^{m(\sigma)}\right) \exp\left(-\left(\frac{\lambda}{\lambda_b}\right)^{m(\lambda)}\right) \exp\left(-\left(\frac{t}{t_b a_T}\right)^{m(t)}\right) \\ &= R_\sigma * R_\lambda * R_t \end{aligned} \qquad (4\text{-}27)$$

where σ_b, λ_b, and t_b are respectively described as the nominal strength, extension ratio, and load endurance time where $R_W = 1/e = 0.37$ for each process acting independently. The parameters σ, λ, and t represent the applied stress, extension ratio, and load dwell time. The presence of a_T from Eq. (4-22) in the last term of Eq. (4-27) introduces the reduced variables criteria into the time dependent failure term. The Weibull factors $m(\sigma)$, $m(\lambda)$, and $m(t)$ define respective distribution shape factors. Since R_W is the product of the three exponential decay terms, Eq. (4-27) assumes that these mechanisms interact in series to give a conservative estimate of structural reliability.

The model polymer described by the curves of Fig. 4-3 is a high molecular weight polyvinyl chloride polymer with the following chemical structure properties: $V_p = 0.71$ cc/gm, $\delta_p^2 = 94.1$ cal/cc, $T_g = 348$ K, Me = 6490 gm/mole, $M_p = 8.32 \cdot 10^5$ gm/mole and $M_c = 8.32 \cdot 10^5$ gm/mole to reflect no crosslinking. The computed curves of nominal tensile stress and strain are shown in the upper view of Fig. 4-4. The stress-strain curve number and the associated computation temperature are listed in Table 4-5. The lower curve of Fig. 4-4 shows the fracture energy rise from brittle failure values at 220 K which maximize near room temperature and diminish toward zero at the flow temperature $T_m \simeq 458$ K. The experimental deformation and failure properties of an

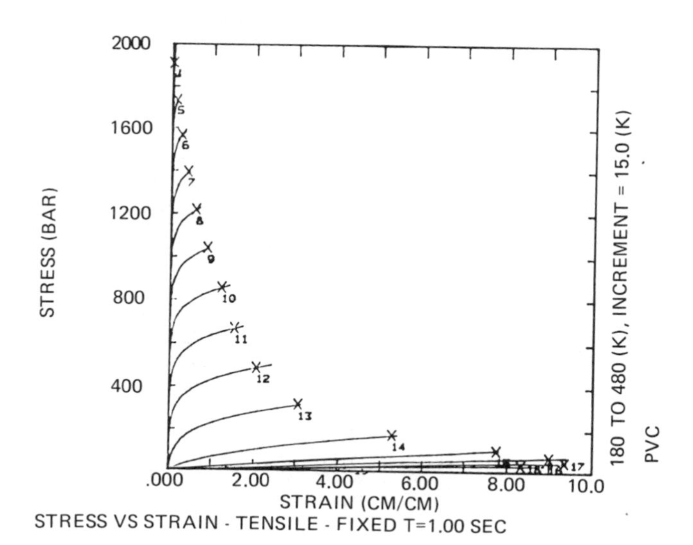

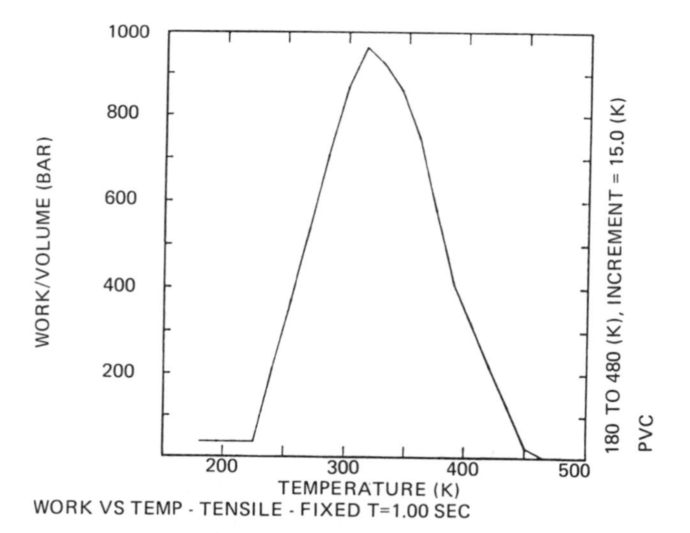

FIG. 4-4 Computed estimates of nominal tensile stress vs strain response and failure (indicated by X in upper curves) and fracture energy (lower curve) of linear polyvinyl chloride (T_g = 348 K, M_n = 8.53E5 gm/mole).

TABLE 4-5 Relation of Stress-Strain Curve Number to Test Temperature

Curve No.	Temp (K)	Curve No.	Temp (K)	Curve No.	Temp (K)
1	180	8	285	15	390
2	195	9	300	16	405
3	210	10	315	17	420
4	225	11	330	18	435
5	240	12	345	19	450
6	255	13	360	20	465
7	270	14	375	21	480

unplasticized PVC polymer of T_g = 346 K and weight average molecular weight $\bar{M}_w = 1.16 \cdot 10^6$ gm/mole have been previously reported[10,23] and display reasonable agreement with the computed results shown in Fig. 4-4. In the experimental study the tensile strength S_b and extensibility displayed a well defined failure envelope as shown in the lower portion of Fig. 4-5. One notes in comparing upper Fig. 4-4 and Fig. 4-5 that the strength $S_b \simeq 50$ bar at maximum extensibility $\varepsilon_b \simeq 9.0$ is in good agreement with experiment. Experimental strength and extensibility data for temperatures below T_g are shown in Fig. 4-5 to be widely scattered. These low temperature data can be converted to true tensile stress $S_b\lambda_b$ where $\lambda_b = \varepsilon_b + 1$ is the extension ratio at break.

In Fig. 4-6 the 26 experimental true tensile stress values fall on a modified Weibull type distribution function where the shape factor $m(\sigma)$ changes from 5.26 to 1.65 as $R(\sigma)$ decreases. This type of bimodal Wiebull function has been observed and discussed previously with regard to polymer cavitation processes.[24] By utilizing the experimental relation of $R(\sigma)$ versus true tensile strength $S_b\lambda_b$ curves of constant $R(\sigma)$ and $S_b\lambda_b$ can be defined for the failure data of Fig. 4-5. Inspection shows that these functions satisfactorily enclose the upper branch of the failure envelope of Fig. 4-5 which is dominated by the stress criteria of failure described by the $R\sigma$ mechanism of Eq. (4-27).

A single example illustrates the use by which the computer model permits investigation of changes in chemical structure properties on mechanical response. The high molecular weight thermoplastic PVC response shown in Fig. 4-4 can be modified by crosslinking. The curves of Fig. 4-7 show the same model PVC polymer of Fig. 4-4 which is modified by a single molecular structure change which is a low molecular weight between crosslinks of M_c = 500 gm/mole.

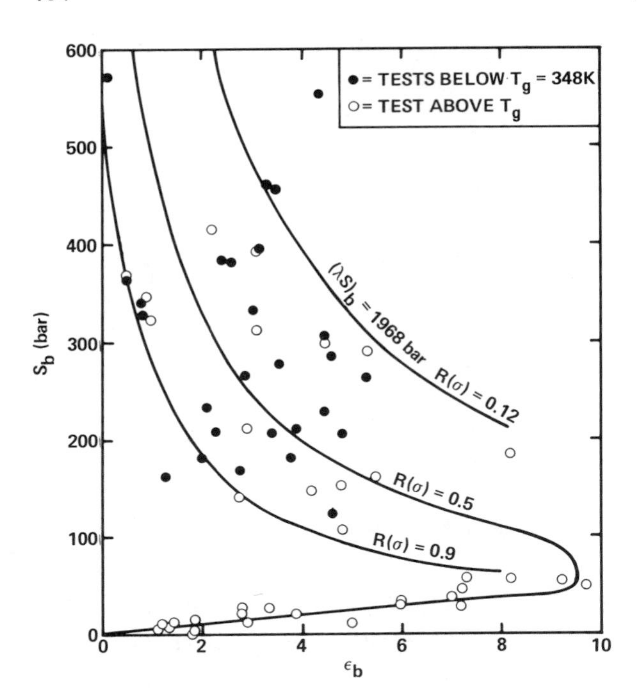

FIG. 4-5 Experimental values of nominal tensile stress S_b vs extensibility ε_b for polyvinyl chloride film (T_g = 346 K, M_w = 1.16E6 gm/mole).

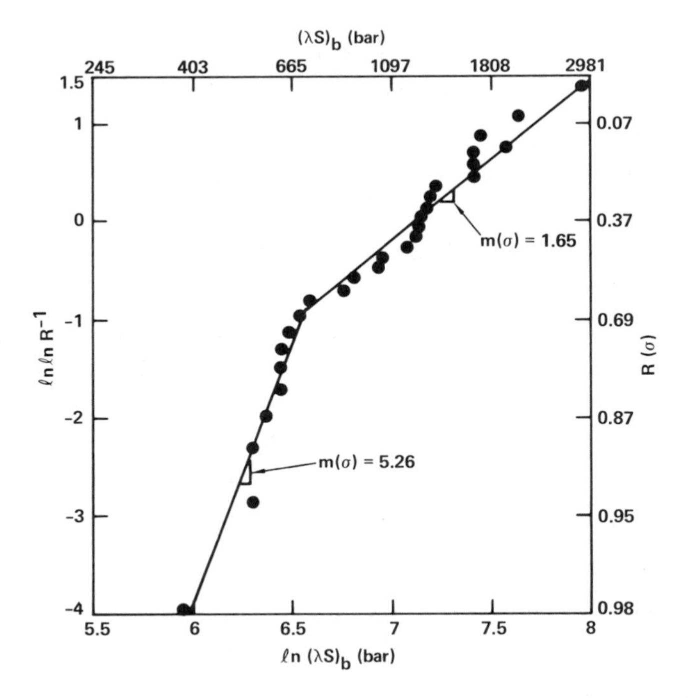

FIG. 4-6 Experimental reliability distribution $R(\sigma)$ for true tensile strength $(\lambda S)_b$ of polyvinyl chloride below T_g.

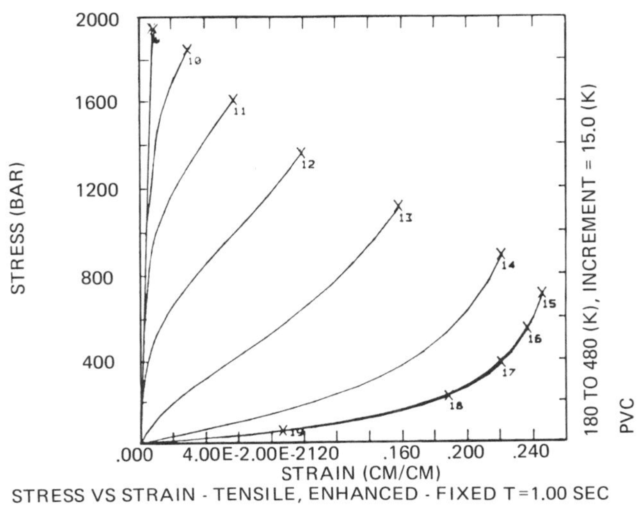

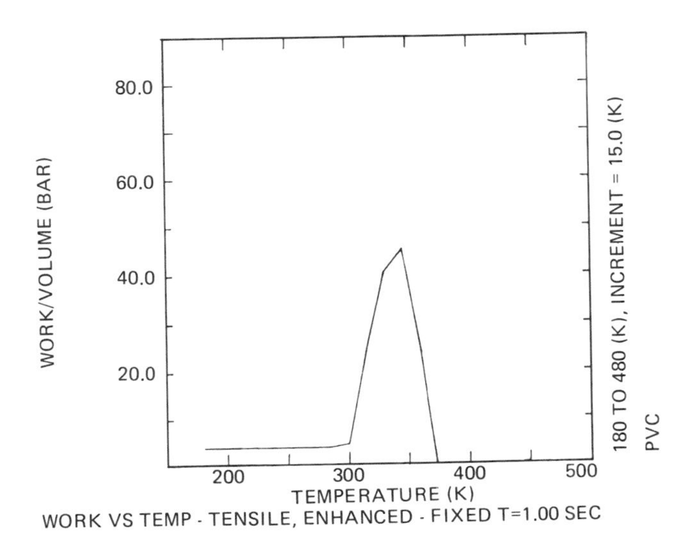

FIG. 4-7 Computed estimates of nominal tensile stress vs strain response (upper curves) and fracture energy (lower curve) for crosslinked PVC.

This high crosslink density of $(1/V_D M_C) = 2.82 \cdot 10^{-3}$(moles/cc) is typical of fully crosslinked thermoset resins such as epoxies. The effect of this simulated dense crosslinking is to raise the stress and lower the polymer extensibility as shown in the upper curves of Fig. 4-7. The effects of chain orientation are dominant at temperatures well above T_g (see curves 15-19) where the stress strain curves and failure points are controlled by the equilibrium network modulus. The lower curve of Fig. 4-7 shows the narrow temperature span and lower magnitude of fracture energy produced by dense crosslinking. These same results are displayed in intercomparisons of experimental deformation and fracture master curves for thermoplastic PVC and thermoset epoxy resins at corresponding free volume states where effects of chemical structure dominate mechanical properties.[10,23]

4-5. Computer Aided Analysis of Acetylene Terminated Resin

Generalized approaches to the experimental characterization of small quantities of newly developed structural polymers have been outlined and demonstrated for acetylene terminated sulfone (ATS) polymers.[25] Further characterization of the effects of cure state upon dynamic mechanical and fracture response of ATS has been presented in a companion paper.[26] Recently developed computer based models[27,28] for estimating polymer physical and mechanical response from chemical structure are applied in this discussion. These computer models are based on a detailed molecular theory of polymer mechanical properties and cohesion developed by Kaelble.[10]

The chemical structure of a model ATS oligomer unit is shown in Fig. 4-8. The polymerization and crosslinking reactions of ATS polymers are now known to be quite complex and can yield combinations of simple chain extension, branching, and crosslinking during cure with only about 30 percent of the ethynyl groups undergoing trimerization.[29] Initial calculation of four important physical properties for linear ATS polymer are developed from the functional group properties listed in Table 4-6. For the infinite linear ATS chain below T_g the computed values of polymer specific volume V_P, cohesive energy density (solubility parameter squared) δ_p^2, glass temperature T_g and molecular weight between entanglements M_e are summarized in Table 4-7.

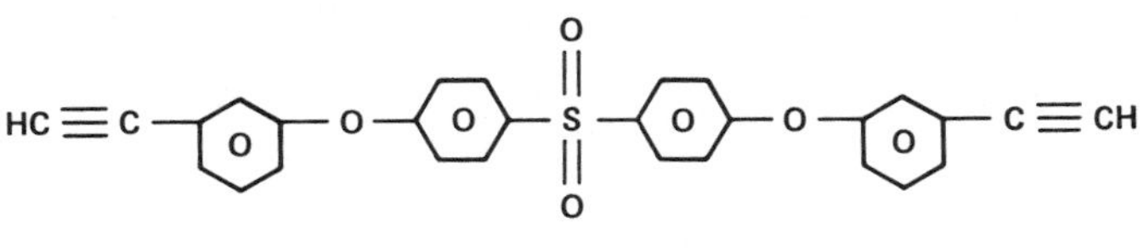

FIG. 4-8 Model ATS oligomer structure.

TABLE 4-6 Functional Group Properties for Polymers

Structure Group	U (J/mole)	H	N	V (m^3/mole)	M (kg/mole)	Reference Polymer
-M-C6H4-	2.58E4	10	3	8.86E-5	7.6E-2	isophthalate
-P-C6H4-	2.38E4	5	4	8.86E-5	7.6E-2	terepthalate
-O-	6.82E3	6	1	1.06E-5	1.6E-2	ether
-CHCH-	7.49E3	8	2	3.70E-5	2.6E-2	1, 4-butadiene
-S((O)2)-	4.54E4	23	1	4.04E-5	6.4E-2	sulfone

The summed functional group properties listed in Table 4-6 are defined as follows:

U = the cohesive energy per group

H = group contribution to internal degrees of freedom related to rotational motion

N = number of main chain atoms per group

V = group contribution to molar volume

M = molar weight per group.

The uncured ATS oligomer of this study displayed a T_g = 40°C as shown on the left side of Fig. 4-9. The fully cured ATS resin displays a measured T_g = 344°C shown on the right side of Fig. 4-9. Intermediate values of T_g representing different degrees of partial cure under isothermal cure history are shown distributed along the lower curve of Fig. 4-9.

The solid curves of Fig. 4-9 are calculated from a theoretical expression for T_g which is defined in terms of varied degree of chain extension, r, and ratio of crosslinks per chain unit, c/r, as shown in Eq. (4-5), where Z = 10 defines the local lattice coordination number and $T_{g\infty}$ = 540K (267°C) is computed for the linear polymer in Table 4-7. Inserting the experimental T_g = 313K (40°C) into Eq. (4-5) for the case c/r = 0 defines an r_0 = 3.45 for the uncured ATS resin. Inserting the experimental T_g = 617K (344°C) for the fully cured ATS resin and r = ∞ into Eq. (4-5) yields a computed value of c/r = 0.5 for the fully cured ATS resin.

The standard relations for step polymerizations (see Eq. 4-5 through Eq. 4-21) can be exploited to generate the T_g versus cure curves shown in Fig. 4-9. For linear or branched polymerization without crosslinking we assume the following relation:

$$r = r_0 \left(\frac{1}{1 - F^Z}\right) \tag{4-28}$$

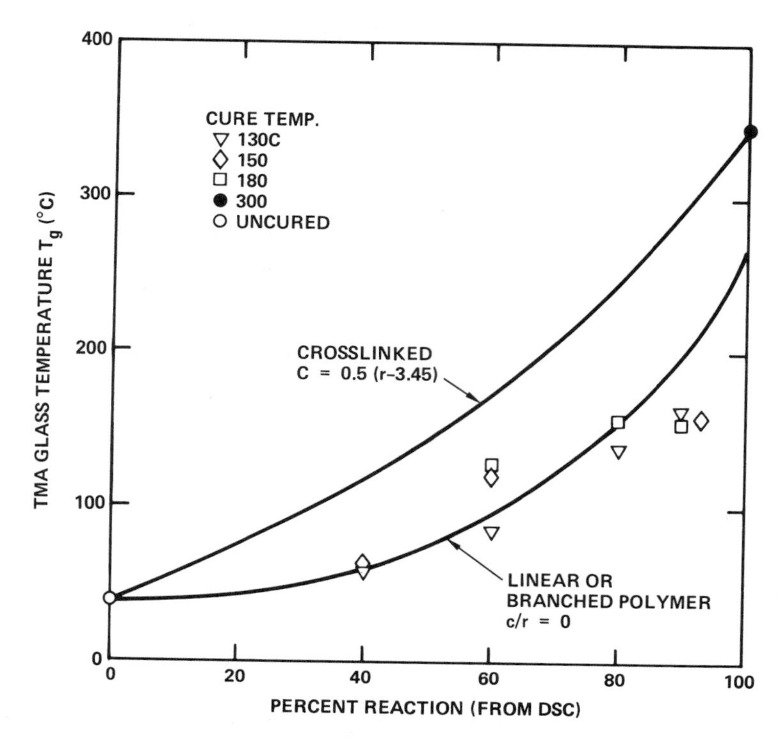

FIG. 4-9 Experimental and theoretical (solid curves) values of T_g for ATS as a function of cure path.

TABLE 4-7 Computed Estimates of ATS Physical Properties from Chemical Structure

Unit No.	Moles	Structures Unit	Polymer Reference
6	2	-M-C6H4	Isophthalate
5	2	-P-C6H4	Terephthalate
20	2	-O-	Ether
8	2	-CHCH-	1-4-Butadiene
33	1	-S((O)2)-	Sulfone

Glass Spec. Vol (M*M*M/KG) = 7 48009E-04 (CC/G) = .748009
Glass C.E.O. (J/M*M*M) = 5.09055E+08 (CAL/CC) = 121.664
Glass Temp. (K) 540.383 (C) = 267.183
Entang. MW (KG/Mole) = 3.39525 (G/MOLE) = 3395.25
U,H,V,M,N 173220 81 4.9E-04 452 21 (Summed Values)

where r is the number of chain repeat units and F is the fractional degree of cure displayed in the abscissa of Fig. 4-9. Combining Eq. (4-5) and Eq. (4-28) for c/r = 0 produces the lower curve of Fig. 4-9. When the reactive groups have functionality greater than two so that crosslinks accompany polymerization the appropriate equations for r and c are as follows:

$$r = r_0 \left(\frac{1}{1 - F}\right) \tag{4-29}$$

$$c = 0.5\ (r - r_0) \tag{4-30}$$

where $r_0 = 3.45$. Combination of Eqs. (4-5), (4-29), and (4-30) produces the upper curve of Fig. 4-9. The upper and lower curves of Fig. 4-9 define two cure paths. The experimental data points of Fig. 4-9 appear to follow the linear polymerization path up to 80 percent reaction. Between 90 percent reaction and full cure the T_g rises abruptly from 150°C to 344°C by a path evidently dominated by a crosslinking reaction.

The ATS properties described in Table 4-6 and Table 4-7 have been combined with calculated values of polymer molecular weight M_p and molecular weight between crosslinks M_c to numerically estimate mechanical responses and cohesive failure for cured ATS resins. The molecular weight of the model ATS repeat unit of Fig. 4-8 is M_0 = 452 gm/ mole. Curing along the upper curve for crosslinking in Fig. 4-8 leads to a model ATS network with a molecular weight between crosslinks of M_c = 1817 gm/mole and T_g = 344°C. As discussed earlier a computer model can be directly applied to describe the tensile stress-strain and failure response over a preassigned range of time and temperature. It is convenient in computing stress-strain response to set time constant at t = 1.0 sec to represent a standard nominal performance condition. The temperature range and step increments define an expected use condition as shown in Table 4-8.

Nineteen test temperatures are defined in Table 4-8 ranging from -50°C to 375°C at 25°C intervals. These test temperatures are identified by curve number in Fig. 4-10 which displays the computed curves of nominal tensile stress versus strain. The computed tensile failure point is graphically indicated by the circled point on each curve. The curves of Fig. 4-10 represent the calculated mechanical response of a high molecular weight linear ATS polymer formed by chain extension along the lower cure curve of Fig. 4-9 to a final r = 1725.8 with a number average molecular weight $M_p = 2.26 \cdot 10^5$ gm/mole and a T_g = 267°C. The curves of Fig. 4-10 show that this linear ATS response is predicted to display brittle response below 0°C (see curve 3).

TABLE 4-8 Relation of Stress-Strain Curve Number to Test Temperature at Constant Time t = 1.0 s

Curve No.	Temperature °C	Temperature K	Curve No.	Temperature °C	Temperature K
1	-50	223	11	200	473
2	-25	248	12	225	498
3	0	273	13	250	523
4	25	248	14	275	548
5	50	323	15	300	573
6	75	348	16	325	598
7	100	373	17	350	623
8	125	398	18	375	648
9	150	423	19	400	673
10	175	448			

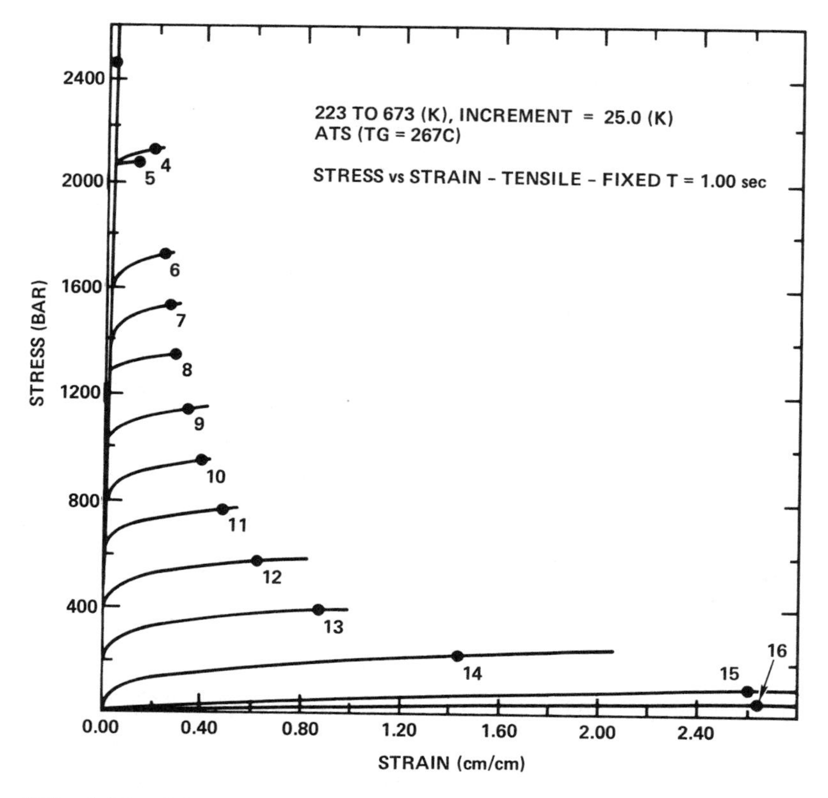

FIG. 4-10 Calculated curves of nominal tensile stress vs strain for linear ATS polymer with T_g = 267°C and M_P = 2.26E5 gm/mole (see Table 4-8 for temperatures).

The appearance of a yield stress and combined elastic-plastic response is displayed at higher temperatures through 250°C (curve 13). Above the computed glass temperature T_g = 267°C the tensile response of linear ATS is dominated by a flow response as shown by curves 14 - 16. The nominal fracture energy W_T per unit volume of unnotched specimen is defined by the area beneath each curve of Fig. 4-10. The curves of Fig. 4-11 plot this calculated tensile work of fracture W_T versus temperature and show a broad range of high W_T => 250 bar from T = 298K (25°C) to 548K (275°C).

The calculated mechanical response for linear and crosslinked ATS can be readily compared. The cure path proceeding along the upper path of Fig. 4-9 is proposed to produce an equivalent chain extension with a high molecular weight $M_p = 2.26 \cdot 10^5$ gm/mole. Additionally, the concurrent crosslinking process reduces the calculated molecular weight between crosslinks at M_c = 1817 gm/mole

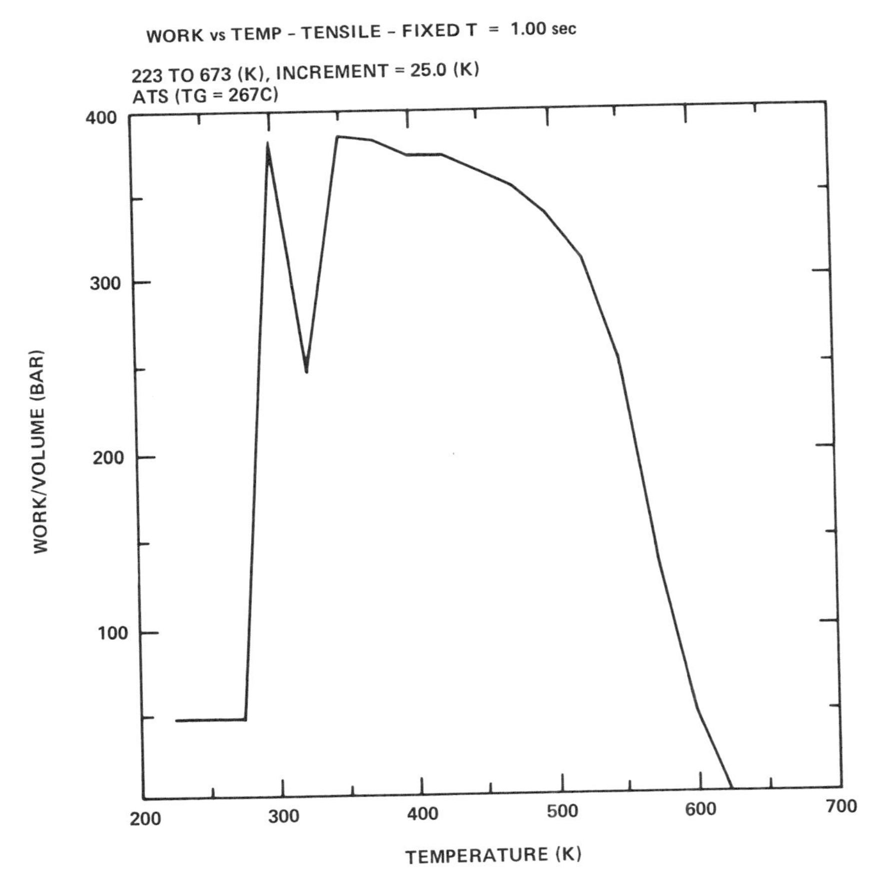

FIG. 4-11 Calculated temperature dependence of tensile fracture energy W_T per unit volume of unnotched linear ATS polymer.

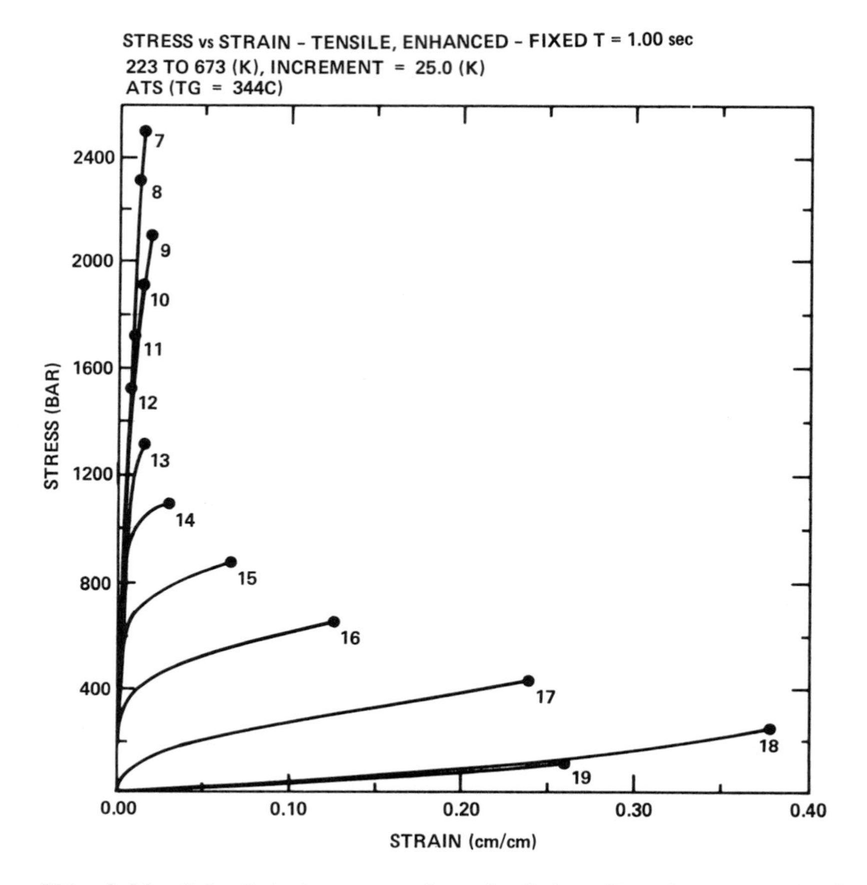

FIG. 4-12 Calculated curves of nominal tensile stress vs strain for crosslinked ATS with T_g = 344°C and Mp = 2.26E5 and M_c = 1817 gm/mole (see Table 4-8 for temperatures).

for the cured thermoset resin with T_g = 344°C as represented by the experimental data for crosslinked ATS. The computed tensile stress versus strain curves shown in Fig. 4-12 for the model crosslinked ATS show that crosslinking produces higher yield stress and strength over the entire range of test temperatures described by Table 4-8. Also shown in Fig. 4-12 is the dramatic reduction of extensibility due to crosslinking. Curve 18 of Fig. 4-12 displays a maximum extensibility of ε_b = 38 percent slightly above T_g at T = 375°C. This combination of high yield stress and strength combined with low extensibility is displayed in the much lower tensile fracture energy shown in the calculated W_T versus T curve of Fig. 4-13 for the model crosslinked ATS.

An experimental study of fracture energy, γ (defined as work per unit area of crack extension), of notched ATS specimens is described in a companion study[26] and the results plotted in Fig. 4-14. The fracture energy specimens

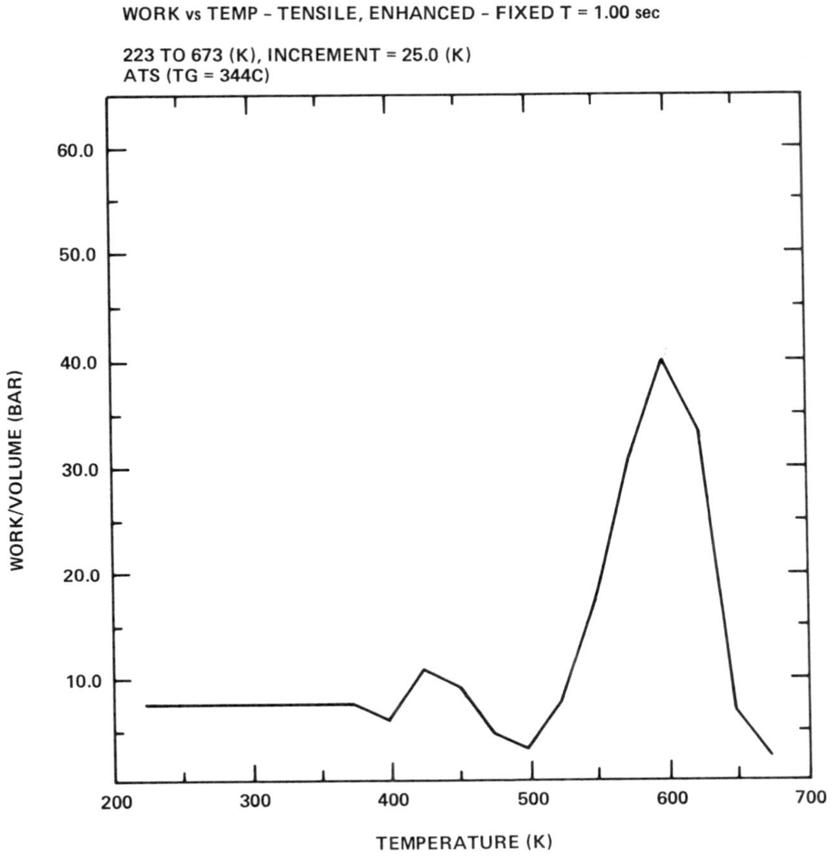
WORK vs TEMP - TENSILE, ENHANCED - FIXED T = 1.00 sec
223 TO 673 (K), INCREMENT = 25.0 (K)
ATS (TG = 344C)
60.0
50.0
40.0
30.0
20.0
10.0
WORK/VOLUME (BAR)
200
300
400
500
600
700
TEMPERATURE (K)

FIG. 4-13 Calculated temperature dependence of tensile fracture energy W_T per unit volume of unnotched crosslinked ATS polymer.

in the experimental study were notched with a single edge razor blade to provide uniform conditions for crack initiation. The data of Fig. 4-14 represent three temperature levels of isothermal cure: 130, 150, and 180°C to 90 percent reaction as shown by the lower curve of Fig. 4-9. All samples were post cured at 300°C for 2.0 h to ensure full crosslinking and a final T_g = 344°C as shown in Fig. 4-9. The experimental values of γ in Fig. 4-14 show that higher cure temperature appears to provide higher fracture energy over the whole range of test temperatures, from 25°C to 375°C. All the experimental γ values are temperature insensitive until about 100°C below T_g and then to display a maximum at T_g. The temperature dependence of experimental notched fracture energy γ is closely modelled by the calculated unnotched fracture energy W_T (see Fig. 4-13) which is shown as the dashed curve in Fig. 4-14.

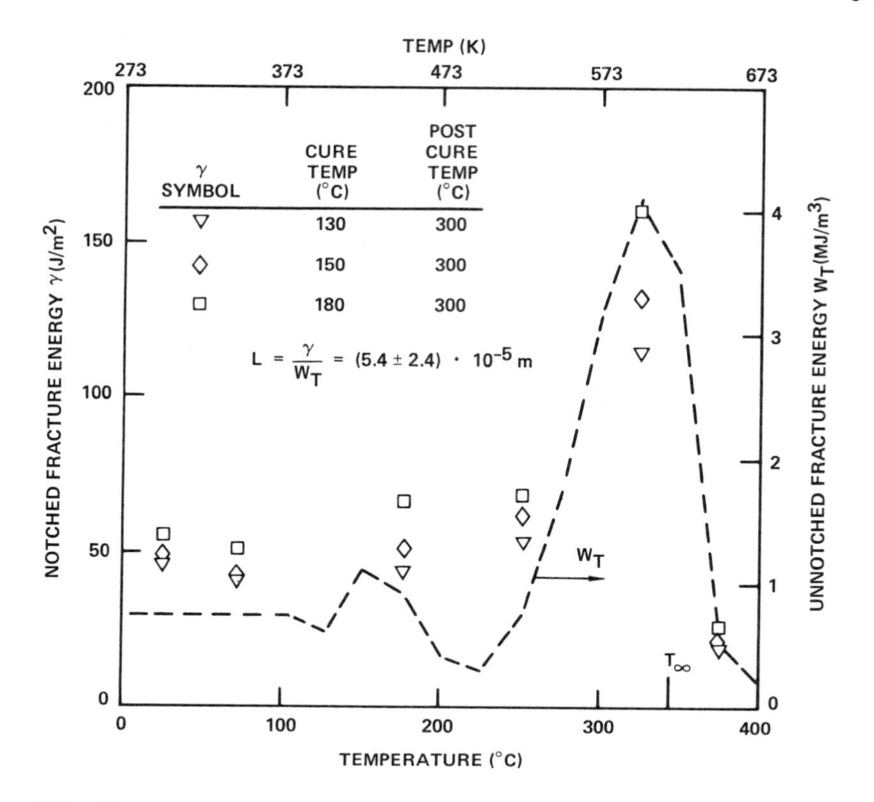

FIG. 4-14 Comparison of experimental notched fracture energy (γ) with unnotched fracture energy W_T from 25°C to 375°C.

The essential shape correspondance between temperature curves of γ and W_T as shown in Fig. 4-14 suggests a simple and perhaps meaningful correlation in the following relation:

$$L = \frac{\gamma}{W_T} \tag{4-31}$$

where L describes the thickness of the stress concentration zone for notched crack propagation. As shown in Fig. 4-14 an average value of L = (5.4 ± 2.4)* 10^{-5}m correlates the experimental values of γ with the calculated values of W_T for temperatures from 25°C to 375°C.

This combined computer analysis is shown to clarify the effects of cure state upon the mechanical response of ATS resin. A polymerization model indicates that ATS may cure by dominantly chain extension and branching until final high temperature post cure which produces dominant crosslinking. The measured notched fracture energy γ of the fully cured ATS closely follows calculated fracture energy values for model crosslinked ATS. New balances between chain extension and crosslinking of ATS resin are suggested by this analysis which may optimize new balances between stiffness, strength, and fracture energy.

4-6. A Generalized Williams-Landel-Ferry (WLF) Equation

The well-known form of the WLF equation which defines the molecular mobility of amorphous polymers above the glass temperature T_g is given as follows:[3]

$$\log a_T = \log \frac{\tau(T)}{\tau(T_0)} = -\frac{A(T-T_0)}{B+T-T_0} \qquad (4\text{-}32)$$

where (T) and (T_0) are polymer relaxation times at a temperature T and a reference temperature T_0, A and B are constants, and a_T is the time-temperature superposition factor. Normally, T_0 is chosen as the dilatometric glass temperature so that $T_0 = T_g$ and the A and B display nearly constant values A = 17.4 and B = 51.6 when temperature is expressed in kelvin (or centigrade) units. Several more general statements of modified WLF equations have been recently introduced to describe polymer relaxation responses, both above and below T_g by a single relation. Representative examples of these types of generalized WLF expressions have been discussed by Rusch and Beck,[31,32] Kaelble,[33] Vrentas and Duda,[34] and Meissner.[35]

The simple change suggested here for the mathematical form of the WLF equation is presented as follows:

$$\log a_T = \log \frac{\tau(T)}{\tau(T_0)} = -\frac{A(T-T_0)}{B + |T-T_0|} \qquad (4\text{-}33)$$

Inclusion of the absolute value $|T-T_0|$ in the denominator of Eq. (4-33) permits application of this relation both above and below T_g. Above T_g the mathematical statements of Eq. (4-32) and Eq. (4-33) are identical where $T_0 \equiv T_g$. Below T_g, Eq. (4-32) will predict too rapid a rise in log a_T culminating in a predicted infinite positive value when $T = T_g - B$. Conversely, Eq. (4-33) predicts realistic variations in log a_T below T_g in a fashion consistent with the modified WLF relations.[31-35] The apparent activation energy of flow, E*, can be defined by differentiation of Eq. (4-33) to obtain the following relation:

$$E^* = 2.303 \frac{R\,d(\log a_T)}{d(1/T)} = \frac{2.303(AB)T^2}{(B + |T-T_g|)^2} = \frac{4.12\ T^2}{(51.6 + |T-T_g|)^2} \qquad (4\text{-}34)$$

where A = 17.4, B = 51.6, R is the gas constant, and E* has units of kcal/mole. A maximum value for E* is predicted at $T = T_g$ which conforms with experi-

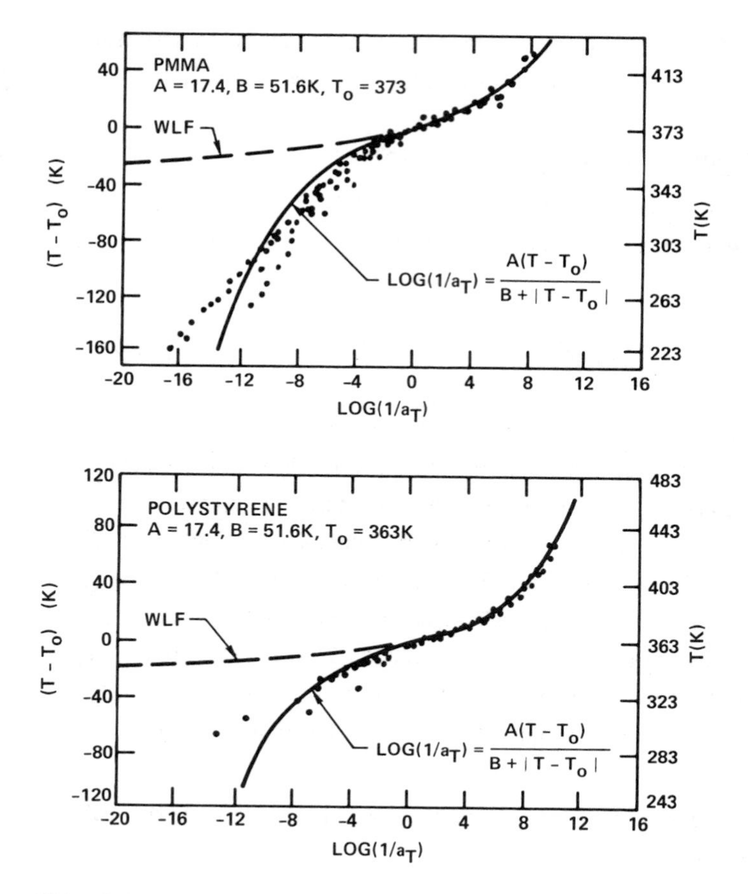

FIG. 4-15 Comparison of standard (dashed) and revised (solid curve) forms of WLF equations. (For summary of experimental time-temperature shift factors a_T, see Refs. 31, 32).

ence.[32-35] Far from T_g, where $|T-T_g| >> B \approx 50K$, the form of Eq. (4-33) shows $|\log a_T| \approx A$. Near T_g, where $|T-T_g| << B$, Eq. (4-33) predicts that $\log a_T \approx (-A/B)(T-T_g)$.

The generalized WLF equations[31-35] and Eq. (4-33) are based upon extensions of the free volume concept to temperatures well below T_g. The assumption is made that the available free volume controls molecular mobility, both in the liquid region above T_g and also in the glass state well below T_g with a rapidly changing free volume state at T_g. These generalized WLF equations thus implicitly contain an empirical or theoretical statement of the rheological state changes which accompany the glass transition. Equation (4-33) and Eq. (4-34) satisfy these general requirements, yet provide the mathematical form of the original WLF equation, Eq. (4-32) above T_g. Plots of experimental values of log (a/a_T) for polymethylmethacrylate (PMMA) and polystyrene as collected by

Rusch[31] from eleven other rheological studies from below T_g to above T_g are shown in Fig. 4-15. The dashed curve plotted through these data is the standard WLF expression of Eq. (4-32) and the solid curve is the modified WLF expression of Eq. (4-33). The ability of Eq. (4-33) to accommodate the sigmoidal form of the experimental log $(1/a_T)$ data both below and above T_g is evident while Eq. (4-32) is seen to apply only above T_g.

References

1. D.H. Kaelble, "Physical Chemistry of Adhesion, Wiley-Interscience, New York, (1971).

2. R.A. Hayes, J. Appl. Polymer Sci, 5, 318 (1961).

3. J.D. Ferry, Viscoelastic Properties of Polymers, 3rd Ed., Wiley, New York, Chap. 11 (1980).

4. L.A. Wood, J. Poly, Sci. 28, 319 (1958).

5. N.W. Johnston, J. Macromol. Sci. Rev. Macromol. Chem. C14(2), p. 215 (1976).

6. R.F. Boyer, J. Poly. Sci. Symposiums 50, p. 189 (1975).

7. J.B. Ennes and J.K. Gillham, ACS Polymer Preprints 22(1), 123 (1981).

8. C.W. Macosko and D.R. Miller, ACS Coatings and Platics Preprints, 35, 38 (1976), E.M. Valks and Macosko, Ibid, 35, 44, 1976.

9. D.H. Kaelble and C. Leung, ACS Org. Coat. and Appl. Poly. Sci. Proc., 46, 328 (1982).

10. D.H. Kaelble, Physical Chemistry of Adhesion, Wiley-Interscience, New York (1971), Chap. 11.

11. F.W. Billmeyer, Textbook of Polymer Chemistry, Interscience, New York (1957), Chap. 20.

12. M. Gordon, Proc. Roy. Soc. (London) A268, 240 (1962).

13. M. Gordon and G.R. Scantlebury, J. Poly. Sci. POart C, No. 16, Part 7, 3933 (1968).

14. P. Hedvig, Dielectric Spectroscopy of Polymers, Wiley, New York (1974), Chap. 2.

15. Proceedings of Composites in Manufacturing, Society of Manufacturing Engineers, Anaheim, Calif., Jan. 12-14, 1982.

16. D.H. Kaelble and P.J. Dynes, "Preventive Nondestructive Evaluation of Graphite Epoxy Composites," in Ceramic Engineering and Science Proceedings, 1, No. 7-8 (1980), p. 458.

17. J.D. Ferry, Viscoelastic Properties of Polymers, 3rd Ed, Wiley, New York (1980).

18. I.M. Ward, Mechanical Propeties of Solid Polymers, Wiley-Interscience, London, (1971).

19. D.H. Kaelble, "Relations Between Polymer Chemistry and Physical Properties," Amer. Chem. Soc. Div. of Plastics and Coatings Preprint Vol. 46 (April, 1982), p. 241.

20. W. Weibull, J. Appl. Mech. 18, (1951), p. 293.

21. N.R. Mann, R.E. Shafer, and N.D. Singpurwalla, "Methods of Statistical Analysis of Reliability and Life Data," Wiley, New York, (1974).

22. D.H. Kaelble, "Polymer Composite Reliability," U.S. Air Force Technical Report AFML-TRR-78-205, pp. 193-207.

23. D.H. Kaelble, Treatise on Adhesion and Cohesion, (R.L. Patrick, Editor) Vol. 1, Chap. 5, Dekker, New York (1967).

24. D.H. Kaelble and E.H. Cirlin, J. Poly. Sci: Symposium No. 43 (1973), p. 131.

25. S. Eddy, M. Lycarelli, T. Helminiak, W. Jones, and L. Picklesimer, "An Evaluation of Acetylene Termianted Sulfone Oligomer," Org. Coatings and Plastics Chemistry Preprint, 42, Amer. Chem. Soc., New York, (March, 1980), pp. 502-508.

26. C. Leung, "Acetylene Terminated Resin Mechanical Characterization Part I: Dynamic Properties and Fracture at Different Cure States," Ibid, Vol. 46, (1982), pp. 322-327.

27. D.H. Kaelble, "Relations Between Polymer Chemistry and Physical Properties,": Ibid, Vol. 46, (1982), pp. 241-245.

28. D.H. Kaelble, "Relations Between Polymer Chemistry and Mechanical Properties," Ibid. Vol. 46, (1982), pp. 246-253.

29. P.M. Hergenrother, "Acetylene Containing Precursor Polymers," J. Macromol. Sci. Rev. Macromol. Chem. C19(1), (1980): pp. 1-34.

30. D.H. Kaelble, E.H. Cirlin, J. Poly. Sci. Symposium 35 (1971) pp. 79-100.

31. K.D. Rusch, J. Macromol. Sci. - Phys., B2(2), 179 (1968).

32. K.C. Rusch and R.H. Beck, Jr., Ibid., B3(3), 365 (1969); B4(3), 621 (1970).

33. D.H. Kaelble, Physical Chemistry of Adhesion, Chap. 9, Wiley-Interscience, New York (1971).

34. J.S. Vrentas and J.L. Duda, J. Appl. Poly. Sci., 22, 2325 (1978).

35. B. Meissner, J. Appl. Poly. Sci.: Polymer Letters, 19, 137 (1981).

36. D.H. Kaelble, "Computer Aided Cure Management," ACS Organic Coatings and Applied Polymer Science Proc., 47, (Sept. 1982), pp. 630-634.

5
Polymer CAD/CAM Models

5-1. The Computer Models

CAD/CAM is the acronym formed by the words "Computer Aided Design and Manufacture." In this chapter we discuss the computer models on which CAD/CAM can be applied to a wide range of polymers. Three CAD/CAM models are introduced and demonstrated for the computer aided design and manufacture of polymers on a TRS-80™ (TANDY CORP.) Model 1- Level-2 microcomputer. The program listings are provided in the Appendix. An excellent reference to the Level-2 BASIC language for these programs is provided by a self-teaching guide by Albrecht, Inman and Zamora.[1] Additional detailed references in BASIC programming techniques are provided by Dwyer and Critchfield[2] and by Borden.[3] The programs discussed in this chapter are developed to operate without computer graphics or a line-printer and also require less than 16K bytes of RAM (random access memory). The outputs are developed in tabular form and the author has utilized a TRS-80™ Screen Printer to record in snapshot fashion both the program input and output as it appears on the video-monitor of the TRS-80.™ The interested reader familiar with personal computer programming will find the advanced graphics methods of Borden[3] and the data file programming methods for floppy-disks by Finkel and Brown[4] useful in modifying and enhancing these programs. The general physical chemistry reference for all three programs is provided by a molecular lattice model for polymer properties developed and discussed by Kaelble.[5]

The previous chapters 3 and 4 have already introduced the concepts and an overview of the contents of the three models discussed here which are:

1) Atomic to Molecular Properties
2) Polymer Chemistry to Physical Properties
3) Polymerization and Crosslinking

The first of the above models is discussed in Chapter 3 and the polymer models are presented in Chapter 4. It should also be mentioned that the models are intended to be used in the sequence of their presentation. Very often models 2 and 3 for polymers can be used without reference to model 1. When considerations of polymer chemical stability and an analysis of acid-base interactions at interfaces are required the chemical bonding analysis of model 1 becomes an important supplement to the polymer physics analysis provided by models 2 and 3. As the discussion of this chapter suggests the inputs and outputs of these three models may be directly interlinked to generate more highly integrated programs with very specific objectives with regard to numeric analysis and polymer performance prediction.

In order to illustrate the generality of the models discussed here we will formulate example problems in which extensive documentation of polymer properties and composite response is available from detailed experimental studies. The first example is taken from the epoxy resin chemistry summarized in Table 1-6 which forms an important part of the current technology in advanced composites using combinations of epoxy resin matrix and graphite fiber reinforcement. Standard epoxy resin chemistry and suggested cure mechanisms for 177°C (350°F) service temperature epoxy resins are illustrated in Fig. 5-1. As shown in Fig. 5-1, both the TGMDA epoxy and DDS curative are tetrafunctional and if reacted stoichiometrically would involve a coreaction of 62.7% by weight TGMDA epoxy and 37.3% by weight DDS curative. Two curing mechanisms are shown in Fig. 5-1 which involve homopolymerization of the epoxy to form the crosslink shown in the lower left of Fig. 5-1 or coreaction of TGMDA and DDS to form the crosslink shown in the lower right of Fig. 5-1.

A contrasting second example involves the alkyl acrylate copolymer shown in Fig. 5-2 which is noncrosslinked and elastomeric in its technical application as a pressure sensitive tape adhesive. Pressure sensitive tapes present many of the properties of uncured composite prepregs. The pressure sensitive tape thus serves as an important model for composite response during cutting and lamination. This second example has received detailed study in relation to the correlations between adhesive chemical structure, physical bonding and fracture energy as measured by peel adhesion.[5]

EPOXY (E): TETRAGLYCIDYL METHYLENE DIANILINE (TGMDA);
M. W. ≅ 422 gm/MOLE

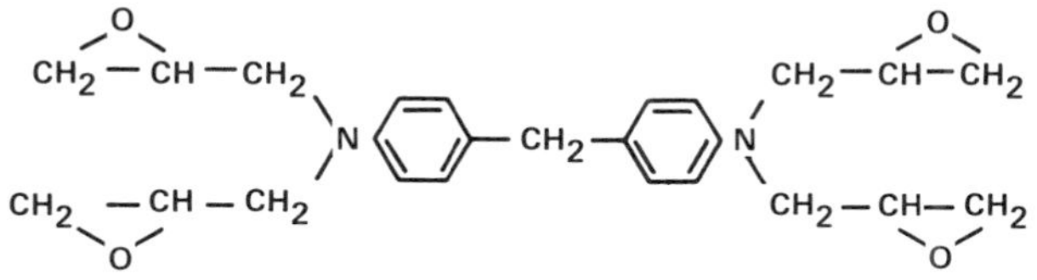

O O
CH_2 — CH — CH_2 CH_2 — CH — CH_2
N — CH_2 — N
CH_2 — CH — CH_2 CH_2 — CH — CH_2
O O

CURATIVE (C): DIAMINODIPHENYLSULFONE (DDS);
M. W. ≅ 248 GM/MOLE

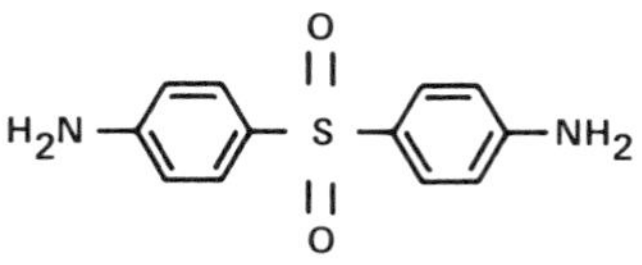

O
||
H_2N — S — NH_2
||
O

CROSSLINK REACTION 1: (100% BY WEIGHT E)

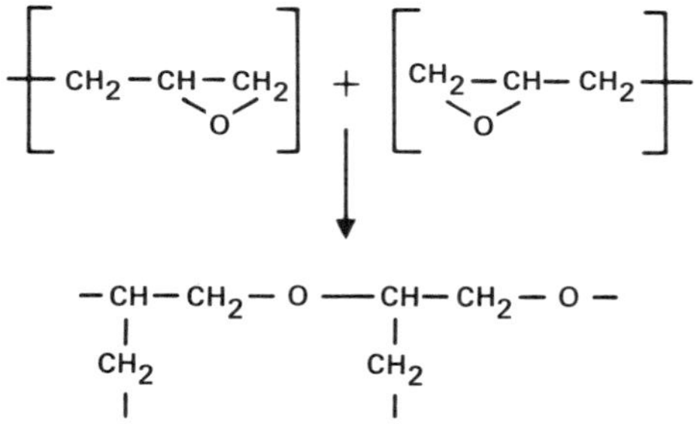

[CH_2 — CH — CH_2] + [CH_2 — CH — CH_2]
O O

— CH — CH_2 — O — CH — CH_2 — O —
| |
CH_2 CH_2
| |

CROSSLINK REACTION 2: (63% BY WEIGHT E + 37% BY WEIGHT C)

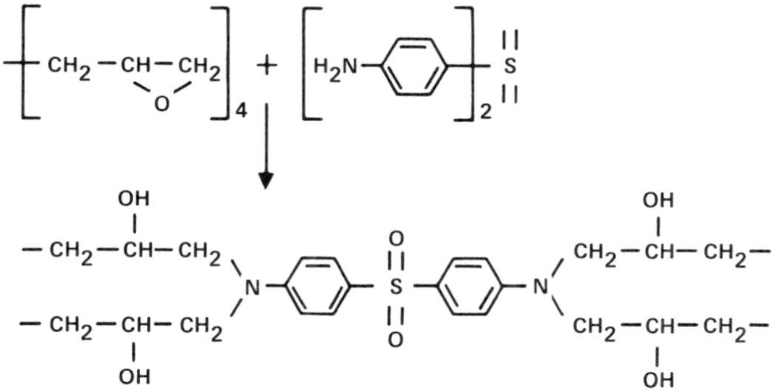

[CH_2 — CH — CH_2]$_4$ + [H_2N —]$_2$ S
O ||

OH OH
| |
— CH_2 — CH — CH_2 O CH_2 — CH — CH_2 —
N — S — N
— CH_2 — CH — CH_2 O CH_2 — CH — CH_2 —
| |
OH OH

FIG. 5-1 Composition and suggested curing mechanisms for 177°C (350°F) service temperature epoxy resins.

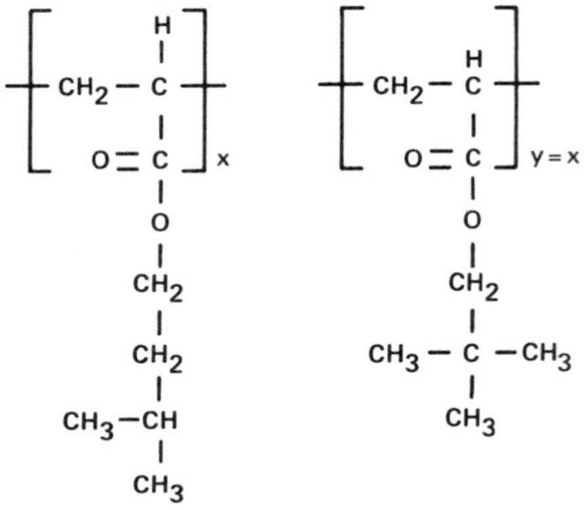

```
 ┌      H  ┐           ┌         ┐
 │      |  │           │     H   │
─┼ CH2— C ─┼─         ─┼ CH2— C ─┼─
 │      |  │           │      |  │
 └  O═  C  ┘x          └  O═  C  ┘y=x
        |                     |
        O                     O
        |                     |
       CH2                   CH2
        |                     |
       CH2             CH3 — C — CH3
        |                     |
   CH3—CH                    CH3
        |
       CH3
```

FIG. 5-2 Repeat structure for 50:50 mole % isoamyl acrylate = Neopentyl acrylate of number average molecular weight M_n = 1.03E6 g/mol, V_p (230K) = 1.01 cc/gm, T_g = 230K, and M_e = 21,000 gm/mole.

In addition to these detailed examples of Fig. 5-1 and Fig. 5-2 a variety of miscellaneous example calculations on a diverse range of polymers will illustrate the operation of these three polymer models. Composites involve the development of new sets of physical responses by the interaction of fiber and matrix through the adhesive bonded interface. When a molecular mixing occurs through interdiffusion a new interfacial phase may form termed an "interphase." Both of these subjects have been analyzed by Kaelble[5] and will not be extensively reviewed here. An excellent recent review of fiber reinforced composites by Delmonte is recommended as a supplement to this text in providing a technology overview of the entire subject of material science for composites.[6] These reviews of adhesion (and cohesion) theory by Kaelble[5] and composite technology by Delmonte[6] will serve as basic references for this chapter on polymer CAD/CAM and the following chapter on composite CAD/CAM. In order to avoid reprinting already well developed material this discussion will refer directly to these two referencs by page number within the text.

5-2. Atomic to Molecular Properties

The properties of reinforcing fibers in composites reflect the optimization of high tensile stiffness combined with high strength. Table 5-1 summarizes the properties of currently important commercial reinforcing fibers which include graphite, boron, polyparabenzamide (aramide-49), and E-glass. Each fiber type represents a special chemical technology and a unique optimization of the balance of specific volume, tensile modulus E, tensile strength S_b, tensile extensibility $\varepsilon_b = S_b/E$, and tensile fracture energy per unit volume $W_b = S_b^2/2E = S_b\varepsilon_b/2$. Fiber modulus and tensile strength reflect

TABLE 5-1 Properties of Commercial Reinforcing Fibers (from Ref. 5, p. 47)

Fiber	Spec. Vol. (cc/g)	Tensile Properties			
		E (GPa)	S_b (GPa)	ϵ_b (%)	W_b (MPa)
Graphite (UHM-S)	0.510	500	1.86	0.37	3.44
(HM-S)	0.523	360	2.34	0.65	7.60
(HT-S)	0.565	244	2.82	1.16	16.36
(A-S)	0.571	208	2.82	1.36	19.18
Boron (W-core)	0.377	386	3.41	0.88	15.00
Aramid-49	0.690	138	2.76	2.00	27.60
E-glass	0.394	72.5	3.44	4.74	81.53

strong contributions from chemical bond energy and crystalline structure and therefore one would expect the atomic to molecular properties model to correlate with these fiber properties.

The atomic to molecular property model, as presented in Listing No.1 of the Appendix, contains Table 3-1 and Eq. (3-3) through Eq. (3-10) in its structure. The computer model therefore recognizes the atomic properties listed in Table 3-1 and computes estimates of the chemical bond energies, the molecular weight, and the intramolecular specific volumes for four idealized crystal forms as described in the discussion of Chapter 3. Three examples of this model computation are presented in the input-output formats of Tables 5-2 through 5-4 for carbon (C) silica (SiO_2) and Aramid-49 (C_6H_4NHCO), respectively.

The simplest computation is shown in Table 5-2 for carbon. The upper portion of Table 5-2 shows the manner in which the computer model requests to know the number of elements in the chemical structure and the moles of each element which is equivalent to writing the empirical formula for carbon. The computer model then requests to know the number of chemical bond types followed by the elements and moles of each bond type. As shown in Table 5-2 this input describes carbon of valence 4 bonded with two chemical bonds per atom into a molecular carbon structure. The lower part of Table 5-2 provides the computed estimates of bond energies, molecular weight, and chemical specific volume. This calculation is not representative of graphite, as well described by Delmonte (Ref. 6, pp. 47-49) but for carbon in the single bonded diamond structure with crystal coordination number Z = 4. As shown in lower Table 4-2 the molecular specific volume V_s = 0.282 cc/gm is in reasonable agreement with V_s = 0.285 for diamond. Since crystalline graphite is a two-dimensional sheet

TABLE 5-2 Estimated Elemental Properties of Carbon

```
HOW MANY ELEMENTS? 1
ELEMENT CODE NO.=? 5
MOLES OF ELEMENT=? 2
NUMBER OF CHEMICAL BOND TYPES=? 1
FOR A-B BOND, ELEMENT A CODE NO.=? 5
ELEMENT B CODE NO.=? 5
MOLES OF A-B BONDS=? 4_


ELEMENTARY PROPERTIES
Z, SY, W, D/1E5, X, R/1E-10, V, PH =
 6 C 12.01  3.48  2.55  .77  4  5
TO CONTINUE PRESS ENTER
?
CHEMICAL ANALYSIS:

BONDING            BOND        % IONIC    BOND        MOLES
ELEMENTS           ENERGY      ENERGY     LENGTH
A     B            (J/MOLE)               (M*1E-10)
C     C            348000       0          1.54        4
TOTAL              1.392E+06                           4
TO CONTINUE PRESS ENTER
?
PHYSICAL ANALYSIS:
ELEMENTS           MOLES
C                  2
MOLECULAR WT. (KG/MOLE)= .02402
              (G/MOLE)= 24.02
SPECIFIC VOLUME (CC/G) :
(Z=12)         (Z=8)          (Z=6)           (Z=4)
 .129547        .140348        .18319          .232413
TO CONTINUE TYPE RUN AND PRESS ENTER
READY
>_
```

structure the higher specific volumes shown in Table 4-1 reflect the intermolecular volume between the molecular sheets.

The estimated intramolecular properties of silica as representing the principal constituent of E-glass are shown by the computer model results in Table 5-3. Here the input table reflects the slightly more complicated composition and chemical bond structure of SiO_2. The chemical analysis shows the strong ionic character calculated for the Si-O bond and an estimate of molecular specific volumes V_s = 0.228 cc/g for Z = 4 which is lower than the typical values 0.38 to 0.45 cc/g for natural silica and fiber glass. The computer model estimates of V_s will always predict the minimum physical value of a perfect crystal structure. Silica which crystallizes in irregular crystals and amorphous solid forms would be expected to display these higher specific volumes.

The third example of the atomic to molecular model is shown in the computation of Aramids-49 intramolecular properties as shown in Table 4-4. One

TABLE 5-3 Estimated Intramolecular Properties of Silica

```
HOW MANY ELEMENTS? 2
ELEMENT CODE NO.=? 12
MOLES OF ELEMENT=? 1
ELEMENT CODE NO.=? 7
MOLES OF ELEMENT=? 2
NUMBER OF CHEMICAL BOND TYPES=? 1
FOR A-B BOND, ELEMENT A CODE NO.=? 12
ELEMENT B CODE NO.=? 7
MOLES OF A-B BONDS=? 4_

ELEMENTARY PROPERTIES:
Z, SY, W, D/1E5, X, R/1E-10, V, PH =
 14 SI 28.09  1.77  1.9  1.11  4  7
 8 O 16  1.39  3.44  .73  2  2
TO CONTINUE PRESS ENTER
?
CHEMICAL ANALYSIS:

BONDING              BOND         % IONIC    BOND         MOLES
ELEMENTS             ENERGY       ENERGY     LENGTH
A     B              (J/MOLE)                (M*1E-10)
SI    O               386360     59.1583    1.7014       4
TOTAL                 1.54744E+06                        4
TO CONTINUE PRESS ENTER
?
PHYSICAL ANALYSIS:
ELEMENTS             MOLES
SI                    1
O                     2
MOLECULAR WT. (KG/MOLE)= .06009
               (G/MOLE)= 60.09
SPECIFIC VOLUME (CC/G) :
(Z=12)          (Z=8)             (Z=6)            (Z=4)
 .104749         .113965           .148114          .228352
TO CONTINUE TYPE RUN AND PRESS ENTER
READY
>_
```

proposed repeat structure is given by the structure formula as follows for polyparabenzamide:

```
   H     H
 -<=====>-N-C -
   H     H H ||
             O
```

The upper portion of Table 5-4 describes the complete empirical formula H_5C_7NO while the chemical bond description describes only the chemical bonding of the main chain structure as 10 moles of C-C bonds (counting the bond between repeat units) and 2 moles of C-N bonds. In this example of partial chemical bond structure calculation, the molecular specific volume V_s has no physical significance. This physical analysis does provide a correct description of

TABLE 5-4 Estimated Intramolecular Properties of Aramid

```
HOW MANY ELEMENTS? 4
ELEMENT CODE NO.=? 1
MOLES OF ELEMENT=? 5
ELEMENT CODE NO.=? 5
MOLES OF ELEMENT=? 7
ELEMENT CODE NO.=? 6
MOLES OF ELEMENT=? 1
ELEMENT CODE NO.=? 7
MOLES OF ELEMENT=? 1
NUMBER OF CHEMICAL BOND TYPES=? 2
FOR A-B BOND, ELEMENT A CODE NO.=? 5
ELEMENT B CODE NO.=? 5
MOLES OF A-B BONDS=? 10
FOR A-B BOND, ELEMENT A CODE NO.=? 6
ELEMENT B CODE NO.=? 5
MOLES OF A-B BONDS=? 2_


 1 H 1.008  4.35  2.2   .32  1  7
 6 C 12.01  3.48  2.55  .77  4  5
 7 N 14.01  1.61  3.04  .75  3  2
 8 O 16   1.39  3.44  .73  2  2
TO CONTINUE PRESS ENTER
?
CHEMICAL ANALYSIS:

BONDING         BOND      % IONIC   BOND       MOLES
ELEMENTS        ENERGY    ENERGY    LENGTH
A     B         (J/MOLE)            (M*1E-10)
C     C          348000    0          1.54       10
N     C          277670    8.34433    1.4759     2
TOTAL            4.03534E+06                     12

TO CONTINUE PRESS ENTER
?
PHYSICAL ANALYSIS:
ELEMENTS          MOLES
H                  5
C                  7
N                  1
O                  1
MOLECULAR WT. (KG/MOLE)= .11912
              (G/MOLE)= 119.12
SPECIFIC VOLUME (CC/G) :
(Z=12)         (Z=8)          (Z=6)          (Z=4)
 .179209        .134979        .253402        .390675
TO CONTINUE TYPE RUN AND PRESS ENTER
READY
>_
```

repeat unit molecular weight due to the input of the correct empirical formula. Assuming that Aramid-49 fiber strength and tensile modulus correlate with the main chain bond energies of Table 5-4 a direct comparison can be made between this polymeric fiber and the other compositions shown in Table 5-1.

We can rapidly explore the simple hypothesis that the specific strength V_s*S_b or specific modulus V_s*E of reinforcing fibers can be correlated with bond energy/molecular weight as calculated by the atomic to molecular model as demonstrated in the above examples. The products of V_s*S_b and V_s*E have the dimensions of (energy/mass) and therefore the appropriate ratio from the atomic to molecular model is the specific bond energy = total bond energy/molecular weight. Summarized in Table 5-5 is a set of fiber compositions ranked in their order of decreasing values of calculated specific bond energies.

The ranking of Table 5-5 correlates well with experience in that carbon and boron are currently considered the fibers with the highest specific modulus and strength. Aramid ranks high in Table 5-5 and is clearly of higher

TABLE 5-5 Calculated Specific Bond Energy for Fibers (Chemical energy/unit mass)

Fibers	Molar Composition	Molar of Bonds	Total Bond Energy (J/mol)	Mol. Wt. (g/mol)	Spec. Bond Energy (J/g)
Commercial					
Carbon	(C_2)	4	1.39E6	24.02	5.79E4
Boron	(B_2)	3	7.59E6	21.62	3.51E4
Aramid-49		12	4.04E6	119.1	3.39E4
Alumina	(Al_2O_3)	6	2.97E6	102.0	2.91E4
Silica	(SiO_2)	4	1.55E6	60.1	2.58E4
Aluminum	(Al_2)	3	6.18E5	53.96	1.15E4
Titanium	$(Ti)_2$	4	1.06E6	95.8	1.11E4
Iron	(Fe_2)	3	6.09E5	111.7	5.45E3
Candidates					
Boron Nitride	$(B_{12}N_5)$	3	9.11E5	24.82	3.67E4
Silicon Carbide	(SiC)	4	1.21E6	40.1	3.02E4
Polyethylene	$(-CH_2-)$	1	3.48E5	14.0	2.48E4
Carbon Precursor					
PAN	$[CH_2-CH(CN)]$	2	6.96E5	51.1	1.36E4

performance than silica (or E-glass) and the lower performing metal fibers such as aluminum, titanium, and iron (or steel). This result is in good agreement with commercial experience (see Ref. 6, p. 46).

The candidate fiber compositions listed in lower Table 5-5 are shown to have intermediate values of specific bond energy which is in agreement with their intermediate performance. The three polymer examples in Table 5-5, Aramid, polyethylene and polyacrylonitrile (PAN) are, of course, presumed to have fully oriented chains parallel to the fiber in order that these chemical calculations are displayed in fiber tensile response. Graphite fibers are now commonly developed from PAN precursor. Delmonte provides an excellent review of the present methods for converting oriented PAN fibers to graphite fibers by pyrolytic graphitization (see Ref. 6, p. 55). As shown in lower Table 5-5 the specific bond energy of PAN fiber is low due to the nitrile side group which is not included in the bond energy. Both nitrogen and hydrogen are lost during pyrolysis to produce the more efficient energy density of carbon or graphite.

Before leaving this section on chemical bonding and properties it is appropriate to review briefly the definitions of the polymer lattice model which will appear repeatedly in the forthcoming sections and are already applied in Eq. (4-5). The basic equations of the polymer lattice model are developed and discussed in detail by Kaelble (Ref. 5, p. 56) and result in a general equation for the intermolecular coordination number q as follows:

$$q = Z - 2 + \frac{2}{r} - \frac{2C}{r} \tag{5-1}$$

where Z is the total number of lattice nearest neighbors (both intra and intermolecular), r is the number average of repeating units per molecule, and c/r is the number average of crosslinked ring structures per repeat unit. For the simplest molecules, for example monatomic metal atoms in the vapor state it follows that $c/r = 0$, $r = 1.0$ and $q = Z$. For an infinite linear polymer chain it follows that $c/r = 0$, $r = \infty$ and $q = Z - 2$.

The discussion of this section on pure intramolecular bonding deals with the special case of the polymer lattice where the intermolecular coordination number $q = 0$ and $r = \infty$ which defines a special relation between c/r and Z as follows:

$$\frac{c}{r} = \frac{Z - 2}{2} \tag{5-2}$$

Thus for the tetrahedral diamond lattice with $Z = 4$ it follows that $c/r = 1.0$. For a fully chemically bonded face centered cubic lattice with $q = 0$ and $Z = 12$

it follows that $c/r = 5.0$. The lattice specific volumes calculated by the atomic to molecular model are thus related to the special limiting case defined by $q = 0$ where Eq. (5-2) applies. The discussion of the next section on polymer chemistry to physical properties describes the special limiting case of infinite linear chains where $q = Z - 2$ and many polymer properties are dominated by the intermolecular forces and free volume associated with molecular segment motion. In later sections the whole proposition of variable r and c/r is dealt with in terms of the polymerization and crosslinking model and the mechanical properties model.

5-3. Polymer Chemistry to Physical Properties

As shown in the previous section and also Chapter 3 the atomic to molecular model deals exclusively with properties of matter dominated by chemical bond structure. As discussed in Chapter 4 many important physical properties of polymers are dominated by the intermolecular forces and the motion of the random coil chain molecule. These subjects are extensively discussed by Kaelble and developed in mathematical relations by use of the polymer lattice model (Ref. 5, p. 56).

The polymer chemistry to physical property model discussed here is presented in Listing No. 2 of the Appendix. This model contains Table 4-1 and Eqs. (4-1) through (4-4) as discussed in Chapter 4-2. The present program listing provides a slight addition to the input-output example shown in Table 4-2 in that the summed values for the five structure unit properties which describe the polymer repeat unit are included in the output table of physical properties as shown in Table 4-7 for acetylene terminated polysulfone (ATS) resin.

In this section we will further illustrate the versatility of this computer model for more complicated polymer structures such as presented in Fig. 5-1 and Fig. 5-2. A model chain extended polymer formed by reaction of equimolar amounts of DGMBA and DDS is schematically shown in Fig. 5-3. The reaction product represented in Fig. 5-3 is an idealized epoxy linear polymer formed by pure chain extension of reactive DDS and TGMDA molecules which normally crosslink and chain extend simultaneously in a process termed "curing" to form a thermoset resin. The advantage of the computer model is that one can write an exact (but hypothetical) polymer structure and immediately make a numeric estimation of the consequent physical properties based upon the known properties of structural groups contained within the computer model. The computer modelling of the linear epoxy polymer of Fig. 5-3 is shown in detail in terms of input format in the upper portion of Table 5-6 and the output table in

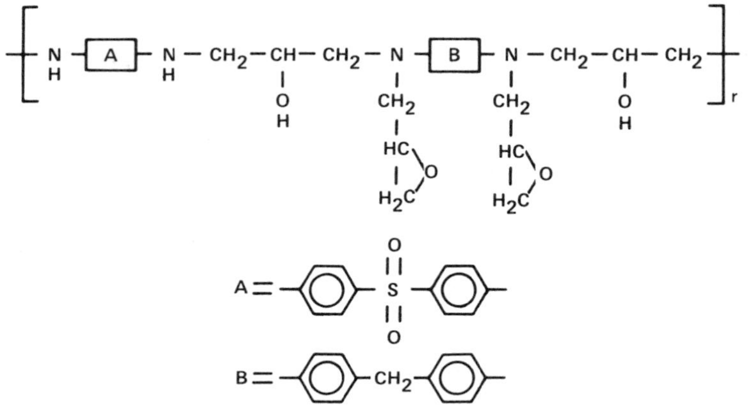

FIG. 5-3 Molecular structure of equimolar amounts of TGMDA and (37 wt%) DDS polymerized by chain extension.

the lower portion of Table 5-6. Some of the subtle features of the use of this polymer chemistry to physical property model are demonstrated in this example and ones that follow.

Input data to this model is entered to the computer model in the upper portion of Table 5-1 and consists of first identifying the number of functional groups from Table 4-1 which are required to describe the main chain structure. Inspection of Fig. 5-3 and Table 4-1 shows that this epoxy structure displays two secondary amines with structure (-NH-) and no group of this type is shown in the 33 functional groups of Table 4-1. The input table of Table 5-6 shows how readily this problem is solved by inputting two moles of amide (-NH C(O)-) and subtracting two moles of carbonyl (-C(O)-) to the main chain structure to create two moles of secondary amine groups. This step requires some chemical judgment as to validity, and experience with the computed consequence but represents one of the very creative aspects of utilizing this computer model.

In the side chain structure described in Table 5-6 the same sort of input procedure is followed by first describing the number of required structure groups from Table 4-1 followed by the group type and moles of each group in the repeat unit. The two epoxy groups (-CH(O)CH-) of the side chains of Fig. 5-3 are developed by adding the properties of a vinyl group (-CH = CH-) and ether (-O-).

The calculated properties in lower Table 5-6 refer to the glass state of the amorphous infinite linear polymer. The chemical characterization and physical property analysis of the TGMDA and DDS thermosets is extensive and this subject is briefly introduced in Chapter 1 through Table 1-6 and Figs. 1-11

TABLE 5-6 Estimated Physical Properties of Equimolar TGMDA and DDS Linear Polymer (see Fig. 5-3)

```
MONOMER-POLYMER PREDICTION PART-1,D.H KAELBLE MAY,81
HOW MANY MAIN CHAIN UNITS?? 6
STRUCTURE UNIT NO.=? 5
MOLES OF STRUCTURE UNIT=? 4
STRUCTURE UNIT NO.=? 33
MOLES OF STRUCTURE UNIT=? 1
STRUCTURE UNIT NO.=? 1
MOLES OF STRUCTURE UNIT=? 5
STRUCTURE UNIT NO.=? 18
MOLES OF STRUCTURE UNIT=? 2
STRUCTURE UNIT NO.=? 21
MOLES OF STRUCTURE UNIT=? 4
STRUCTURE UNIT NO.=? 16
MOLES OF STRUCTURE UNIT=? -4
HOW MANY SIDE GROUPS? (NONE=0)? 3
STRUCTURE UNIT NO.=? 1
MOLES OF STRUCTURE UNIT=? 2
STRUCTURE UNIT NO.=? 8
MOLES OF STRUCTURE UNIT=? 2
STRUCTURE UNIT NO.=? 20
MOLES OF STRUCTURE UNIT=? 2_

I. MAIN CHAIN UNITS:
UNIT NO. MOLES STRUCTURE                                  POLYMER
               UNIT                                       REFERENCE
 5        4    -P-C6H4-                                   TEREPHTHALATE
 33       1    -S((O)2)-                                  SULFONE
 1        5    -CH2-                                      ETHYLENE
 18       2    -CH(OH)-                                   VINYL ALCOHOL
 21       4    -NHC(O)-                                   AMIDE
 16      -4    -C(O)-                                     KETONE
II  SIDE CHAIN UNITS:
 1        2    -CH2-                                      ETHYLENE
 8        2    -CHCH-                                     1-4-BUTADIENE
 20       2    -O-                                        ETHER
GLASS SPEC. VOL.(M*M*M/KG)=  8.30266E-04 (CC/G)= .830266
GLASS C.E.D. (J/M*M*M)=  7.13963E+09 (CAL/CC)= 170.637
GLASS TEMP. (K)=  551.407 (C)= 278.207
ENTANG. MW. (KG/MOLE)=  4.33352 (G/MOLE)= 4833.52
U,H,V,M,N 399720  183  8.062E-04  .67  28
TO CONTINUE TYPE RUN AND PRESS ENTER
READY
>_
```

through 1-13. A recent book edited by May[7] documents the extensive number of studies of this state-of-the-art epoxy resin system. Suffice it to say that the calculated properties shown in Table 5-6 are consistent with experience for this model resin chemistry.

The effect of changing the reaction ratio of DDS to TGMDA is shown in the model linear structure of 1 moles DDS to two moles TGMDA is shown in Fig. 5-4 and the computer simulation of Table 5-7. In this example the mole fractions of computer simulated groups are adjusted to reflect the change in the chemical composition shown in Fig. 5-4. Comparing Table 5-7 and Table 5-8 shows that lowering the DDS concentration lowers both the glass state cohesive energy density δ^2 and the glass transition T_g of the model linear epoxy polymer while leaving both glass state specific volume V_s and entanglement molecular weight M_e relatively unchanged. Appropriate changes in the summed values of U, H, V, M, and N are also reflective of the new composition derived by reducing the DDS concentration. It should be noted that the chemical ratio of 2 moles TGMDA and 1 mole DDS is shown in Table 1-6 for NARMCO 5208 epoxy and therefore this composition is of considerable practical interest.

The model linear homopolymer of TGMDA is shown in the chemical repeat structure of Fig. 5-5 and the computer model calculations of Table 5-8. Dynes and Kaelble[8] have shown that pure TGMDA will polymerize and crosslink at high temperatures (between 525K and 625K). As shown in Table 5-8 the calculated values of both cohesive energy density δ^2 and T_g are markedly lowered by the exclusion of DDS from this epoxy homopolymer. The values of specific volume V_s and entanglement molecular weight M_e are shown to be similar to the two DDS containing compositions already examined. The time required to complete all three sample computations shown in Table 5-6 through 5-8 is roughly equal to the time required to manually write the structural formulas shown in Figs. 5-3 through 5-5 which is typically less than 10 minutes. These examples will be further defined by model calculations of epoxy curing and mechanical response.

As a change in pace let us return to the elastomeric acrylate polymer described in Fig. 5-2. The experimental values of V_s (at 230K) = 1.01 cc/gm, T_g = 230K, and M_e = 21000 gm/mole offer an interesting contrast to thermoset resins. Two computer simulations of the model structure of Fig. 5-2 are shown in Tables 5-9 and 5-10. In Table 5-9 the main chain is synthesized from the repeat properties of four methylene ($-CH_2-$) groups from Table 4-1 and one sees in the estimated properties that the calculated T_g = 205K is farther below the experimental value than one would like. In the second computer simulation of Table 5-10 two of the methylene ($-CH_2-$) groups in the main chain are replaced by methyl methylene ($-CH(CH_3)-$) and appropriate changes in side chain structure

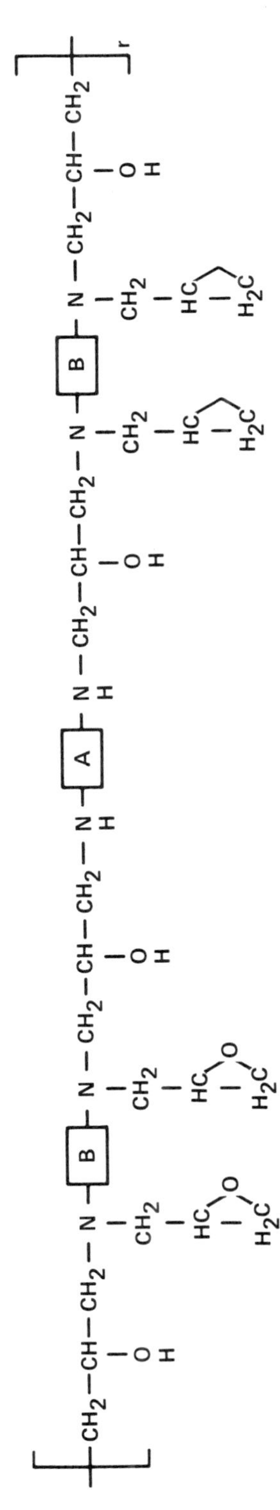

FIG. 5-4 Molecular structure of 2 moles TGMDA and 1 mole (22.7 wt%) DDS polymerized by chain extension.

TABLE 5-7 Estimated Physical Properties of 2 Moles of TGMDA and 1 Mole of DDS Linear Polymer (see Fig. 5-4)

```
MONOMER-POLYMER PREDICTION PART-1,D.H.KAELBLE MAY,81
HOW MANY MAIN CHAIN UNITS?? 6
STRUCTURE UNIT NO.=? 5
MOLES OF STRUCTURE UNIT=? 6
STRUCTURE UNIT NO.=? 33
MOLES OF STRUCTURE UNIT=? 1
STRUCTURE UNIT NO.=? 1
MOLES OF STRUCTURE UNIT=? 10
STRUCTURE UNIT NO.=? 19
MOLES OF STRUCTURE UNIT=? 4
STRUCTURE UNIT NO.=? 21
MOLES OF STRUCTURE UNIT=? 6
STRUCTURE UNIT NO.=? 16
MOLES OF STRUCTURE UNIT=? -6
HOW MANY SIDE GROUPS? (NONE=0)? 3
STRUCTURE UNIT NO.=? 1
MOLES OF STRUCTURE UNIT=? 4
STRUCTURE UNIT NO.=? 8
MOLES OF STRUCTURE UNIT=? 4
STRUCTURE UNIT NO.=? 20
MOLES OF STRUCTURE UNIT=? 4_

I. MAIN CHAIN UNITS:
UNIT NO. MOLES STRUCTURE                      POLYMER
               UNIT                           REFERENCE
 5       6     -P-C6H4-                       TEREPHTHALATE
 33      1     -S((O)2)-                      SULFONE
 1       10    -CH2-                          ETHYLENE
 18      4     -CH(OH)-                       VINYL ALCOHOL
 21      6     -NHC(O)-                       AMIDE
 16     -6     -C(O)-                         KETONE
II. SIDE CHAIN UNITS:
 1       4     -CH2-                          ETHYLENE
 8       4     -CHCH-                         1-4-BUTADIENE
 20      4     -O-                            ETHER
GLASS SPEC. VOL.(M*M*M/KG)=  8.59914E-04 (CC/G)= .859914
GLASS C.E.D. (J/M*M*M)=  6.67803E+08 (CAL/CC)= 159 605
GLASS TEMP. (K)=  502.679 (C)= 229.479
ENTANG. MW. (KG/MOLE)=  5.16813 (G/MOLE)= 5168.13
U,H,V,M,N 632290  319  1.3634E-03  1.094  45
TO CONTINUE TYPE RUN AND PRESS ENTER
READY
>_
```

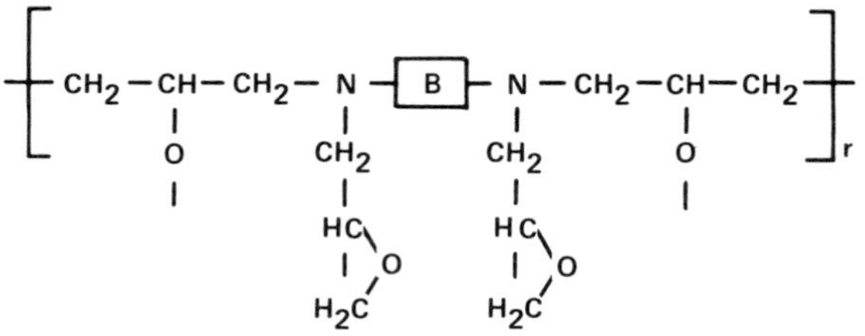

FIG. 5-5 Molecular structure of chain extended TGMDA homopolymer.

TABLE 5-8 Estimated Physical Properties of TGMDA Linear Homopolymer (see Fig. 5-5)

```
MONOMER-POLYMER PREDICTION PART-1,D.H.KAELBLE MAY,81
HOW MANY MAIN CHAIN UNITS?? 5
STRUCTURE UNIT NO.=? 5
MOLES OF STRUCTURE UNIT=? 2
STRUCTURE UNIT NO.=? 1
MOLES OF STRUCTURE UNIT=? 5
STRUCTURE UNIT NO.=? 8
MOLES OF STRUCTURE UNIT=? 1
STRUCTURE UNIT NO.=? 21
MOLES OF STRUCTURE UNIT=? 2
STRUCTURE UNIT NO.=? 16
MOLES OF STRUCTURE UNIT=? -2
HOW MANY SIDE GROUPS? (NONE=0)? 3
STRUCTURE UNIT NO.=? 20
MOLES OF STRUCTURE UNIT=? 4
STRUCTURE UNIT NO.=? 1
MOLES OF STRUCTURE UNIT=? 2
STRUCTURE UNIT NO.=? 8
MOLES OF STRUCTURE UNIT=? 2_


1. MAIN CHAIN UNITS:
UNIT NO. MOLES STRUCTURE                       POLYMER
               UNIT                            REFERENCE
 5        2    -P-C6H4-                        TEREPHTHALATE
 1        5    -CH2-                           ETHYLENE
 8        1    -CHCH-                          1-4-BUTADIENE
 21       2    -NHC(O)-                        AMIDE
 16      -2    -C(O)-                          KETONE
II  SIDE CHAIN UNITS:
 20       4    -O-                             ETHER
 1        2    -CH2-                           ETHYLENE
 8        2    -CHCH-                          1-4-BUTADIENE
GLASS SPEC. VOL.(M*M*M/KG)=  8.45986E-04 (CC/G)= .845985
GLASS C.E.D. (J/M*M*M)=  5.57993E+08 (CAL/CC)= 133.36
GLASS TEMP. (K)=  402.495 (C)= 129.295
ENTANG. MW. (KG/MOLE)=  5.30042 (G/MOLE)= 5300.42
U,H,V,M,N 209490  128  5.174E-04  .422  17
TO CONTINUE TYPE RUN AND PRESS ENTER
READY
>_
```

groups made to reproduce an equivalent overall composition for the acrylate repeat unit. The second computation provides a higher T_g = 240K which indicates an improved and acceptable agreement with experiment. This example shows that alternate modeling of the same repeat unit can illuminate specific features of molecular motion and thermal response as related by the U and H contributions to T_g as expressed by Eq. (4-2). Inspection of Tables 5-9 and 5-10 shows the V_s and M_e are not sensitive to these internal structure variations.

The ability of the polymer chemistry to physical property model to represent the effects of chemical composition changes is illustrated by larger

TABLE 5-9 First Estimate of the Physical Properties of Equimolar Isoamyl-Neopentyl Acrylate Copolymer

```
MONOMER-POLYMER PREDICTION PART-1,D.H.KAELBLE MAY,81
HOW MANY MAIN CHAIN UNITS?? 1
STRUCTURE UNIT NO.=? 1
MOLES OF STRUCTURE UNIT=? 4
HOW MANY SIDE GROUPS? (NONE=0)? 4
STRUCTURE UNIT NO.=? 14
MOLES OF STRUCTURE UNIT=? 2
STRUCTURE UNIT NO.=? 1
MOLES OF STRUCTURE UNIT=? 5
STRUCTURE UNIT NO.=? 2
MOLES OF STRUCTURE UNIT=? 1
STRUCTURE UNIT NO.=? 3
MOLES OF STRUCTURE UNIT=? 1_


UNIT NO. MOLES STRUCTURE                    POLYMER
               UNIT                         REFERENCE
 1         4   -CH2-                        ETHYLENE
II  SIDE CHAIN UNITS:
 14        2   -C(O)O-                      ETHYLENE ADIPATE
 1         5   -CH2-                        ETHYLENE
 2         1   -CH(CH3)-                    PROPYLENE
 3         1   -C((CH3)2)-                  ISOBUTYLENE
GLASS SPEC. VOL.(M*M*M/KG)=  9.16437E-04 (CC/G)= .916437
GLASS C.E.D. (J/M*M*M)=  3.44195E+08 (CAL/CC)= 82.2626
GLASS TEMP. (K)=  204.575 (C)=-68.6251
ENTANG. MW. (KG/MOLE)=  34.185 (G/MOLE)= 34185
U,H,V,M,N 90159  121  3.772E-04  .284  4
TO CONTINUE TYPE RUN AND PRESS ENTER
READY
>_
```

samplings of single physical properties of polymers. In Fig. 5-6 is shown a comparison of calculated and experimental values of density for a diverse group of polymers which include semicrystalline high density polymers such as polytetrafluorethylen ($-CF_2-$) and vinylidine bromide ($-CH_2CBR_2-$). The computer model calculates specific volume since high values of V_s correlate with high specific performance, other properties equal.

In Fig. 5-7 is provided a direct comparison of calculated and experimental solubility parameters δ (the square root of cohesive energy density) for 12 polymers ranging from fluoropolymers to polar polyamide polymers. The experimental solubility parameters are expressed in range values which define the horizontal bars of Fig. 5-7.

The effect of chemical composition of the repeat unit of side chains and main chains versus polymer T_g is plotted relative to the calculated curves from the computer model in Fig. 5-8 and the agreement is shown to be adequate. While this model defines T_g for the infinite linear polymer the effects of molecular weight on T_g of atactic polystyrene is shown in preliminary fashion in

TABLE 5-10 Second Estimate of the Physical Properties of Equimolar Isoamyl-Neopentyl Acrylate Copolymer

```
MONOMER-POLYMER PREDICTION PART-1,D.H.KAELBLE MAY,81
HOW MANY MAIN CHAIN UNITS?? 2
STRUCTURE UNIT NO =? 1
MOLES OF STRUCTURE UNIT=? 2
STRUCTURE UNIT NO.=? 2
MOLES OF STRUCTURE UNIT=? 2
HOW MANY SIDE GROUPS? (NONE=0)? 4
STRUCTURE UNIT NO.=? 14
MOLES OF STRUCTURE UNIT=? 2
STRUCTURE UNIT NO.=? 1
MOLES OF STRUCTURE UNIT=? 3
STRUCTURE UNIT NO.=? 2
MOLES OF STRUCTURE UNIT=? 1
STRUCTURE UNIT NO.=? 3
MOLES OF STRUCTURE UNIT=? 1_


I. MAIN CHAIN UNITS:
UNIT NO. MOLES STRUCTURE                         POLYMER
                UNIT                              REFERENCE
  1        2     -CH2-                            ETHYLENE
  2        2     -CH(CH3)-                        PROPYLENE
II. SIDE CHAIN UNITS:
 14        2     -C(O)O-                          ETHYLENE ADIPATE
  1        3     -CH2-                            ETHYLENE
  2        1     -CH(CH3)-                        PROPYLENE
  3        1     -C((CH3)2)-                      ISOBUTYLENE
GLASS SPEC. VOL.(M*M*M/KG)=  9.16437E-04 (CC/G)= .916437
GLASS C.E.D. (J/M*M*M)=  3.78706E+08 (CAL/CC)= 90.5108
GLASS TEMP. (K)=  240.33 (C)=-32.8198
ENTANG. MW. (KG/MOLE)=  34.185 (G/MOLE)= 34185
U,H,V,M,N 99200  111  3.772E-04  .284  4
TO CONTINUE TYPE RUN AND PRESS ENTER
READY
>_
```

Fig. 5-9. The more extensive data on this same polymer shown in Fig. 5-10 agrees with the polymer lattice equation for T_g which is derived by Kaelble[5] and introduced earlier as Eq. (4-5) in Chapter 4-2. In Fig. 5-10 the equation of the polymer chemistry to physical property model is shown as the calculated upper bound of the T_g versus reciprocal number average molecular weight of atactic polystyrene whose lower bound is styrene monomer. The generality of both Eq. (4-1) and Eq. (4-5) is demonstrated in the data of Fig. 5-10.

Experimental estimates of the molecular weight between entanglements introduces several sources of experimental error in both measurement and interpretation of data. In Fig. 5-11 is shown the quite limited data for relatively non-polar linear polymers which was applied in the early development of Eq. (4-4) for estimating M_e. In Fig. 5-12 is a bilogarithmic plot comparing calculated and experimental values of M_e for forty polymers of known composi-

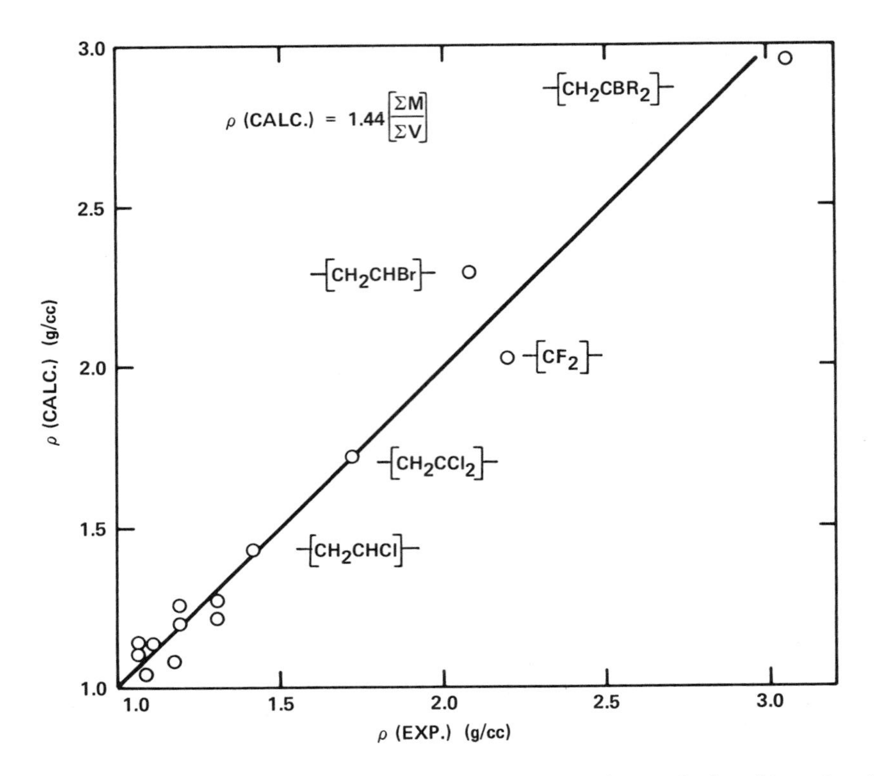

FIG. 5-6 Comparison of calculated and experimental density of solid polymers at 298K (data from Ref. 5).

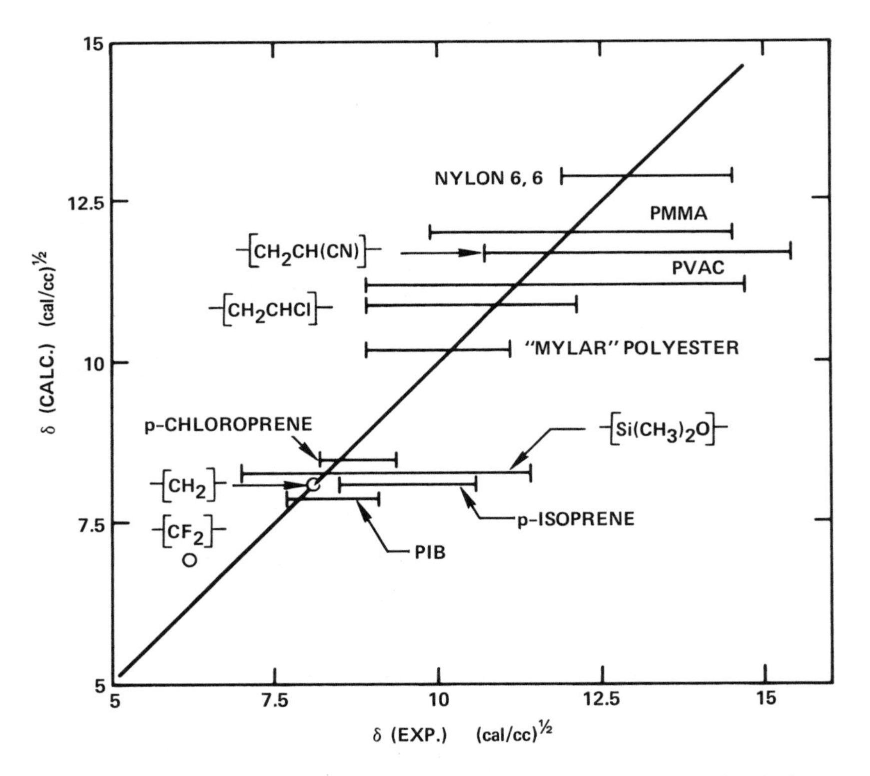

FIG. 5-7 Comparison of calculated and experimental solubility parameter (data from Refs. 9, 10).

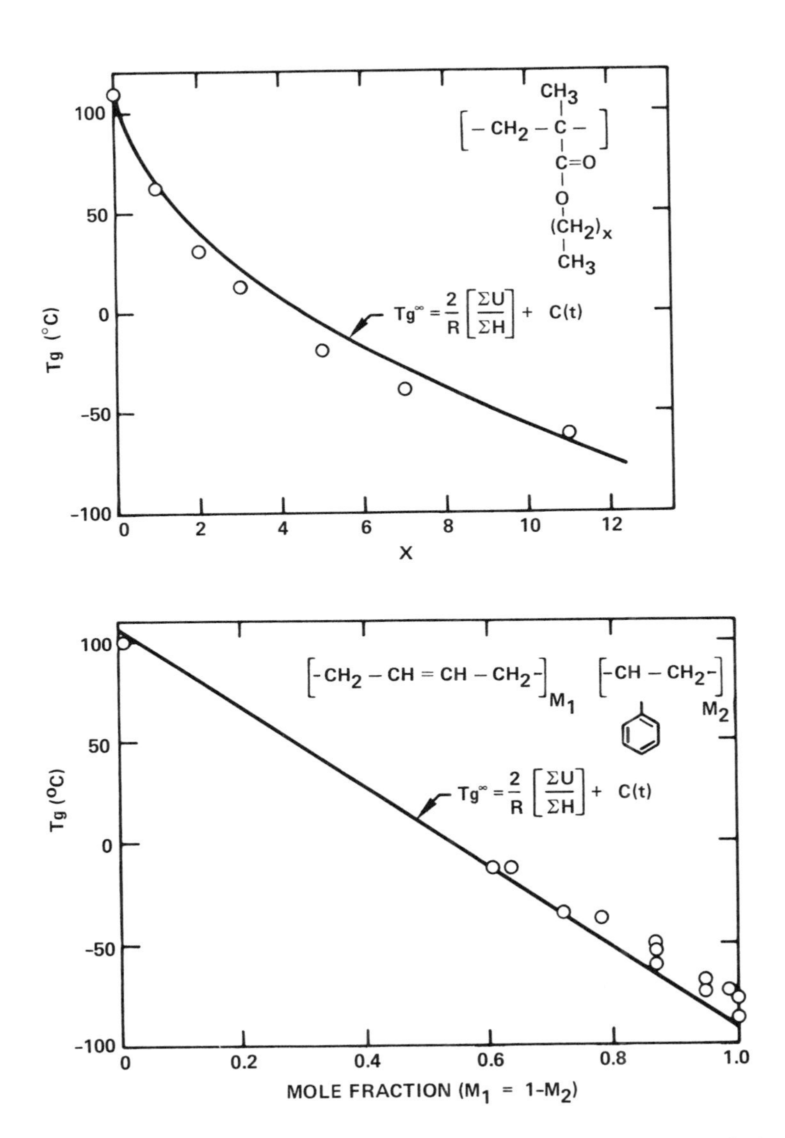

FIG. 5-8 Comparison of calculated and experimental glass temperatures for polyacrylates (upper curve) and butadiene-styrene copolymers (Refs. 11, 12).

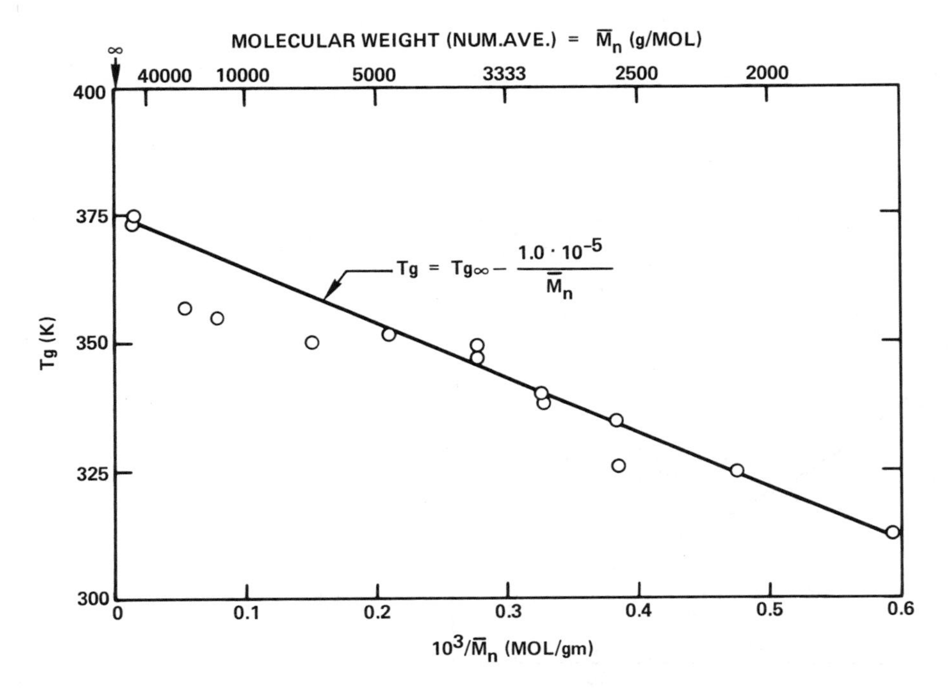

FIG. 5-9 M_n vs T_g for atactic polystyrene (data from Refs. 13, 14).

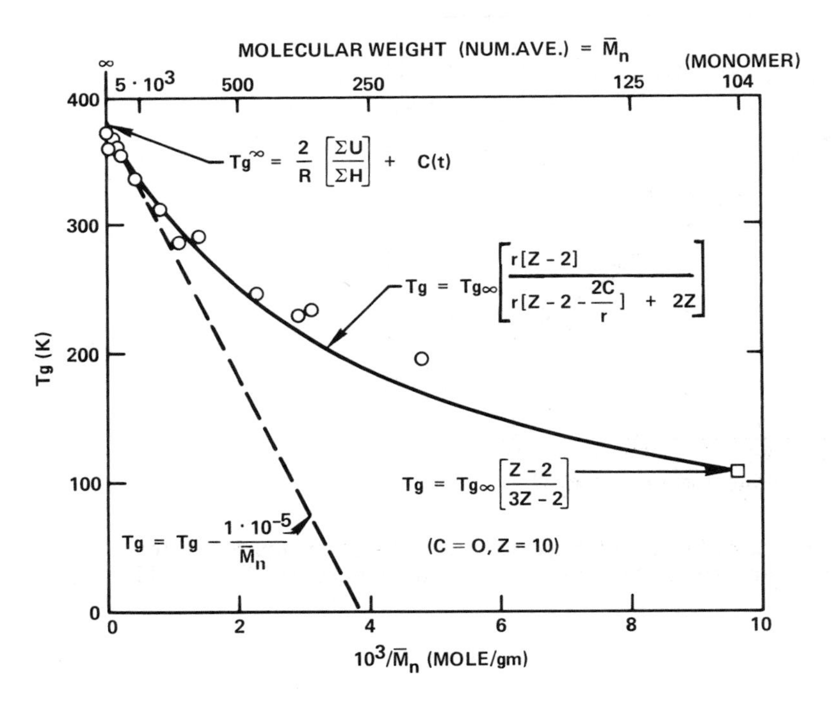

FIG. 5-10 M_n vs T_g for atactic polystyrene (data from Refs. 15, 16).

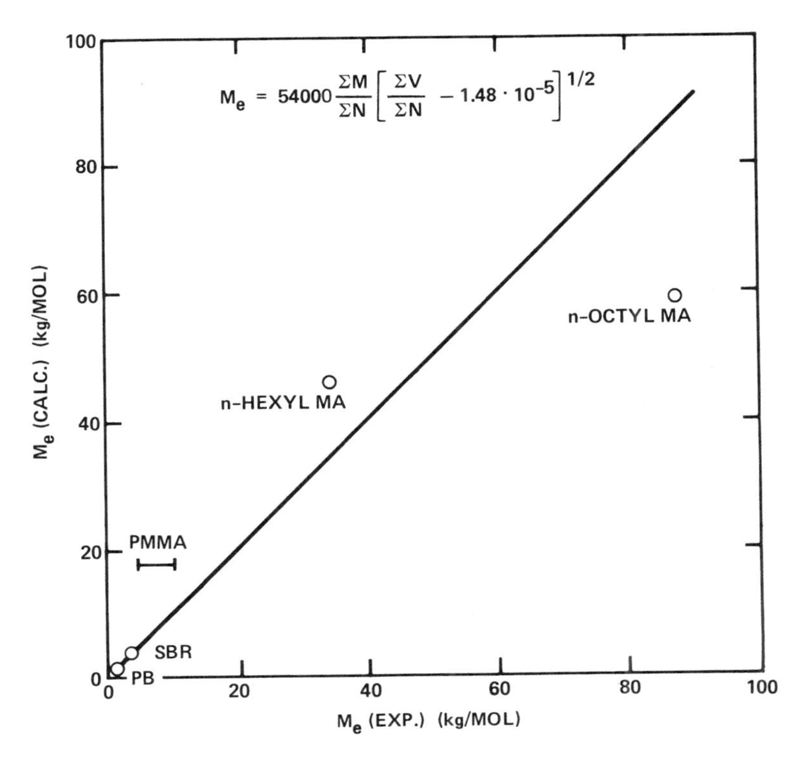

FIG. 5-11 Comparison of calculated and experimental entanglement molecular weight (data from Ref. 17).

tion. In this data display it is evident that Eq. (4-4) gives a fair estimate of the upper value of M_e where strong polar interchain interaction or extreme main chain stiffness due to steric constraints does not operate.

The computer model properties for these forty diverse polymers are summarized in Table 5-11 with regard to the calculated physical properties. Table 5-12 provides the summed group properties for each of the repeat units described by the corresponding polymer number in Table 5-11. The polymer lattice model provides a direct relation for calculation of the temperature dependence of shear $(\sigma_{12})_y$ and tensile $(\sigma_{11})_y$ yield stresses as follows (see Ref. 5, p. 363):

$$\frac{1}{2}\left(\frac{-\sigma_{11}}{\partial T_g}\right)_y = \left(\frac{-\partial\sigma_{12}}{\partial T_g}\right)_y = \frac{R}{4}\frac{q_L}{Z_g}\left(\frac{\Sigma H}{0.69\Sigma V}\right) \qquad (5\text{-}3)$$

and the shear yield stress at room temperature (293K) as follows:

$$(\sigma_{12})_y = (T_g - 293)\left(\frac{-\partial\sigma_{12}}{\partial T_g}\right)_y \tag{5-4}$$

where R = gas constant (= 87 bar * cc/mol * K), q_L = liquid state intermolecular nearest neighbors (≃ 9.0) and Z_g = total number of glass state nearest neighbors (≃ 10) and the remaining terms are defined by Eq. (4-1) through Eq. (4-4). Inspection of the stress-temperature functions of lower Fig. 4-3 shows that Eq. (5-3) and Eq. (5-4) defines the upper stress bound of the glass state above the brittle temperature T_b and the intercept of $(\sigma_{12})_y = 0$ at $T = T_g$. These calculated mechanical property values from Eq. (5-3) and Eq. (5-4) are listed for the forty polymers in the right columns of Table 5-12.

In composite materials the role of the reactive silane coupling agents and primer coatings applied to reinforcing fibers is a matter of much conjec-

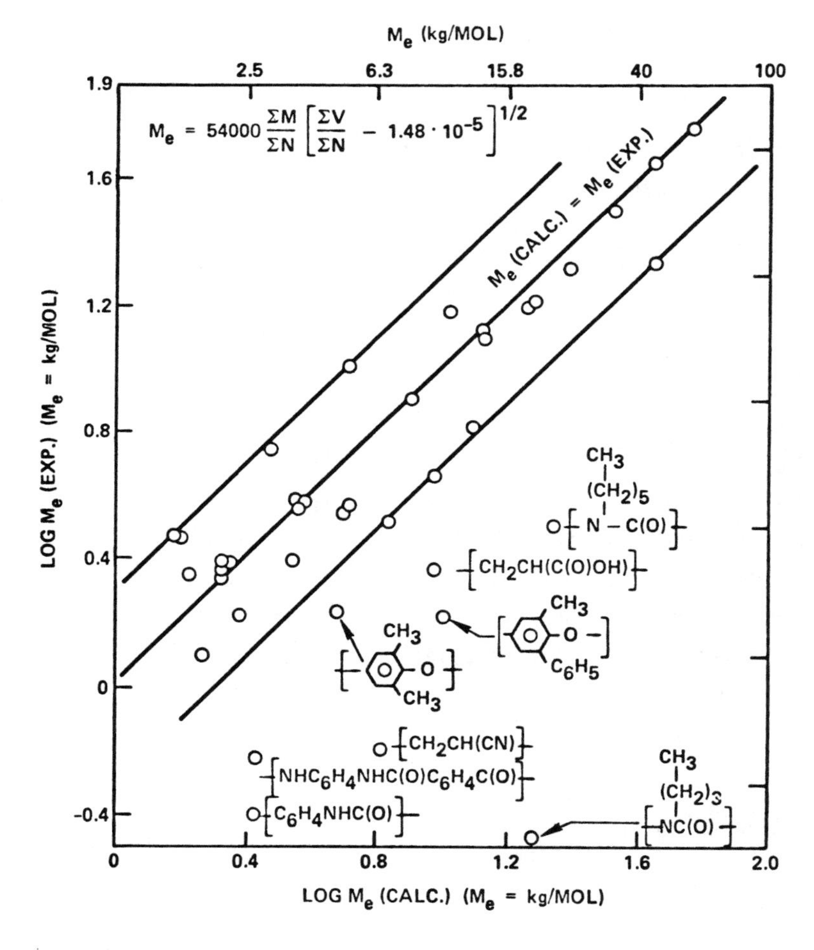

FIG. 5-12 Comparison of calculated and experimental entanglement molecular weight (data from Ref. 18).

TABLE 5-11 Relations Between Polymer Chemistry and Physical Properties

Polymer Number	Polymer	Calc. V_s $(\frac{cc}{g})$	Calc. δ^2 $(\frac{cal}{cc})$	Calc. T_g (K)	Calc. M_e $(\frac{kg}{mol})$	Exp.[18] M_e $(\frac{kg}{mol})$
1	p-dimethyl siloxane	0.81	68.7	163	10.6	12.2
2	p-isobutylene	1.09	62.1	201	8.2	8.0
3	p-cisisoprene	1.05	65.7	201	3.1	3.8
4	p-cis-transbutadiene	1.04	66.7	183	1.7	2.2
5	p-cisbutadiene	1.04	66.7	183	1.7	3.0
6	p-ethylene	1.09	64.2	150	2.1	2.5
7	p-propylene	1.09	87.5	240	4.9	3.5
8	p-styrene	0.88	88.5	384	20.2	17.5
9	p-α-methyl styrene	0.91	90.3	434	25.2	20.4
10	p-ethyleneoxide	0.86	94.5	190	1.5	3.0
11	p-propyleneoxide	0.92	105.9	254	3.5	3.9
12	p-tetramethylene oxide	0.95	81.0	173	1.8	1.3
13	p-methylacrylate	0.79	113.3	276	13.6	12.1
14	p-methylmethacrylate	0.83	143.7	380	18.1	15.8
15	p-n-butylmethacrylate	0.91	115.3	285	34.0	30.2
16	p-n-hexylmethacrylate	0.94	105.4	256	46.1	45.9
17	p-n-octylmethacrylate	0.96	98.8	237	59.3	57.0
18	p-2 ethylbutylmethacrylate	0.94	112.2	288	46.1	21.4
19	p-vinylacetate	0.79	125.9	304	13.6	12.3
20	p-vinylalchohol	1.12	148.6	362	5.4	3.8
21	p-vinylchloride	0.69	118.4	351	6.9	3.2
22	p-decamethyleneadipate	0.92	78.6	177	2.1	2.2
23	p-decamethylenesebecate	0.95	75.8	172	2.1	2.4
24	p-decamethylenesuccinate	0.90	80.5	181	2.1	2.3
25	p-ethyleneterephthalate	0.72	104.0	348	2.4	1.7
26	p-ethyleneisophthalate	0.72	104.0	348	2.4	1.7
27	p-bisphenol-A-carbonate	0.70	96.4	400	3.5	2.5
28	p-bisphenol-A-diphenyl-sulfone	0.75	118.4	533	3.6	3.6
29	p-2-methyl-6 phenyl-1,4-phenylene oxide	0.80	96.0	613	10.3	1.7
30	p-2, 6-dimethyl-1, 4-phenylene oxide	0.83	93.2	501	4.8	1.7
31	p-caprolactam	0.91	150.5	321	2.2	2.5
32	p-propylene sulfide	0.87	93.1	250	5.4	10.0
33	p-acrylic acid	0.75	171.8	363	9.6	2.4
34	p-acrylonitrile	0.93	136.7	450	6.5	0.65
35	p-tetrafluoroethylene	0.48	47.6	169	12.0	6.6
36	p-acrylamide	0.80	220.3	463	9.8	4.6
37	p-phenyleneterephthalamide	0.73	185.5	938	2.7	0.6
38	p-benzamide	0.73	185.5	938	2.7	0.4
39	p-n-hexylisocyanate	0.93	139.2	299	28.8	3.7
40	p-n-butylisocyanate	0.88	165.5	351	18.6	0.35

TABLE 5-12 Summed Properties of Functional Groups

Polymer Number	U $(\frac{J}{mol})$	H	V $(\frac{cc}{mol})$	M $(\frac{g}{mol})$	N	$-(\frac{\partial\sigma_{12}}{\partial T_g})$ $(\frac{bar}{deg})$	Shear Yield σ_{12} (at 293K) bar
1	17200	30	86.2	74	2	9.9	0
2	16040	22	88.8	56	2	7.0	0
3	19780	27	103.6	68	4	7.4	0
4	15770	24	81.4	54	4	8.3	0
5	15770	24	81.4	54	4	8.3	0
6	4140	8	22.2	14	1	10.2	0
7	16940	19	66.6	42	2	8.0	0
8	34240	23	133.0	104	2	4.9	446
9	40740	24	155.0	118	2	5.7	804
10	15100	22	55.0	44	3	11.3	0
11	23760	25	77.2	53	3	9.1	0
12	23380	30	99.4	72	5	11.1	0
13	32240	31	97.9	86	2	8.9	0
14	50140	34	120.1	100	2	8.0	696
15	62560	58	186.7	142	2	8.8	0
16	70840	74	231.1	170	2	9.1	0
17	79120	90	275.5	198	2	9.2	0
18	75360	60	231.1	170	2	8.4	0
19	35840	31	97.9	86	2	8.9	98
20	30740	22	71.2	44	2	8.7	600
21	21640	16	62.9	62.5	2	7.2	418
22	86160	136	377.2	284	18	10.2	0
23	102720	168	466.0	340	22	10.2	0
24	77280	120	332.8	256	16	10.2	0
25	60280	45	199.4	192	10	6.4	352
26	62280	50	199.4	192	9	7.1	227
27	80980	52	287.2	254	12	5.1	546
28	166660	79	482.6	442	20	4.6	1104
29	58560	24	210.0	182	5	3.3	947
30	43420	22	143.6	120	5	5.3	1102
31	65100	53	148.9	113	7	10.1	283
32	25200	27	93.1	74	3	8.2	0
33	39240	28	78.6	72	2	9.7	679
34	28240	16	71.1	53	2	6.4	1005
35	4810	8	34.8	50	1	6.5	0
36	52680	29	82.3	71	2	9.9	1105
37	68200	18	126.5	119	6	4.03	2599
38	68200	18	126.5	119	6	4.03	2599
39	69240	61	171.1	127	2	10.1	61
40	60960	45	126.7	99	2	8.5	493

ture. Delmonte (Ref. 6, p. 185) provides a discussion of the varied success of coating treatments on glass and carbon fiber reinforcement. As discussed in Chapter 2.4 surface energy analysis (SEA) has been extensively applied to describe the surface properties and adhesion of the fiber sizes and primer coatings. One example of such detailed surface analysis is summarized for polymerized silane coatings on glass for eight commercial silane coupling agents is summarized in Table 5-13. The data of Table 5-13 were generated by studies of Lee[19] and Kaelble.[20] The results summarized in Table 5-13 show that polymerized silane coatings on glass display a wide range of surface properties as expressed in the SEA model of Eq. (1-12) through Eq. (1-18). The remaining important question is the physical state and mechanical properties of these coatings. Very few if any experimental studies provide direct answers to this important question. The chemical structure to physical property model can provide computational estimates of these physical properties.

The process suggested by Lee[19] for the polymerization of reactive silane coatings on hydroxylated surfaces such as glass proposes the formation of hydroxylated polysiloxanes. A list of chemical structures for commercially important reactive silane monomers is presented in Table 5-14. Reaction of these trimethoxy or triethoxy silanes by acidic hydroxylation followed by condensation polymerization to polysiloxanes would lead to linear (or crosslinked) polymers with chemical structures as described in Table 5-15. These infinite linear siloxane polymers are amenable to computer analysis of physical properties and these results are summarized in Table 5-16 for summed structure group properties and Table 5-17 for calculated physical properties. Composition No. 41 representing polydimethyl siloxane is introduced as a reference material. The range of calculated physical properties shown in Table 5-17 is as diverse as the range of surface properties shown for different reactive silanes in Table 5-13. These calculations would indicate a wide range of solubility parameter, glass temperature T_g and entanglement molecular weight M_e for polymers of Table 5-12 while polymer specific volume V_s does not change markedly with composition.

There is strong spectroscopic evidence in a number of recent studies that surface hydroxyl concentration and the acidity (pH) during silane hydroxylation have important influences upon the orientation of the siloxane polymer side chain, the conformation of the siloxane polymer chains, and the thickness and crosslink network structure of the 20 to 50 nm thickness coatings formed.[21-26] Many of these recent studies shed light upon the important technological function of reactive silanes and particularly upon the role of silane coating rheology as recently outlined by Pleuddeman.[27] As will be shown in

TABLE 5-13 Polymerized Silane Chemistry and Surface Properties [19,20]

Test Liquid				H_2O	Glycerol	Eth. Glycol	PG E-200	PG 15-200	PB 1200
γ_{LV} (dyn/cm)				7.28	64.0	48.3	43.5	36.6	31.3
$2\alpha_L$ $(dyn/cm)^{1/2}$				9.34	12.16	10.70	10.62	10.20	9.90
β_L/α_L				1.54	0.94	0.81	0.74	0.64	0.53
R-Structure	Source	γ_{SV}^d dyn/cm	γ_{SV}^p dyn/cm	$W_{SL}/2\alpha_L$ $(dyn/cm)^{1/2}$					
$H_2N(CH_2)_2NH(CH_2)_3$-	DC Z-6020	30.0	4.6		7.54	7.21	6.99	6.92	
$CH_2{=}C(CH_3){-}C(=O){-}O{-}(CH_2)_3$-	CD Z-6030	8.4	41.7		9.08	8.03	7.41	7.42	6.21
CH_2-CH-$CH_2O(CH_2)_3$- (epoxide, O bridging CH_2-CH) (catalyzed)	DC Z-6040	10.2	43.6	13.23	9.69	8.57	8.13	7.12	
CH_2-CH-CH_2-O-$(CH_2)_3$- (epoxide, O bridging CH_2-CH) (noncatalyzed)	DC Z-6040	17.6	25.4	11.94	9.01	8.29	7.87		
Cl-$(CH_2)_3$-	DC XZ-8-0999	36.5	3.8	9.09	7.80	7.21	7.90		
NH_2-CH_2-$)_3$	UC A-1100	17.9	19.8	11.35	7.63	7.76	7.41	7.63	
HS-$(CH_2$-$)_3$	DC XZ-8-0951	67.4	0.0		8.06	8.33	8.02		
CH_2=CH-		28.5	2.1	7.72	6.25	6.35	6.56	6.81	5.90

Analysis of vapor-liquid-solid interactions for polymerized coatings of reactive silane coupling agents with structure R-$S_i(OCH_3)_3$.

TABLE 5-14 Reactive Silane Monomers (Refs. 19,20)

Number	Reactive Silane
41	(dimethyl)(dimethoxy)silane - model compound
42	tetraethoxy silane
43	(vinyl)(triethoxy)silane
44	(γ-chloropropyl)(trimethoxy)silane
45	(γ-mercaptopropyl)(trimethoxy)silane
46	(methacryloxypropyl)(trimethoxy)silane
47	(γ-glycidoxypropyl)(trimethoxy)silane
48	(β-3, 4-epoxycyclohexylethyl)(trimethoxy)silane
49	(γ-aminopropyl)(trimethoxy)silane
50	(γ-aminopropyl)(trimethoxy)silane
51	N-β-aminoethyl-γ-aminopropyl(trimethoxy)silane
52	(4-styryl-methylene-β-aminoethyl-γ-aminopropyl)(trimethoxy)silane

TABLE 5-15 Linear Hydroxy Polymers of Reactive Silane Primers

Number	Linear Polymer
41	(dimethyl)siloxane
42	(dihydroxy)siloxane
43	(vinyl)(hydroxy)siloxane
44	(γ-chloropropyl)(hydroxy)siloxane
45	(γ-mercaptogropyl)(hydroxy)siloxane
46	(methacryloxypropyl)(hydroxy)siloxane
47	(γ-glycidoxypropyl)(hydroxy)siloxane
48	(β-3, 4-epoxycyclohexylethyl)(hydroxy)siloxane
49	(γ-aminopropyl)(hydroxy)siloxane
50	(γ-aminopropyl)(hydroxy)siloxane
51	N-β-aminoethyl-γ-aminopropyl(hydroxy)siloxane
52	(4-styryl-methylene-β-aminoethyl-γ-aminopropyl)(hydroxy)siloxane

TABLE 5-16 Summed Values of Monomer Group Properties for Linear Hydroxy Polymers of Reactive Silane Primers

Number	U J/mole	H	V (cc/mole)	M	N
41	17200	30	86.2	74	2
42	44800	36	95.4	78	2
43	34350	33	105.6	88	2
44	52640	49	153.7	138.5	2
45	47540	57	161.7	136	2
46	66280	72	227.4	188	2
47	64550	77	215.6	176	2
48	62630	52	242	186	2
49	76360	56	151	119	2
50	76360	56	151	119	2
51	121720	79	211	162	2
52	157150	100	359	278	2

TABLE 5-17 Calculated Properties of Linear Hydroxy Polymers of Reactive Silanes at T_g

Number	V_s (cc/g)	δ^2 (cal/cc)	T_g (K)	M_e (kg/mole)
41	0.80	69	163	10.6
42	0.84	162	325	12.1
43	0.83	112	276	14.6
44	0.77	118	284	29.4
45	0.82	101	226	29.8
46	0.83	100	246	50.5
47	0.85	103	227	45.8
48	0.90	89	315	51.8
49	0.87	174	354	25.0
50	0.87	174	354	25.0
51	0.90	198	396	41.7
52	0.89	151	404	96.3

following sections the mechanical properties and environmental response of the silane primer should depend upon both the molecular chemistry of the monomer and the macromolecular structure of the polymerized film which controls the rheological and mechanical response.

The molecular weight between entanglements M_e and the molecular weight between chemical crosslinks M_c have significant influences upon both the rheology and extensibility of polymers. In Table 5-18 is summarized the correlation between experimental values of M_c and the maximum extension ratio $\lambda_b(max)$ obtained from evaluation of the complete tensile failure envelope above the glass temperature. These data are collected from several reviews and represent a wide variety of polymer types including two important classes of thermoset epoxy resins. In chemically crosslinked polymers there is a well defined relation between crosslink density (ρ/M_c) and maximum extension ratio λ_m which is approximated by the following semiempirical relation (Ref. 5, p. 385):

$$\lambda_m = K(M_c/\rho)^{1/2} \qquad (5\text{-}5)$$

where ρ is polymer density and M_c is the number average molecular weight between chemical crosslinks. The data summary of Table 5-17 shows a proportionality factor in Eq. (5-5) of K = 0.0515 with a standard deviation of about ± 30%. This new result is in close agreement with an earlier value of K = 0.057 developed by Kaelble based on data for 10 crosslinked polymers (Ref. 5, p 385). The ± 30% scatter in K values reflects experimental uncertainties in the precise evaluation of both λ_m and M_c.

For non-crosslinked thermoplastic polymers, Kramer[32] has proposed a molecular theory of fracture toughness based upon chain extensibility between non-slipping entanglements. More recently, Donald and Kramer,[33] present the result of detailed studies which show that chain orientation and breakage are significant limiting features of craze formation below the glass temperature. The data summary of Table 5-19 shows that a simple relation for maximum craze extension ratio λ_c and the entanglement crosslink density (ρ/M_e) can be defined as follows:

$$\lambda_c = K_c\,(M_e/\rho)^{1/2} \qquad (5\text{-}6)$$

where ρ is polymer density and M_e is the measured number average molecular weight between entanglements determined from melt viscosity. As shown in Table 5-19 for 11 polymers, the value of K_c = 0.032 ± 0.008 in Eq. (5-6) is only 60% of the value of K = 0.052 for chemical crosslinked networks. The lower value of

TABLE 5-18 Correlation Between Crosslink Density (ρ/M_c) and Maximum Network Extensibility $\lambda_b(\max) \approx K(M_c/\rho)^{1/2}$

Polymer	T_g (°C)	$(M_c/\rho)^{1/2}$ $(cc/mole)^{1/2}$	$\lambda_b(\max)$	K	Ref.
Silicone elastomer	-123	153	6.85	0.045	28
SBR elastomer	-61	113	7.20	0.064	28
Polybutadiene	-86	112	5.09	0.045	28
EPR elastomer	-55	105	6.85	0.065	28
Butyl elastomer	-70	104	6.17	0.059	28
Viton - b elastomer		93	5.19	0.056	28
Butyl elastomer	-70	92	7.20	0.078	28
Epoxy thermosett	115	31	1.59	0.051	28
Epoxy thermosett	72	23.2	1.27	0.055	29
Epoxy - polyamide	45	34.6	1.46	0.042	29
Epoxy - polyamide	20	56.4	1.95	0.035	29
Epoxy - polyamide	6	72	2.50	0.035	29
Viton - B elastomer		466	19.10	0.041	30
Viton - B elastomer		245	15.5	0.063	30
Viton - B elastomer		187	12.6	0.067	30
Viton - B elastomer		143	8.9	0.062	30
Viton - B elastomer		128	7.9	0.062	30
Viton - B elastomer		92	5.7	0.062	30
Epoxy - CTBN (50%)	-50	52.4	2.78	0.053	31
Epoxy - CTBN (39%)		47.0	2.41	0.051	31
Epoxy - CTBN (29%)		34.6	1.56	0.045	31
Epoxy - CTBN (17%)		30.6	1.32	0.044	31
Epoxy	100	29.2	1.35	0.046	31
			Ave. =	0.0515	
			Std. dev. =	±0.0150	

TABLE 5-19 Correlation Between Entanglement Crosslink Density (ρ/M_e) and Maximum Craze Extensibility $\lambda_c = K_c(M_e/\rho)^{1/2}$

Polymer	M_e (gm/mole)	$(M_e/\rho)^{1/2}$	λ_c	K_c
p-tert.-butylstyrene	4.3E4	203	7.2	0.035
p-para vinyltoluene	2.5E4	151	4.5	0.030
p-styrene	1.9E4	129	3.8	0.029
p-styrene-maleicanhydride (9 wt%)	1.9E4	128	4.2	0.033
p-styrene-acrylonitrile (24 wt%)	1.2E4	103	2.7	0.026
p-methylmethacrylate	9.2E3	87	2.0	0.023
p-styrene-methylmethacrylate (65 wt%)	9.0E3	87	2.0	0.023
p-styrene-acrylonitrile (66 wt%)	6.4E3	76	2.0	0.026
p-2,6 dimethyl-1,4-phenylene oxide (-E)	4.3E3	60	2.6	0.043
p-2,6 dimethyl-1,4-phenylene oxide (-M)	7.4E3	78	2.6	0.033
p-bisphenol-A carbonate	2.5E3	42	2.0	0.048
			Ave. =	0.032
			Std. Dev. =	±0.008

Note: M_e and λ_c data generated by experiments of Donald and Kramer (Ref. 7) and ρ is calculated from molecular structure.

K_c may simply reflect the greater steric hindrance to chain orientation below T_g where this value is determined. Considering that M_e can be reasonably calculated from molecular structure as defined by Eq. (4-4) and that M_c can be calculated from polymer cure models as defined by Eq. (4-19), an important chemical basis for predicting mechanical response is provided by Eq. (5-5) for crosslinked polymers and Eq. (5-6) for non-crosslinked glassy polymers.

The combined data of Table 5-18 and Table 5-19 are presented on a bilogarithmic plot of maximum extension ratio versus entanglement or crosslink density in Fig. 5-13 to illustrate the broad ranges of M_c or M_e over which the correlations of Eq. (5-5) and Eq. (5-6) are shown to apply. A further important prediction of non-extensibility, where $\lambda_m \leq 1.0$, $\lambda_c \leq 1.0$, at low values of M_c/ρ or M_e/ρ, respectively, implying chemically produced inextensibility due to high crosslink or entanglement density.

In the limit of short time or low temperature a brittle fracture without induced molecular motion or molecular orientation is predicted for all polymers. Gardon[34] developed a simple relation between cohesive energy density as defined by Eq. (4-2) and the maximum tensile strength for brittle-

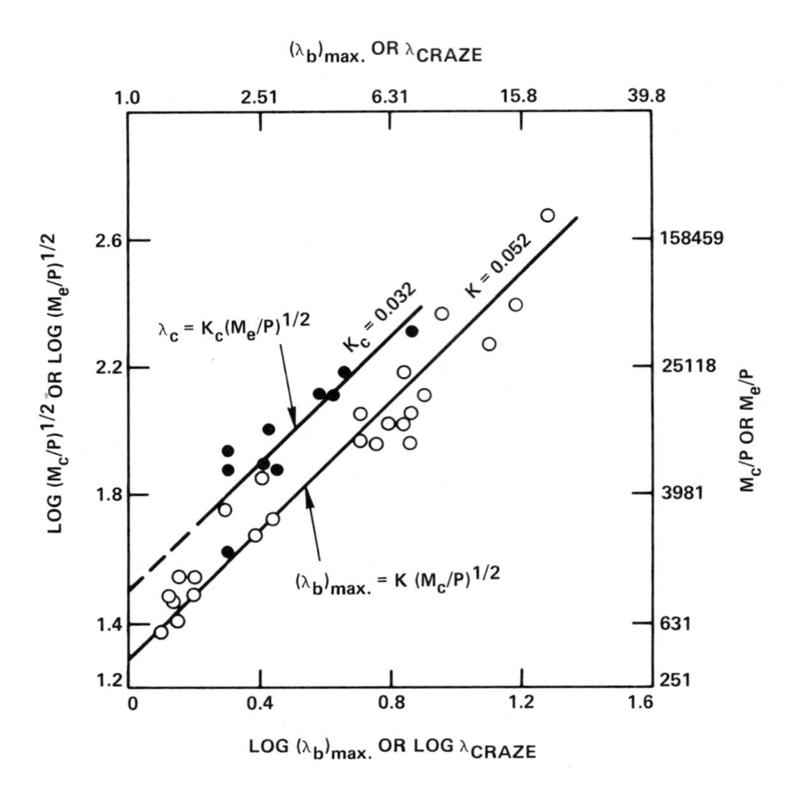

FIG. 5-13 Correlation between entanglement density (ρ/M_e) or chemical crosslink density (ρ/M_c) and maximum extension ratio λ_b or maximum craze extensibility λ_{craze}.

fracture of an unnotched specimen. This relation of brittle strength to cohesive energy density was rederived by the polymer lattice model and given the following form by Kaelble for nonoriented polymers (Ref. 5, p. 382):

$$(\sigma_{11})_b = \delta_p^2(0.25 + 0.68\ (\frac{Z - q}{q})) + \frac{P}{3} \tag{5-7}$$

where δ_p^2 is the polymer cohesive energy density defined by Eq. (4-2) and Z and q are defined by the polymer lattice model through Eq. (5-1). The external hydrostatic pressure P is included in Eq. (5-7) but has a negligible contribution at ambient pressure. For high polymers with typical lattice coordination numbers Z = 10 and q = 8 characteristic of flexible linear chains it follows that (Z - q)/q = 0.25 and neglecting pressure effects Eq. (5-7) becomes:

$$(\sigma_{11})_b = 0.42\ \delta_p^2 \quad . \tag{5-8}$$

Conversely for linear high polymer with stiff rod-like chains a lower value of

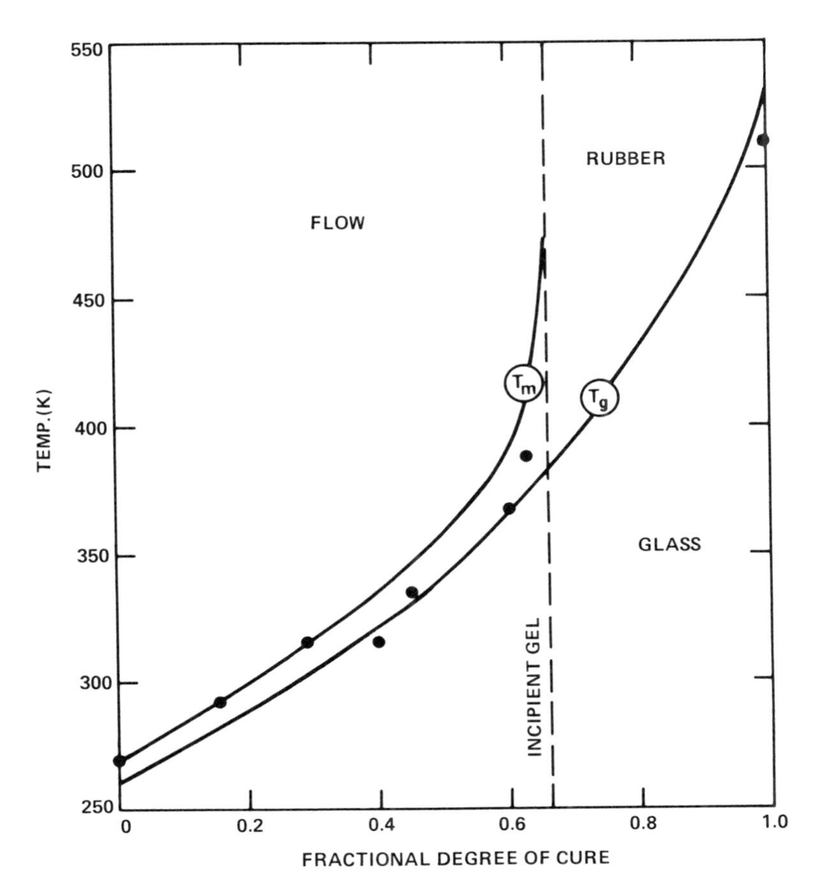

FIG. 5-14 Comparison of computed T_g and T_m curves for nonstoichiometric TGMDA/DDS (see Table 5-22) and measured T_g (●) for Hercules 35-1-5 epoxy resin (see Table 1-6 and Ref. 44).

Z = 6 and q = 4 provides a higher value of (Z - q)/q = 0.50 and Eq. (5-7) predicts:

$$(\sigma_{11})_b = 0.59\ \delta_p^2 \tag{5-9}$$

For nonpolymeric solids where Z = q and pressure effects are negligible, Eq. (5-7) simplifies to become equivalent to the Gardon relation:[34]

$$(\sigma_{11})_b = 0.25\ \delta_p^2 \tag{5-10}$$

An experimental test of Eq. (5-7) was conducted using microtensile specimens of nonoriented polymer films and special clamping techniques suited for cryogenic measurements as described by Kaelble and Cirlin.[35] The maximum measured

TABLE 5-20 Correlation of Polymer Cohesive Energy δ_p^2 Density and Maximum Tensile Strength $(\sigma_{11})_b$ at 80 to 130K

	Chemical Composition	Calc. δ_p^2 (bar)	Meas. $(\sigma_{11})_b$ (bar)	$(\sigma_{11})_b/\delta_p^2$
1)	Fluorocopolymer $(C_2F_4)_{1.0}\ (C_3F_6)_{0.136}$	1667	980	0.59
2)	C_2F_4 Homopolymer	1735	794	0.46
3)	Fluorocopolymer $(CF_2CFCl)_{1.0}(CF_2CH_2)_{0.031}$	2608	1147	0.44
4)	Bisphenol-A Carbonate $(OC_6H_4C(CH_3)_2C_6H_4OC(O))$	4088	1333	0.33
5	Polyethylene Terephthalate	5225	2108	0.40
6	Polyimide $(N(CO)_2C_6H_2(CO_2)NC_6H_4OC_6H_4)$	6186	2157	0.35
	Average			0.43
	Standard dev. ±			0.09

tensile strengths occurred at 80K to 130K for polymers of widely varied values of δ_p^2 and these results are summarized in Table 5-20. As shown by Table 5-20 the experimental value of maximum tensile strength of $(\sigma_{11})_b = 0.43\ \delta_p^2$ agrees well with the polymer lattice model and particularly with Eq. (5-8) for flexible linear chains. This information on the maximum strength of nonoriented polymers can be reasonably calculated from the computer model for physical response.

5-4. Polymerization and Crosslinking

The chemical process of curing for many classes of polymers can be described by the step growth polymerization mechanism in which the chains grow by progressive attachment of reactive monomer. The classical and recent approaches to calculating molecular weight distributions and crosslink network properties for this class of reactions is presented in a group of papers by Macosko, Miller and coworkers.[36-42] These papers provide new derivations and test methods for quantitative studies of step growth polymerization. The polymerization and crosslinking model utilized here is presented in Listing No.3 of the Appendix and incorporates Eq. (4-5) through Eq. (4-19) of Chapter 4.

While the glass transition temperature T_g is readily identified with polymers it is a much less well known property of low molecular weight liquids and monomers. An extensive listing of crystalline melting T_c and glass temperatures for over 60 hydrocarbons, alchohols, halogenated, and other miscellaneous low molecular weight compounds are collected in a recent review by Fedors.[16] For all 65 of the nonpolymeric liquids the ratio of T_g to T_c, in Kelvin units, is given by a simple ratio as follows:

$$T_g/T_c = 0.60 \pm 0.07 \tag{5-11}$$

where water is experimentally shown to display a T_g = 137K and a T_c = 273K with a ratio T_g/T_c = 0.51 which falls reasonably close to the norm displayed by Eq. (5-11). The effects of water of solution on polymer response is an important subject already discussed in earlier chapters.

The polymerization and crosslinking model computes both the glass temperature T_g and the rheological flow temperature T_m as a function of the fractional extent of chemical reaction. Four examples of computer inputs for this model are given in the upper portions of Tables 5-21 through 5-25. This computer model describes a coreaction between molecules Type A and B in which molecules Type B can be in excess of stoichiometry. The computer acquires information on the moles, the reactive functionality, and the molecular weight of both A and B molecules.

The next input to the computation defines the fraction of all molecules with functionality greater than 2 which can participate in crosslinking as described in Eq. (4-14). When other input factors are fixed, the adjustment of this variable to existing experimental data can often identify steric or other molecular restrictions to crosslinking. Entanglement molecular weight M_e is taken from the polymer chemistry to physical property model as is the T_g of the linear polymer if experimental data are not available. This model also asks for the monomer T_g and experimental data as summarized by Eq. (5-11) is applicable. More often, however, in the case of thermoset resins some polymerization has occurred as part of the supplier manufacturing process. In this case it is of value to obtain a measured value of T_g from thermal analysis of the uncured material.

The computer model takes the above information and makes an initial calculation of gel point in terms of % A and B reacted and an estimate of the initial number average degree of polymerization from the input T_g data. Based upon these results the user may want to revise the inputs or proceed with polymerization analysis.

TABLE 5-21 Estimated Cure Path Properties of Equimolar TGMDA and DDS (see Table 5-6)

```
A AND B COREACTION-MOL. WT. DIST -THERMAL TRANS.-D.H.KAELBLE-OCT
. 27,1982
IF NONSTOICHIOMETRIC REACTION HAVE MOLES OF B IN EXCESS
MOLES OF TYPE A (MOLE)=? 1
TYPE A FUNTIONALITY(=>2)=? 4
MOL. WT. OF TYPE A (G/MOLE)=? 248
MOLES OF TYPE B (MOLE)=? 1
TYPE B FUNCTIONALITY (=>2)=? 4
MOL. WT. OF TYPE B (G/MOLE)=? 422
FRACTION OF MOLECULES OF FUNCTIONALITY >2=? 1
NUMBER OF A AND B MAIN CHAIN ATOMS (A1,A2)=? 11,17
MOL. WT. BETWEEN ENTANGLEMENTS (G/MOLE)=? 4833
GLASS COORDINATION NUMBER (3<Z<10)=? 10
MONOMER AND LINEAR POLYMER GLASS TEMPERATURES (T1,T2) IN DEG. K=
? 260,551
GEL POINT (% A REACTED)= 33.3333
GEL POINT (% B REACTED)= 33.3333
INITIAL NUM. AVE. DEG. OF POLYMERIZATION= 2.23368
TO ANALYSE POLYMERIZATION PRESS ENTER?
% A         BRANCH      NUM. AVE. WT. AVE.  GLASS      FLOW
REACTED     COEF.       MW(G/MOL) MW(G/MOL) TEMP(K)    TEMP(K)
 0           0           335       335       260        267.829
 3.32667     .0332667    358.877   384.519   269.469    277.773
 6.65334     .0665334    385.42    446.388   279.653    288.48
 9.98001     .0998001    418.541   525.882   290.638    300.049
 13.3067     .133067     456.486   631.787   302.52     312.599
 16.6333     .166333     501.998   779.384   315.416    326.28
 19.96       .1996       557.59    1001.66   329.46     341.283
 23.2867     .232867     627.029   1370.31   344.813    357.88
 26.6134     .266134     716.221   2103.95   361.567    376.516
 29.94       .2994       934.995   4276.06   380.253    398.276
 33.2667     .332667     1001      223324    400.853    491.121
TO ANALYSE CROSSLINKING PRESS ENTER? _

%A          BRANCH      WT. FR.    NUM. AVE. X-LINK MW     GLASS
REACTED     COEF.       GEL        MW(G/MOL) (G/MOL)       TEMP(K)
 33.4169     .334169     .0100001   1010.07   2.67125E+07   401.837
 34.2996     .342996     .109       1066.85   212041        407.83
 35.2944     .352944     .208       1139.02   54751.9       415.919
 36.4317     .364317     .307       1234.5    23479.1       423.796
 37.7555     .377555     .406       1367.97   12439.1       434.764
 39.3337     .393337     .505       1570.37   7370.28       448.917
 41.2786     .412786     .604       1920.57   4653.78       468.049
 43.7981     .437981     .703       2700.79   3035.76       495.843
 47.3477     .473477     .802       6315.3    1987.04       541.671
 53.317      .53317      .901       3.35E+08  1240.37       598.397
 98.087      .98087      1          3.35E+08  344.897       727.705
% B REACTED= 98.087   TO CONTINUE TYPE RUN AND PRESS ENTER
READY
>_
```

TABLE 5-22 Estimated Cure Path Properties of 2 Moles of TGMDA and 1 Mole of DDS (see Table 5-7)

```
A AND B COREACTION-MOL. WT  DIST.-THERMAL TRANS.-D.H.KAELBLE-OCT
. 27,1982
IF NONSTOICHIOMETRIC REACTION HAVE MOLES OF B IN EXCESS
MOLES OF TYPE A (MOLE)=? 1
TYPE A FUNTIONALITY(=>2)=? 4
MOL. WT. OF TYPE A (G/MOLE)=? 248
MOLES OF TYPE B (MOLE)=? 2
TYPE B FUNCTIONALITY (=>2)=? 4
MOL. WT. OF TYPE B (G/MOLE)=? 422
FRACTION OF MOLECULES OF FUNCTIONALITY >2=? 1
NUMBER OF A AND B MAIN CHAIN ATOMS (A1,A2)=? 11,17
MOL. WT. BETWEEN ENTANGLEMENTS (G/MOLE)=? 5168
GLASS COORDINATION NUMBER (3<Z<10)=? 10
MONOMER AND LINEAR POLYMER GLASS TEMPERATURES (T1,T2) IN DEG  K=
? 260,503
GEL POINT (% A REACTED)= 66.6667
GEL POINT (% B REACTED)= 33.3333
INITIAL NUM. AVE. DEG  OF POLYMERIZATION= 2.6749
TO ANALYSE POLYMERIZATION PRESS ENTER?
```

% A REACTED	BRANCH COEF.	NUM. AVE. MU(G/MOL)	WT. AVE. MW(G/MOL)	GLASS TEMP(K)	FLOW TEMP(K)
0	0	364	364	260	268.066
6.65334	.0332667	389.944	417.306	268.635	277.179
13.3067	.0665334	419.371	485.03	277.862	286.934
19.96	.0998001	454.773	571.497	287.747	297.407
26.6134	.133067	496.003	686.479	298.36	308.694
33.2667	.166333	545.455	847.397	309.786	320.912
39.92	.1996	605.859	1088.37	322.123	334.215
46.5734	.232867	681.308	1488.93	335.482	348.829
53.2267	.266134	778.222	2296.08	349.998	365.143
59.8801	.2994	907.279	4645.22	365.327	384.175
66.5334	.332667	1092.63	242657	383.155	475.197

```
TO ANALYSE CROSSLINKING PRESS ENTER?
```

%A REACTED	BRANCH COEF.	WT. FR. GEL	NUM. AVE. MW(G/MOL)	X-LINK MW (G/MOL)	GLASS TEMP(K)
66.8339	.334169	.0100001	1097.51	2.90251E+07	383.978
68.5993	.342996	.109	1159.21	230396	388.993
70.5888	.352944	.208	1237.62	59491.5	395.02
72.8634	.364317	.307	1341.36	25511.6	402.332
75.5111	.377555	.406	1486.39	13515.9	411.572
78.6674	.393337	.505	1706.31	8008.31	423.396
82.5573	.412786	.604	2096.83	5056.65	439.391
87.5962	.437981	.703	2934.59	3298.56	462.172
94.6954	.473477	.802	6861.99	2159.05	499.291
99.9265	.499632	.85348	495168	1709.73	530.392

```
% B REACTED= 49.9632   TO CONTINUE TYPE RUN AND PRESS ENTER
READY
>_
```

TABLE 5-23 Estimated Cure Path for TGMDA Homopolymer (see Table 5-8)

```
A AND B COREACTION-MOL. WT. DIST -THERMAL TRANS -D.H.KAELBLE-OCT
. 27,1982
IF NONSTOICHIOMETRIC REACTION HAVE MOLES OF B IN EXCESS
MOLES OF TYPE A (MOLE)=? 1
TYPE A FUNTIONALITY(=>2)=? 4
MOL. WT. OF TYPE A (G/MOLE)=? 422
MOLES OF TYPE B (MOLE)=? 1
TYPE B FUNCTIONALITY (=>2)=? 4
MOL. WT. OF TYPE B (G/MOLE)=? 422
FRACTION OF MOLECULES OF FUNCTIONALITY >2=? 1
NUMBER OF A AND B MAIN CHAIN ATOMS (A1,A2)=? 17,17
MOL. WT. BETWEEN ENTANGLEMENTS (G/MOLE)=? 5300
GLASS COORDINATION NUMBER (8<Z<10)=? 10
MONOMER AND LINEAR POLYMER GLASS TEMPERATURES (T1,T2) IN DEG. K=
? 260,402
GEL POINT (% A REACTED)= 33 3333
GEL POINT (% B REACTED)= 33.3333
INITIAL NUM. AVE. DEG  OF POLYMERIZATION= 4.57747
TO ANALYSE POLYMERIZATION PRESS ENTER?
% A         BRANCH      NUM. AVE.  WT. AVE.    GLASS       FLOW
REACTED     COEF.       MW(G/MOL)  MW(G/MOL)   TEMP(K)     TEMP(K)
 0           0           422.       422         260         268.5
 3.32667     .0332667    452.078    484.38      266.258     275.243
 6.65334     .0665334    486.773    562.315     272.824     282.344
 9.98001     .0998001    527.237    662.455     279.722     289.84
 13.3067     .133067     575.037    795.863     286.978     297.78
 16.6333     .166333     632.358    982.422     294.621     306.226
 19.96       .1996       702.397    1261.79     302.681     315.269
 23.2867     .232867     789.868    1726.18     311.196     325.057
 26.6134     .266134     902.224    2550.35     320.203     335.892
 29.94       .2994       1051.85    5386.56     329.747     348.693
 33.2667     .332667     1260.96    281322      339.877     437.852
TO ANALYSE CROSSLINKING PRESS ENTER? _


%A          BRANCH      WT. FR.     NUM. AVE.  X-LINK MW    GLASS
REACTED     COEF.       GEL         MW(G/MOL)  (G/MOL)      TEMP(K)
 33.4169     .334169     .0100001    1272.38    3.365E+07    340.351
 34.2996     .342996     .109        1343.92    267108       343.267
 35.2944     .352944     .208        1434.83    68970.9      346.821
 36.4317     .364317     .307        1555.1     29576.7      351.202
 37.7555     .377555     .406        1723.23    15669.5      356.595
 39.3337     .393337     .505        1978.19    9284.36      363.753
 41.2786     .412786     .604        2419.35    5862.38      373.166
 43.7981     .437981     .703        3402.19    3824.16      386.481
 47.3477     .473477     .802        7955.39    2503.07      407.368
 53.317      .53317      .901        4.22E+08   1562.5       431.108
 98.087      .98087      1           4.22E+08   434.467      530.921
% B REACTED= 98.087   TO CONTINUE TYPE RUN AND PRESS ENTER
READY
>_
```

TABLE 5-24 Estimated Polymerization Path for Equimolar Isoamyl-Neopentyl Acrylate Copolymer (see Table 5-9)

```
A AND B COREACTION-MOL. WT  DIST -THERMAL TRANS.-D.H.KAELBLE-OCT
. 27,1982
IF NONSTOICHIOMETRIC REACTION HAVE MOLES OF B IN EXCESS
MOLES OF TYPE A (MOLE)=? 1
TYPE A FUNTIONALITY(=>2)=? 2
MOL. WT. OF TYPE A (G/MOLE)=? 142
MOLES OF TYPE B (MOLE)=? 1
TYPE B FUNCTIONALITY (=>2)=? 2
MOL. WT. OF TYPE B (G/MOLE)=? 142
FRACTION OF MOLECULES OF FUNCTIONALITY >2=? 0
NUMBER OF A AND B MAIN CHAIN ATOMS (A1,A2)=? 2,2
MOL. WT. BETWEEN ENTANGLEMENTS (G/MOLE)=? 34185
GLASS COORDINATION NUMBER (8<Z<10)=? 10
MONOMER AND LINEAR POLYMER GLASS TEMPERATURES (T1,T2) IN DEG. K)
? 69,240
GEL POINT (% A REACTED)= 100
GEL POINT (% B REACTED)= 100
INITIAL NUM. AVE. DEG. OF POLYMERIZATION= 1.00877
TO ANALYSE POLYMERIZATION PRESS ENTER? _
```

% A REACTED	BRANCH COEF.	NUM. AVE. MW(G/MOL)	WT. AVE. MW(G/MOL)	GLASS TEMP(K)	FLOW TEMP(K)
0	0	142	142	69	70.8494
9.98001	9.96006E-03	143.429	144.857	69.4932	71.3977
19.96	.0399402	147.892	153.784	71.0159	73.0866
29.94	.0896405	155.982	169.965	73.7076	76.0586
39.92	.159361	168.919	195.838	77.8381	80.591
49.9	.249001	189.082	236.163	83.8818	87.1751
59.8801	.358562	221.379	300.755	92.6766	96.6839
69.8601	.488043	277.367	412.734	105.785	110.753
79.8401	.637444	391.663	641.327	126.415	132.734
89.8201	.806765	734.855	1327.71	162.284	171.113
99.8001	.996006	35552.9	70963.8	237.548	264.813

```
% B REACTED= 99.8001   TO CONTINUE TYPE RUN AND PRESS ENTER
READY
>_
```

The lower portions of Tables 5-21 through 5-25 shows that the polymerization analysis describes 10 degrees of reaction between initial state and the gel point in terms of the chemically defined branching coefficient, and number and weight average molecular weights of the reacting polymer. This calculation also provides both T_g and T_m to define the changing physical states which accompany curing as described earlier in Fig. 2-2. The program again pauses at the gel point and then if reaction is not complete will proceed to compute ten levels of reaction between the gel point and complete cure.

The branching coefficient, weight fraction gel, and molecular weight between crosslinks M_c are the three chemical properties calculated for the crosslinked polymer and T_g the significant physical property. The T_g calculation now incorporates the effects of both chain extension and crosslinking.

TABLE 5-25 Estimated Polymerization Path for Polystyrene

```
A AND B COREACTION-MOL. WT. DIST.-THERMAL TRANS.-D.H.KAELBLE-OCT
  27,1982
IF NONSTOICHIOMETRIC REACTION HAVE MOLES OF B IN EXCESS
MOLES OF TYPE A (MOLE)=? 1
TYPE A FUNTIONALITY(=>2)=? 2
MOL. WT. OF TYPE A (G/MOLE)=? 104
MOLES OF TYPE B (MOLE)=? 1
TYPE B FUNCTIONALITY (=>2)=? 2
MOL. WT. OF TYPE B (G/MOLE)=? 104
FRACTION OF MOLECULES OF FUNCTIONALITY >2=? 0
NUMBER OF A AND B MAIN CHAIN ATOMS (A1,A2)=? 2,2
MOL. WT. BETWEEN ENTANGLEMENTS (G/MOLE)=? 20200
GLASS COORDINATION NUMBER (8<Z<10)=? 10
MONOMER AND LINEAR POLYMER GLASS TEMPERATURES (T1,T2) IN DEG. K=
? 110,384
GEL POINT (% A REACTED)= 100
GEL POINT (% B REACTED)= 100
INITIAL NUM. AVE. DEG. OF POLYMERIZATION= 1.00365
TO ANALYSE POLYMERIZATION PRESS ENTER? _
```

% A REACTED	BRANCH COEF.	NUM. AVE. MW(G/MOL)	WT. AVE. MW(G/MOL)	GLASS TEMP(K)	FLOW TEMP(K)
0	0	104	104	110	111.849
9.98001	9.96006E-03	105 046	106.093	110.787	112.692
19.96	.0398402	108.315	112.631	113.219	115.289
29.94	.0896405	114.241	124.481	117.517	119.868
39.92	.159361	123.715	143.431	124.113	126.866
49.9	.249001	138.482	172.965	133.767	137.06
59.8801	.358562	162.136	220.271	147.919	151.827
69.8601	.488043	203.142	302.284	168.774	173.744
79.8401	.637444	286.852	469.704	201.777	208.145
89.8201	.806765	538.204	972.408	259.226	268.061
99.8001	.996006	26038.7	51973.5	380.217	407.779

```
% B REACTED= 99.8001   TO CONTINUE TYPE RUN AND PRESS ENTER
READY
>_
```

The examples of epoxy resins in Table 5-21 through 5-23 continue the analysis begun by the structural formulas shown earlier in Fig. 5-3 through 5-5. As pointed out earlier the molecular structure shown in Fig. 5-4 closely resembles commercially important epoxy resins as analyzed in Table 1-6. The effect of degree of cure as measured by differential scanning calorimetry (DSC) for Hercules 3501-5 epoxy, which closely models the example of 2 moles TGMDA and 1 mole of DDS, has been reported by Kaelble and Dynes.(44) The comparison of the model calculation of T_g and T_m in Table 5-22 for nonstoichiometric 2 moles TGMDA and 1 mole DDS is shown in Fig. 5-14 to be in good agreement with the experimental values of T_g obtained by DSC measurement.

The correlation of T_g with degree of cure has been reported in a number of prior studies.(45-48) More recent quantitative evaluations of both cure state and specific property correlations with cure state have been undertaken. Adabbo and Williams[49] propose a physical state diagram of degree of cure versus T_g for thermoset resins. The relevance of combining the temperature difference into a single variable $(T - T_g)$ in analyzing curing and aging of thermosets is proposed by Lee and Goldfarb.[50] Based upon viscoelastic measurements for a variety of thermoset systems Tung and Dynes[51] define the gel point of a curing thermoset, well above T_g where $T - T_g = 30K$, as the point where the dynamic loss modulus G" is less than storage modulus G' so that the loss tangent tan δ is less than unity. A detailed molecular analysis of dielectric cure monitoring by Tajima[52] correlates closely with the chemical modeling approach discussed in this section.

Intercomparison of the estimated cure properties in Tables 5-21 through 5-23 shows that the stoichiometric equimolar TGDMA and DDS of Table 5-21 displays a final T_g far above the projected range of epoxy thermal stability, while the pure TGMDA homopolymer of Table 5-23 has a relatively low final T_g. The intermediate composition of Table 5-23 parallels commercial formulas, shown in Table 1-6, where a useful balance of physical properties are obtained by formulation of nonstoichiometric mixtures.

The application of the polymerization and crosslinking model to pure chain extension reactions is shown in Table 5-24 for the chain extension polymerization of the equimolar isoamyl-neopentyl acrylate structure of Fig. 5-2. As shown in Table 5-24 this reaction projects an incipient gel point only at complete polymerization. The ratio of weight to number average molecular weight approaches $M_W/M_N = 2.0$ also at complete cure as expected for pure chain extension. The variation in T_g and T_m also follow conventional predictions for linear polymers.

The last polymerization example for polystyrene in Table 5-25 returns a whole set of calculated properties typical for this polymer. In particular the T_g versus number average molecular weight data of Table 5-25 will correlate with the experimental data of Kanig[15] as shown in the curves of Fig. 5-10.

The three CAD/CAM models discussed in this chapter provide for computer aided analysis of the relation of chemical and physical properties of a wide range of nonpolymeric and polymeric materials. As shown in the extensive examples of this discussion these CAD/CAM models are designed for both separate and combined usage. In the next and final chapter of this discussion we will illustrate the further use of these models to numerically evaluate polymer composite response. The appropriate philosophy in the use of these CAD/CAM models is illustrated in the dual path approach illustrated in Fig. 1-5 where experimental data are combined with molecular analysis. The models presented here and in the forthcoming chapter are intentionally simplified for ease of use on microcomputers.

References

1. B. Albrecht, D. Inman and R. Zamora, "TRS-80 BASIC," John Wiley and Sons Inc., New York, (1980).

2. T. Dwyer and M. Critchfield, "BASIC and the Personal Computer," Addison-Wesley Publishing Co., Reading, Mass. (1978).

3. W. Borden, "Programming Techniques for Level-II BASIC," Tandy Corp. Forth Worth, TX, (1980).

4. L. Finkel and J.R. Brown, "Data File Programming in Basic," John Wiley & Sons Inc., New York, (1981).

5. D.H. Kaelble, "Physical Chemistry of Adhesion," Wiley-Interscience, New York (1971).

6. J. Delmonte, "Technology of Carbon and Graphite Fiber Composites," Van Nostrand - Reinhold Co., New York (1981).

7. C.A. May (Editor), "Resins for Aerospace" ACS Symposium Series, No. 132 (1980).

8. P.J. Dynes and D.H. Kaelble, "Physio-chemical Analysis of Graphite-Epoxy Composite Systems," ASTM Spec. Tech., Pub. 674, ASTM, Philadelphia (1979), p. 566.

9. H. Burrell, Enc. of Poly. Sci. and Tech. 12, Wiley-Interscience, New York (1970), p. 618.

10. J.L. Gardon, Ibid., 3, (1965), p. 833.

11. S.S. Rogers and L. Mandelkern, J. Phys. Chem., 61, (1957), p. 985.

12. L.A. Wood, J. Poly. Sci., 28 (1958), p. 314.

13. T.G. Fox and P.J. Flory, J. Appl. Phys., 21, (1950), p. 581.

14. T.G. Fox and P.J. Flory, J. Poly. Sci., 14, 1954, p. 315.

15. V.G. Kanig, Koll-Zeit., 190, (1963), p. 1.

16. R.F. Fedors, J. Poly. Sci., Letters, 17 (1979), p. 719.

17. J.D. Ferry, Viscoelastic Properties of Polymers, 3rd Ed., Wiley, New York, 1980, Chap 11.

18. S.M. Aharoni, ACS Polymer Preprints, 23(1), (1982), p. 275.

19. L.H. Lee, J. Coll. and Interface Sci., 27, 1968, p. 751.

20. D.H. Kaelble, Proc. 23rd Int. Cong. on PUre and Appl. Chemistry, Butterworths, London (1971), Vol. 8, p. 265.

21. A.C. Zettlemoyer and H.H. Hsing, J. Coll. Interface Sci., 58, (1977), p. 263.

22. C-H. Chiang, H. Ishida and J.L. Koenig, Ibid., (1980), p. 396.

23. C-H. Chiang and J.L. Koenig, Ibid., 83, (1981), p. 361.

24. A.T. DiBennedetto and D.A. Scola, Ibid., 64, (1978), p. 480.

25. F.J. Boerio, L. Armogan and C.Y. Cheng, Ibid., 73, (1980), p. 416.

26. Y. Eckstein and P. Dreyfuss, J. Adhesion, 13, (1982), p. 303.

27. E.P. Plueddeman, "Chemical Bonding Technology for Terrestrial Solar Cell Modules," JPL Report No. 5101-132, Jet Propulsion Laboratory, Calif. Inst. of Technology, Pasadena, September 1980.

28. R.F. Landel and R.F. Fedors, Fracture Process in Polymeric Solids (Editor, B. Rosen), Wiley, New York, (1961), Chap. 3B.

29. K. Kanamaru, T. Arai and E. Wado, "Ultimate Tensile Properties of Epoxy-Polymide Resins, Reports of Progress in Polymer Physics," Japan, 12, (1969), p. 353.

30. R.F. Fedors and R.F. Landel, "Relationship Between Extensibility of Networks and the Degree of Crosslinking and Primary Molecular Weight," J. Appl. Poly. Sci., 19, (1975), p. 2709.

31. D.H. Kaelble, "Block Copolymers as Adhesives," in Adhesion Science and Technology (Editor, L.H. Lee), Vol. 9A, Plenum, New York, (1975), p. 199.

32. E.J. Kramer, "A Molecular Theory for the Fracture Toughness of Low Molecular Weight Polymers," J. Matl. Sci., 14, (1978), p. 1381.

33. E.M. Donald and E.J. Kramer, J. Poly. Sci. Polymer Physics Ed., 20, (1982), p. 899.

34. J.L. Gardon, Treatise on Adhesion and Cohesion, (Editor: R.L. Patrick), Dekker, New York (1966), Chap. 8.

35. D.H. Kaelble and E.H. Cirlin, J. Poly. Sci., Part C, 35, (1971), p. 101.

36. C.W. Macosko and D.R. Miller, "A New Derivation of the Average Molecular Weights of Nonlinear Polymers," Macromolecules, 9, (1976), P. 199.

37. D.R. Miller and C.W. Macosko, "A New Derivation of Post Gel Properties of Network Polymers," Ibid. 9, (1976), p. 206.

38. D.R. Miller, E.M. Valles and C.W. Macosko, "Calculation of Molecular Parametres for Stepwise Polyfunctional Polymerization," Poly, Eng. and Sci., 19, (1979), p. 272.

39. E.M. Valles and C.W. Macosko, "Structure and Properties of Polydimethylsiloxanes with Random Branches," Macromolecules, 12, (1979), p. 521.

40. E.M. Valles and C.W. Macosko, "Properties of Networks Formed by Endlinking Polydimethylsiloxane, Ibid., 12, (1979), p. 673.

41. D.R. Miller and C.W. Macosko, "Substitution Effects in Property Relations for Stepwise Polyfunctional Polymerization," Macromolecules, 13, (1980), p. 1063.

42. M. Gottlieb, C.W. Macosko, G.S. Benjamin, K.O. Meyers, and E.W. Merritt, "Equilibrium Modules of Model Polydimethylsilkoxane Networks," Ibid., 14, (1981), p. 1039.

43. C.A. Angell, J.M. Sare, and E.J. Sare, "Transition Temperatures for Simple Molecular Liquids and Their Binary Solutions," J. Phys. Chem., 82 (1978), p. 2622.

44. D.H. Kaelble and P.J. Dynes, "Preventative Nondestructive Evaluation of Graphite Epoxy Composites," Ceramic Eng. & Sci. Proc., 1, (1980), p. 458.

45. M. Gordon and W. Simpson, Polymer, 2, (1961), p. 383.

46. R.A. Fava, Ibid., 9, (1968), p. 137.

47. K. Horie, H. Hiura, M. Sawada, I. Mita, and H. Kambe, J. Poly Sci., Part A-1, 8 (1970) p. 1357.

48. W.P. Brennan, Amer. Laboratory, Feb. 1975, pp. 75-81.

49. H.E. Adabbo and R.O. Williams, J. Appl. Poly, Sci., 27, (1982), p. 1327.

50. C. Y-C. Lee and I.J. Goldfarb, Poly. Eng. and Sci., 21, (1980), p. 951.

51. C-Y.M. Tung and P.J. Dynes, J. Appl. Poly. Sci., 27, (1982), p. 569.

52. Y.A. Tajima, Polymer Composites, 3, (1982), p. 162.

6
Composite CAD/CAM Models

6-1. Introduction

In this final chapter we introduce three additional CAD/CAM models for polymer composites which project the definitions of chemical properties directly to calculations of stiffness, strength, and fracture toughness of simple composite combinations which form the building blocks of all laminated composite structures. These models extend the series of CAD/CAM models introduced in Chapter 5. The models discussed here have their programs listed in the Appendix by the following program listing numbers:

4. Polymer Chemistry to Mechanical Properties
5. Composite Fracture Energy and Strength
6. Peel Mechanics.

All the program outputs of this chapter relate to mechanical reliability and mechanical durability which is made directly traceable to chemical and molecular structure variables. The polymer chemistry to mechanical properties program No. 4 provides this transformation out of molecular chemistry to mechanical response. The mathematical model for program No. 4 is developed and discussed by Kaelble(1) as part of a detailed analysis of polymer cohesion and adhesion. The interested reader is referred to that extended discussion for details. The mathematical derivations for program listing No. 5 for composite fracture energy and strength are also developed by Kaelble(2) and presented in a separate publication. The peel mechanics model of program listing No. 6 was derived in several early publications and incorporated in an earlier computer model by Kaelble.[3-5] The sequence of use of the models in this discussion is to utilize the polymer mechanical properties program No. 4 to computationally define shear and tensile properties which are required as inputs to programs No. 5 and No. 6. As mentioned in Chapter 5 it is very simple to combine these separate models to directly describe composite fracture or laminate peeling properties based upon polymer chemistry and macromolecular structure inputs.

6-2. Polymer Tensile Response

Before introducing the composite CAD/CAM models it is useful to briefly review the results of tensile property measurements on film forming polymers which are frequently applied as primary mechanical structures in tension skin loaded structures ranging from food packages to the first man powered airplane. The "Gossamer" man powered aircraft designed by Paul Macready utilized a graphite fiber reinforced epoxy as the composite material in the main wing spar and "Mylar" polyester (polyethyleneterephthalate) film as the skin for both wing and fuselage.[6] According to popular accounts a good deal of "Mylar" type pressure sensitive tape was also used to both construct and repair the Gossamer and later versions of this aircraft. Program listing No. 5 relates to the graphite reinforced wing spar of the "Gossamer" and listing No. 6 relates directly to the pressure sensitive tape used in fabrication and repair.

Many of the unique and useful properties of polymer films are displayed by thermal tensile testing which is conducted from cryogenic to elevated temperatures so that all the important rheological transitions are displayed in the tensile deformation and failure response. The results of one such experimental study are shown in the tensile stress-strain curves of Figs. 6-1 through 6-4. Since "Mylar" type polyester film was included in this study as shown in Fig. 6-3 the data will illuminate some features of the utility of this material in aircraft design.

These tensile studies utilized an Instron thermomechanical analysis (TMA) unit with a temperature range of 80K to 620K and a tensile strain range from 0 to 200% strain. The six films tested are identified in Table 5-20 and as mentioned earlier all failures occur in the uniform gage section using the special clamping methods of this TMA measurements.[7] All tests were conducted at a constant tensile strain rate $\dot{e}$ = 2.08 E-3 s^{-1} with multiple tests conducted at 20K temperature intervals. Tests were conducted in a dry nitrogen atmosphere with a maximum exposure time of less than 1 hr with 16 min under tensile loading. The repeat unit chemical structure and thermal transitions of these six film forming polymers are summarized in Table 6-1. The listing of thermal transitions in Table 6-1 is taken principally from reviews of Boyer[8] and the

TABLE 6-1 Chemical Structure and Thermal Transitions of Six Film Forming Polymers

No.	Repeat Unit Structure	Thermal Transition (K) T_Y	T_g	T_c
1	$[CF_2 - CF_2]_x [CF_2 - CF(CF_3)]_{0.14x}$	177	358	555
2	$-CF_2-CF_2-$	177	400	600
3	$[CF_s - CF(Cl)]_x [CF_2 - CH_2]_{0.03x}$	273	323	493
4	$[C(=O)-O-C_6H_4-C(CH_3)_2-C_6H_4-C]$	176	423	538
5	$[O-CH_2-CH_2-O-C(=O)-C_6H_4-C(=O)]$	243	350	536
6	$[N(C(=O))_2C_6H_2(C(=O))_2N-C_6H_4-O-C_6H_4]$	180	530	-

T_λ is an amorphous transition below T_g involving local motion of 2-4 atoms.
T_g is the amorphous glass transition involving cooperative motion of 20-40 atoms in chain segments.
T_c is the crystalline phase melting temperature involving cooperative motion of the entire polymer molecule.

discussion of solid state molecular motion presented by McCrum, Read and Williams.[9]

The two fluorocarbon polymers shown in Fig. 6-1 display a brittle tensile response with maximum tensile strength $(\sigma_{11})_b$ = 800 to 1000 bar. These low cohesive energy density polymers show yielding anud onset of a simple elastic-plastic response at temperatures consistent with the T_γ temperatures reported in Table 6-1. These polymers also display ease of recrystallization which enhances the plastic response and makes them important as dry lubricants. The dashed curve in Figs. 6-1 and 6-2 denote the combination of nominal strength S_b and extensibility ε_b characteristic of about 50% failure probability. This dashed curve is termed the "tensile failure envelope."

The tensile responses shown for the fluorochlorocarbon copolymer No. 3 in Fig. 6-2 shows the appearance of a strong-tough response at temperatures between 200K and 260K wherein a high yield stress combines with strength and

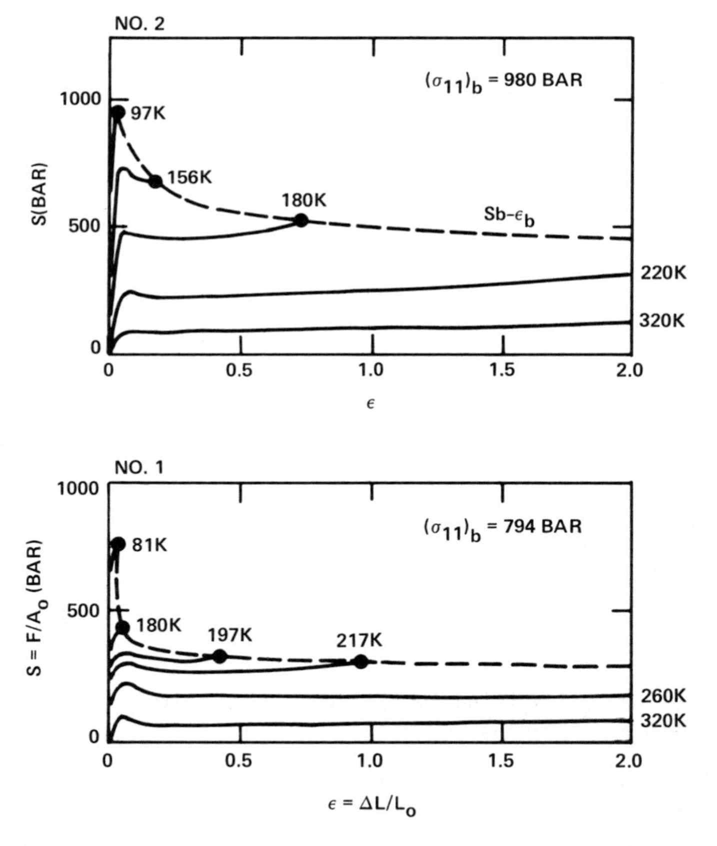

FIG. 6-1 Tensile response of C_2F_4 homopolymer (lower view) and $(C_2F_4)_{1.0}$ $(C_3F_6)_{0.14}$ copolymer (upper view) films.

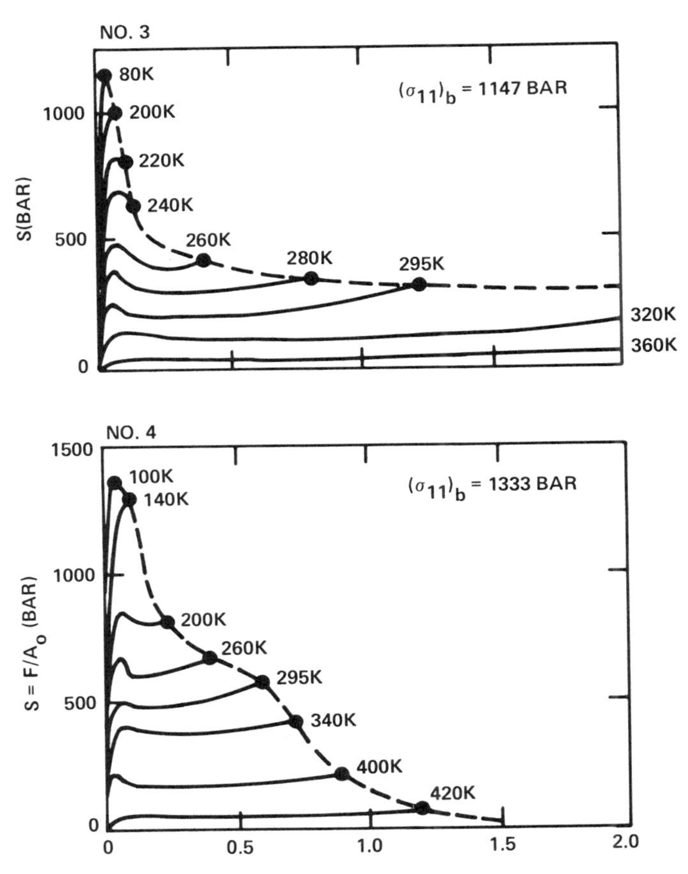

FIG. 6-2 Tensile response of $(CF_2CFCL)_{1.0}$ $(CF_2CH_2)_{0.03}$ copolymer (upper view) and polybisphenol-A carbonate (lower view) films.

extensibility. Above 300K this polymer behaves very similarly to the fluoropolymers in that ideal plasticity permits flow without fracture.

The polycarbonate film data of lower Fig. 6-2 shows a further enhancement of the strong-tough response which extends from 140K to 400K. This strong-tough response is combined with easy processability as indicated by the elastic-plastic response above 420K which explains the importance of this polymer as an engineering thermoplastic.

The tensile response of "Mylar" type polyster film shown in Fig. 6-3 demonstrates the outstanding strong-tough response of this polymer at temperatures from 190K to 475K. The high crystalline melt temperature T_c = 536K limits the processability of this polymer primarily to production of films and fibers.

The tensile response of the polyimide film illustrated in Fig. 6-4 shows the optimization of a strong-tough tensile response which operates over an

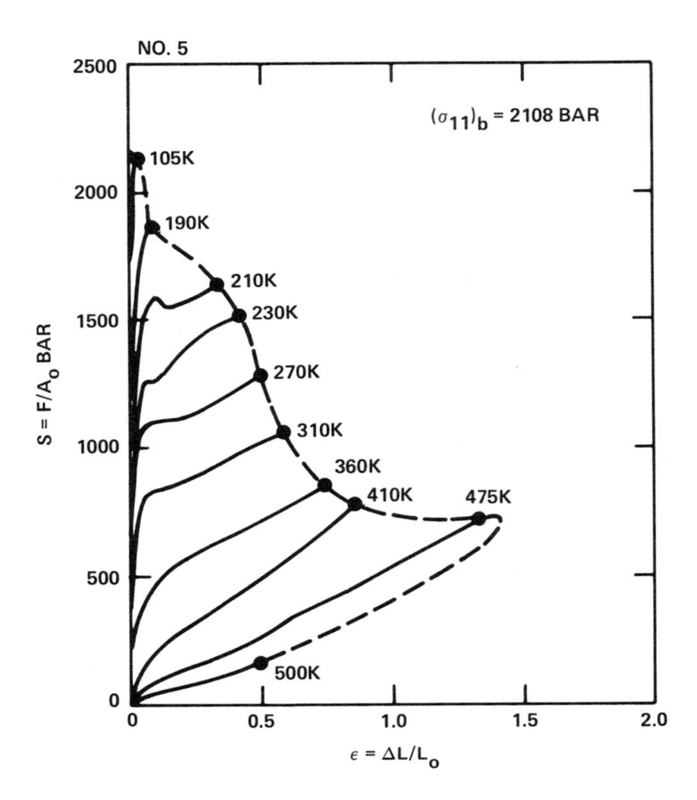

FIG. 6-3 Tensile response of polyethyleneterephthalate film.

extreme temperature range from 155K through 610K. The tensile response of this polymer shows no elastic-plastic yielding which in the other polymers, No. 1 - 5 is indicative of entanglement slippage. As shown in Table 6-1 there is no well defined crystalline melting phase for this polyimide yet the high tensile response at 610K, which is well above T_g = 530K for this polymer, indicates interchain constraints typical of crystallites or light crosslinking.

The polyester film response of Fig. 6-3 shows the most complete failure envelope (dashed curve) with the strong-tough response in the upper quadrant and the maximum extensibility character shown in the lower right quadrant between 475K and 500K. Above 500K both the strength and extensibility diminish due to onset of crystalline melting. It is suspected that the failure envelope of the polyimide film of Fig. 6-4 would show a lower quadrant response similar to Fig. 6-3 if experiments were conducted well above 610K which was the upper measurement limit of this study.

All the tensile responses in Figs. 6-1 through 6-4 can be easily associated with features of the chemical structure and thermal transitions shown in Table 6-1. The range of tensile response as shown by comparing the elastic-

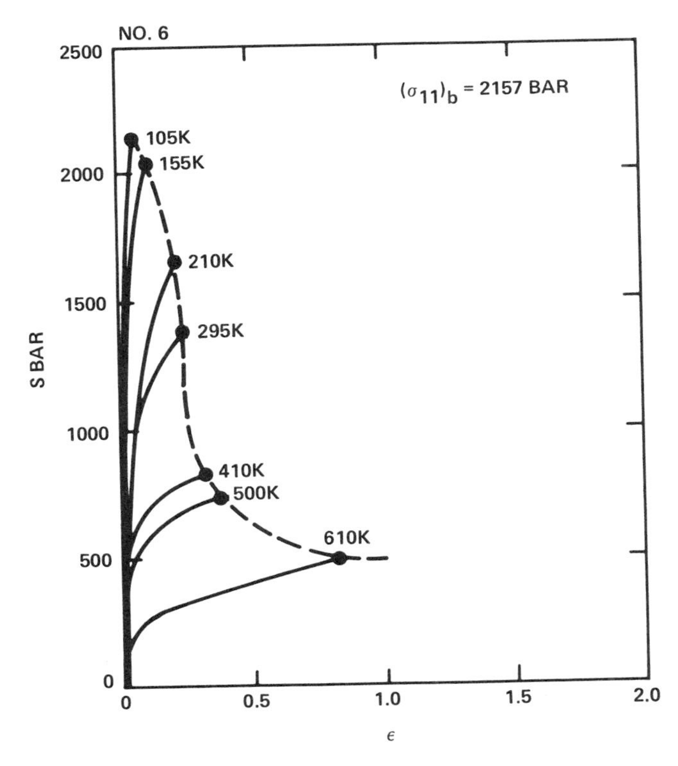

FIG. 6-4 Tensile response of $(N(CO)_2\ C_6H_2(CO)_2NC_6H_4OC_6H_4)$ polyimide film.

plastic responses of Fig. 6-1 with the strong-tough response of Fig. 6-4 is broad and selectable from chemical composition of the polymer. It is this last fundamental point that mechanical response is selectable based upon polymer composition and macromolecular structure which is fundamental to this discussion.

6-3. Polymer Chemistry to Mechanical Properties

The data of the previous Section 6.2 describe experimental tests where one stress-strain curve is produced per hour in a measurement system restricted to temperatures of 80K to 625K and tensile strains of less than 200% extension. A good days effort will produce 8 stress-strain curves and cover 40K to 60K in temperature using three specimens per temperature and 20K steps in temperature.

The computer model of listing No. 4 provides a method for generating both tensile and shear responses based upon the relations and concepts discussed in Chapter 4-3. This chemical to mechanical response model describes the composite response of a polymer-diluent combination. The diluent may be one of three types which are:

1. a plasticizer which lowers the polymer T_g
2. an extender which does not affect T_g
3. an antiplasticizer which raises polymer T_g.

In all cases the diluent is assumed soluble in the polymer and the polymer is assumed amorphous so that a single phase and glass temperature describe the polymer-diluent combination. The output of the program calculates stress-strain response in simple shear and tension for preassigned conditions of constant temperature (isothermal) and constant time (isochronal) which are most useful to the composite stress analysis models presented in listing No. 5 and No. 6.

The input format for this chemical-mechanical property program is shown in Table 6-2 to describe 15 polymer properties, 10 for the diluent and 5 characteristics of use. Experience has shown that it is best to have an input table already made up on a blank form like Table 6-2 before running a computer simulation. In this way data can be readily entered and cross-checked for correctness. The symbols and meaning of these input data are described in Table 6-3. As the subsequent example problems will show, all of these input data for polymer and diluent can be generated by the computer models of Chapter 5.

After the data input is complete, the chemistry-mechanical model calculates and displays immediately below the input report a number of intermediate results whose symbols and definitions are summarized in Table 6-4. Both the input and the intermediate tables serve as useful records in evaluating the calculated mechanical responses. The mathematical definitions of the intermediate results is presented in program listing No. 4 of the Appendix and the derivations provided in the theory of cohesion and adhesion by Kaelble.[1] A rapid and easy way to learn the exact relations between the input variables and the intermediate results is to use the computer to instantaneously translate selective changes in inputs into new intermediate results.

The bottom row of the input table shown in Table 6-2, in row 7, describes the use conditions. The first of these variables is the mechanical strain (or thermal scan) rate defined in the computer program as variable BA. If the rate BA differs from the reference rates AJ and AZ assigned to polymer and diluent, respectively, the computer model automatically adjusts the calculated T_g of both polymer and diluent for the use condition rate BA. The program then defines the stress dependence of T_g for the polymer-diluent solution. The results of these computations are shown in upper Table 6-4. The listing of variables in Table 6-4 follows the sequence of the computer calculations. The volume fractions of polymer and diluent are computed along with the cohesive energy of the polymer-diluent solution. The theory of the model permits a de-

TABLE 6-2 Input Format for Chemical-Mechanical Property Program (Upper symbols found in Ref. 1 and lower symbols in Table 6-3)

		C1	C2	C3	C4	C5
	R1	Polymer or Experiment Name AO$				
Polymer Properties	R2	M_p (g/mol) AA	X_p (mole) AB	Σh_p AC	ρ_p (g/cc) AD	$(T_{gAR})_P$ (K) AE
	R3	Z_{gP} AF	V^o_{gP} (cc/mol) AG	q_{LP} AH	δ_P $(cal/cc)^{1/2}$ AI	AR_P (S^{-1}) AJ
	R4	M_o (g/mol) AK	M_c (g/mol) AL	M_e (g/mol) AM	τ_g (S) AN	Φ AP
Diluent Properties	R5	M_D (g/mol) AQ	X_D (mol) AR	Σh_D AS	ρ_D (g/cc) AT	$(T_{gAR})_D$ (K) AU
	R6	Z_{gD} AV	V^o_{GD} (cc/mol) AW	q_{LD} AX	δ_D $(cal/cc)^{1/2}$ AY	AR_D (S^{-1}) AZ
Use Condition	R7	A (S^{-1}) BA	t_I (S) BB	T (K) BC	ΔT (K) BD	NT BE

tailed definition of the efficiency of polymer-diluent molecular force interactions through the input parameter AP (see Ref. 1, Eq. 11-5 for details) is defined by the following relation:

$$AP = \Phi = d_P d_D + p_P p_D + \Delta_{PD} \tag{6-1}$$

where d_P and d_D are the dispersion fractions and p_P and p_D the polar fractions of the polymer (P) and diluent (D) solubility parameter δ and Δ_{PD} is an excess interaction of chemical type. When AP = 1.0 the polymer and diluent have perfectly matched ratios of dispersion and polar fractions of intermolecular forces. A mismatch in molecular force components can lower polymer-diluent interactions

TABLE 6-3 Input Nomenclature for Chemical-Mechanical Program

Row	Symbol	Meaning
2	AA	Polymer molecular weight (number ave.)
	AB	Moles polymer
	AC	Polymer repeat unit rotational degrees of freedom
	AD	Polymer density
	AE	Polymer glass temperature at reference strain rate AJ
3	AF	Total adjacent lattice (Z) sites for glass (nominally = 10)
	AG	Polymer glass repeat unit molar volume
	AH	Intermolecular lattice sites in polymer liquid (nominally = 9)
	AI	Polymer solubility parameter
	AJ	Strain (or thermal scan) rate for reference polymer glass temperature (nominally = 1.0)
4	AK	Polymer repeat unit molecular weight
	AL	Molecular weight between crosslinks (number ave.)
	AM	Molecular weight between entanglements (number ave.)
	AN	Relaxation time at T_g (nominally = 1.0)
	AP	Polymer-diluent interaction parameters (nominally = 1.0)
5	AQ	Diluent molecular weight
	AR	Moles diluent
	AS	Diluent molecular rotational degrees of freedom
	AT	Diluent density
	AU	Diluent glass temperature at reference rate AZ
6	AV	Total adjacent lattice (Z) sites of diluent glass (nominally = 10)
	AW	Diluent glass molar volume
	AX	Intermolecular lattice (q) sites of diluent liquid (nominally = 9)
	AY	Diluent solubility parameter
	AZ	Strain (or thermal scan) rate for diluent reference glass temperature (nominally = 1.0)
7	BA	Mechanical (or thermal scan) strain rate (nominally = 1.0)
	BB	Constant time for isochronal stress-strain response (nominally = 1.0)
	BC	Starting temperature for family of stress-strain curves
	BD	Temperature increment
	BE	Number of temperatures

TABLE 6-4 Nomenclature for Intermediate Results in Chemical-Mechanical Program

Line No.	Symbol	Meaning
1	BF	Wood constant in T_g calculation
	BG	Rate ratio in polymer T_g calculation
	BH	Log BG
	BI	Polymer T_g at rate BA (K)
	BJ	Polymer T_g change with shear stress (K/bar)
2	BK	Rate ratio in diluent T_g calculation
	BL	Log BK
	BM	Diluent T_g at rate BA (K)
	BN	Diluent T_g change with shear stress (K/bar)
3	TG	T_g of polymer-diluent at zero stress (K)
	UR	Polymer-diluent T_g change with shear stress (K/bar)
4	BO	Volume fraction polymer
	BQ	Volume fraction diluent
	BR	Cohesive energy of polymer-diluent solution (cal/cc)
5	BS	Fraction of effective crosslinked segments
	BT	Fraction of effective entangled segments
6	BU	Glass state shear modulus (bar)
	BV	Rubber state shear modulus at τ_g (bar)
	BX	Rubber state shear modulus at τ_l (bar)
	BY	Rubber state shear modulus at τ_m (bar)
	BZ	Crosslink network shear modulus (bar)
7	SB	Brittle shear strength (bar)
	TL	$\text{Log}_{10}(\tau_m/\tau_g)$
	TM	Melt (or flow) temperature (K)
	NH	Fraction Neohookian versus Hookian tensile response
8	T	Current temperature (K)
	SM	Flow shear strength (bar)
	SS	Current shear strength (bar)

and produce AP < 1.0. Special intermolecular interactions between polymer and diluent which produce complexing or compound formation produce AP > 1.0.

The remaining portion of the intermediate calculations identified in Table 6-4 define the rheological properties of the polymer-diluent combination. Most commonly the diluent will describe the monomer or repeat unit for the polymer. This is the case presented in the example of Table 6-5. The example of Table 6-5 reduces to being pure polymer since the moles of diluent AR = 0.0. By increasing AR > 0.0 the effects of residual monomer on the polymer physical and rheological properties would appear as changes in the intermedaite results. The effects of AR > 0 on mechanical stress-strain response and failure properties would be reported in the lower portion of Table 6-5.

TABLE 6-5 Estimated Mechanical Properties of an Acrylate Copolymer

```
POLYMER-DILUENT=EQUIMOLAR ISOAMYL-NEOPENTYL ACRYLATE
POLYMER PROPERTIES:
AA,AB,AC,AD,AE= 1.03E+06  1  121  1.09  230
AF,AG,AH,AI,AJ= 10  260  9  9.07  1
AK,AL,AM,AN,AP= 234  1.03E+06  34200  1  1
DILUENT PROPERTIES:
AQ,AR,AS,AT,AU= 234  0  121  1.09  59
AV,AW,AX,AY,AZ= 10  260  9  9.07  1
TEST CONDITIONS:
BA,BB,BC,BD,BE= 1  1  100  25  15
FRACTION NEOHOOKIAN TENSILE RESPONSE= 0
PRESS ENTER TO CONTINUE

INTERMEDIATE RESULTS:
BF,BG,BH,BI,BJ= 1  1  0  230  .114881
BK,BL,BM,BN= 1  0  59  .114881
TG,UR= 230  .114881
BO,BQ,BR= 1  0  82.2649
BS,BT= 0  .933592
BU,BV,BX,BY,BZ= 27547.9  27547.9  .609377  .157935  0
SB,TL,TM,NH= 860.491  12.1372  349.802  0
T,SM,SS= 100  2167.48  860.491
SHEAR AND TENSION ANALYSIS:
INPUT NUMBER OF STRESS INCREMENTS? 12_

INPUT NUMBER OF STRESS INCREMENTS? 12
```

SHEAR MODULUS (BAR)	SHEAR STRESS (BAR)	SHEAR STRAIN	TENSILE STRESS (BAR)	TENSILE STRAIN
	0	0	0	0
27547.9	71.7076	2.60301E-03	143.415	1.73534E-03
27547.9	143.415	5.20603E-03	286.83	3.47069E-03
27547.9	215.123	7.80904E-03	430.245	5.20603E-03
27547.9	286.83	.0104121	573.661	6.94137E-03
27547.9	358.538	.0130151	717.076	8.67672E-03
27547.9	430.245	.0156181	860.491	.0104121
27547.9	501.953	.0182211	1003.91	.0121474
27547.9	573.661	.0208241	1147.32	.0138827
27547.9	645.368	.0234271	1290.74	.0156181
27547.9	717.076	.0260301	1434.15	.0173534
27547.9	788.783	.0286332	1577.57	.0190888
27547.9	860.491	.0312362		

```
SHEAR FAILURE PROPERTIES:
STRESS(BAR)= 860.491 STRAIN= .0312362   ENERGY/VOL(BAR)=
 11.2927
E-MODULUS(BAR)= 23147.9           E-WORK(BAR)= 11.2927
E-STRAIN= .0312362                P-WORK(BAR)= 0

TENSILE FAILURE PROPERTIES:
STRESS(BAR)= 1686.56 STRAIN= .0204074   ENERGY/VOL(BAR)=
 17.2091
E-MODULUS(BAR)= 82643.7           E-WORK(BAR)= 17.2091
E-STRAIN= .0204074                P-WORK(BAR)= 0
PRESS ENTER TO CONTINUE:? _
```

In evaluating the mechanical response of the polymer-diluent combination this computer model offers two options. The first option is offered at the end of data input and requests a selection of fractional Neohookian versus Hookian response. A selection of NH = 1.0 defines pure Neohookian response and NH = 0 defines pure Hookian response. Mixtures of these responses are calculated for values of NH between zero and unity. The standard definitions of Hookian response utilized in the computer model are:

$$E(t) = 3\ G(t) \tag{6-2}$$

$$S(t) = 2\ \sigma(t) \tag{6-3}$$

$$\varepsilon = (2/3)\ \gamma = \lambda - 1 \tag{6-4}$$

where E(t) and G(t) are the time dependent tensile and shear modulus, S(t) and $\sigma(t)$ are the nominal tensile and shear stress, ε and γ the nominal tensile and shear strain and λ is the extension ratio. The time dependent shear modulus G(t) is defined by a proportional stress-strain relation:

$$G(t) = \frac{\sigma(t)}{\gamma} \tag{6-5}$$

for both Hookian and modified Neohookian response. For finite strain $\lambda > 1.0$ the Neohookian tensile response is defined as follows:

$$E(t) = G(t)\ \left(\frac{\lambda - \lambda^{-2}}{\lambda - 1}\right) \tag{6-6}$$

$$S(t) = \frac{2}{3}\left(\frac{\lambda - \lambda^{-2}}{\lambda - 1}\right)\ \sigma(t) \tag{6-7}$$

$$\varepsilon = (2/3)\ \gamma = (\lambda - 1) \tag{6-8}$$

By introducing the variable Q as follows:

$$Q = \frac{1}{3}\ \left(\frac{\lambda - \lambda^{-2}}{\lambda - 1}\right)$$

the combined Hookian-Neohookian response can then be defined as follows:

$$S(t) = 2\sigma(t)\ [1 + NH\ (Q - 1)\ ] \tag{6-9}$$

where NH is defined as the fraction of Neohookian response. Neohookian response is shown by Ward[10] to be derivable based upon three general requirements which are material isotropy, incompressability and reduction to Hook's law at small strains.

The polymer lattice theory of Kaelble[1] presumes that stress can induce a rubbery response at constant volume and thereby the above relations satisfy all three of the Neohookian requirements. The effects of material anisotropy are separately defined in this computer model by a network orientation factor R defined as follows:

$$R = \frac{1}{1 - (\lambda/\lambda_m)^2} \tag{6-10}$$

where λ_m is defined by Eq. (5-5). Both shear and tensile properties are modified by Eq. (6-9) as shown on inspection of the program listing No. 4. When a polymer-diluent combination is chemically cross-linked it is presumed the diluent is present and in equilibrium with the polymer prior to crosslinking. Thus crosslinking produces no internal stresses other than described by Eq. (6-10). It is important to point out that Eq. (5-5) which defines λ_m defines in Eq. (6-10) a condition of chemical strain failure at high crosslink density (ρ/M_c) or low crosslink molecular weight M_c when λ_m < = 1.0. The computer program recognizes this condition and asks for new inputs for M_c.

The input section of Table 6-5 describes test conditions for the acrylate copolymer previously discussed in Chapter 5 (see Tables 5-9 and 5-10). In Table 6-5 all T_g defining rates are equivalent, that is BA = AJ = AZ and the isochronal test time BB and polymer T_g relaxation AN are equivalent. This input example also calls, in the test conditions, for calculations to start at temperature BC = 100K and repeat at rising temperature increments BD = 25K for a total BE = 15 temperature increments. At each temperature the program requests the number of stress increments desired for each stress-strain calculation. The example of Table 6-5 shows 12 increments. This number will vary with the polymer-diluent combination and very often requires 500 or more increments to satisfy the criteria for establishing the failure condition. The program automatically calls for an increase in stress increments when needed.

As shown in the lower half of Table 6-5 the program reports the value of shear modulus involved in each stress-strain increment calculation and prints the accompanying tensile and shear stress and strain at each increment. A simple trapezoidal rule is selected for integration of the work of deformation and these results are reported in the failure analysis shown at the bottom of Table 6-5. The units of stress, stength and energy/volume as shown in Table 6-5 and the rest of this chapter are in the following common unit of bars defined in terms of other common SI units as follows:

$$1.0 \text{ bar} = 0.10 \text{ MN/m}^2 = 0.10 \text{ MPa} = 1.0 \text{ E6 dyn/cm}^2 \tag{6-11}$$

This unit is used as a matter of convenience and can, of course, be readily replaced within the computer program in the print instructions.

The failure report in the lower portion of Table 6-5 defines the failure stress, strain and energy/volume for both shear and tension. This report is followed by an elastic (E)-plastic (P) analysis as discussed earlier in Chapter 2-2 and particularly with reference to Fig. 2-6. Three conditions are defined in the model for elastic-plastic analysis. The procedure for tension E-P analysis is described and applies for shear with appropriate changes in terms. When $W_T => S_b\varepsilon_b$ where S_b is nominal tensile strength and ε_b is extensibility no analysis is made and this is an occurrence with Neohookian deformation.

When $(S_b\varepsilon_b/2) < W_T < S_be_b$ the following relations apply:

$$E = \frac{S_b^{\,2}}{2\ S_b\varepsilon_b - W_T} \tag{6-12}$$

$$W_E = \frac{S_b^{\,2}}{2E} \tag{6-13}$$

$$W_p = W_T - W_E \tag{6-14}$$

$$\varepsilon_E = \frac{2\ W_E}{S_b} \tag{6-15}$$

When $W_T <= (S_b\varepsilon_b/2)$ as is common for crosslinked materials the following relations apply:

$$E = \frac{2W_T}{\varepsilon_b^{\,2}} \tag{6-16}$$

$$W_E = W_T \tag{6-17}$$

$$W_p = 0 \tag{6-18}$$

$$\varepsilon_E = \varepsilon_b \tag{6-19}$$

Equivalent relations are written for shear by appropriate substitution of shear modulus, strength, extensibility, and work for their tensile counterparts. These elastic-plastic (E-P) analog terms provide direct inputs to the Dugdale[11] type fracture mechanics models and elastic-plastic stress analysis models as discussed by Hart-Smith[12] and reviewed earlier in Chapter 2-2. This computer model thus furnishes the full stress-strain curve or the E-P analog of the full curvilinear stress-strain response except where conditions $W_T => S_b\varepsilon_b$ prevent the E-P analysis.

The shear stress-strain curves for the acrylate copolymer described by Table 6-5 are graphed in Fig. 6-5 and show an estimated shear strain at failure in excess of λ = 2000 due to entanglement slippage above T_g = 230K and a loss of both strength and extensibility at the approach of the flow temperature T_m = 349K. As shown in the tensile stress-strain curves of Fig. 6-6 the simple effect of cross section area reduction of tensile strain lowers the calculated maximum tensile extensibility of ε = 26 for Hookian response and ε = 47 for Neohookian response. This maximum tensile extensibility also is displayed as due to entanglement slippage above T_g. The initial yielding response of the

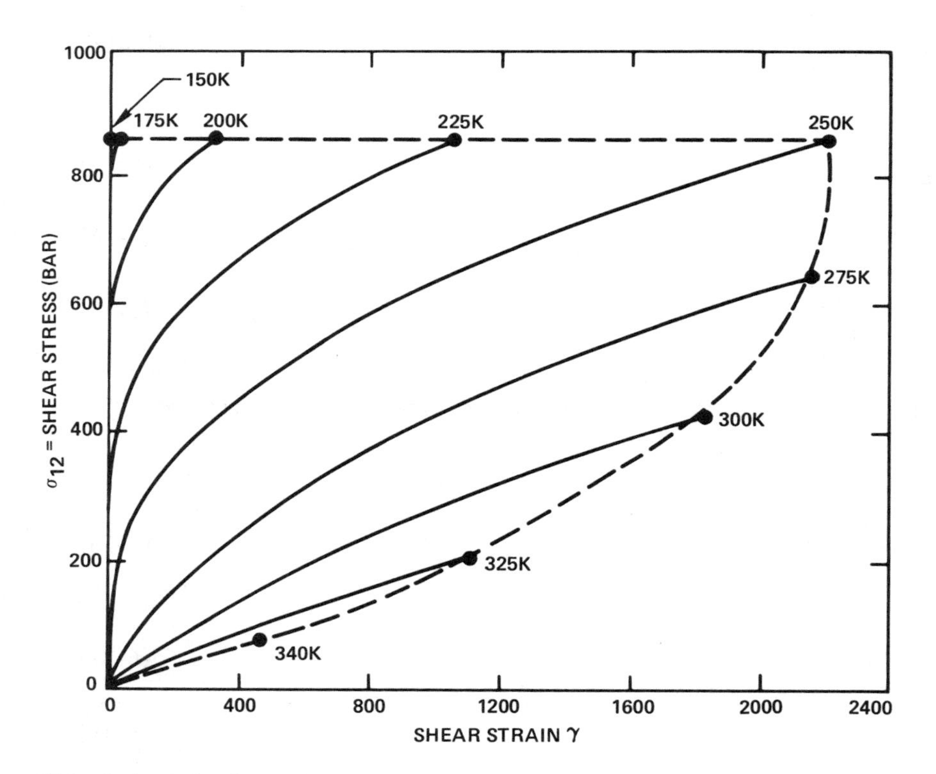

FIG. 6-5 Calculated shear stress vs strain response for equimolar isoamyl-neopentyl acrylate copolymer (M_n = 1.03E6 g/mol, T_g = 230K).

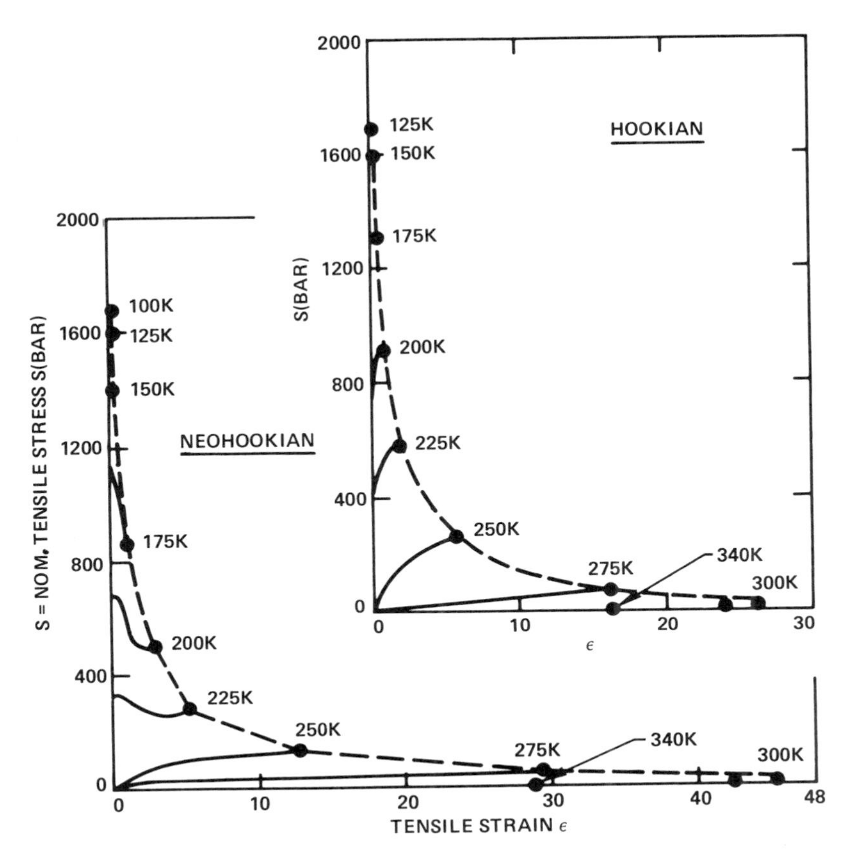

FIG. 6-6 Calculated tensile stress vs strain response for equimolar isoamyl-neopentyl acrylate copolymer (M_n = 1.03E6 g/mol, T_g = 230K).

Neohookian tensile curves of Fig. 6-6 is quite characteristic of the experimental stress strain curves of Figs. 6-1 through 6-3. The calculated Hookian behavior of Fig. 6-6 provides stress-strain curve shapes very characteristic of the experimental curves of polyimide film as shown in Fig. 6-4.

The calculated fracture energies/volume for tension W_T and shear W_S for this linear acrylate copolymer described by Table 6-5 is shown in Fig. 6-7. These curves indicate that maximum values of W_S and W_T are centered at about T_g = 230K and that the calculated brittle temperature, $T_b \approx$ 125 to 150K is well below T_g. The upper curve of Fig. 6-7 plots log W_S to display the tremendous predicted range of change in this property due to the entanglement slippage mechanism. The lower curves of Fig. 6-7 show W_T as varying over a much smaller range and that Neohookian response predicts higher values of W_T at lower temperatures where elastic yielding is more dominant in the tensile stress-strain response. Experience shows that the Hookian calculation gives a

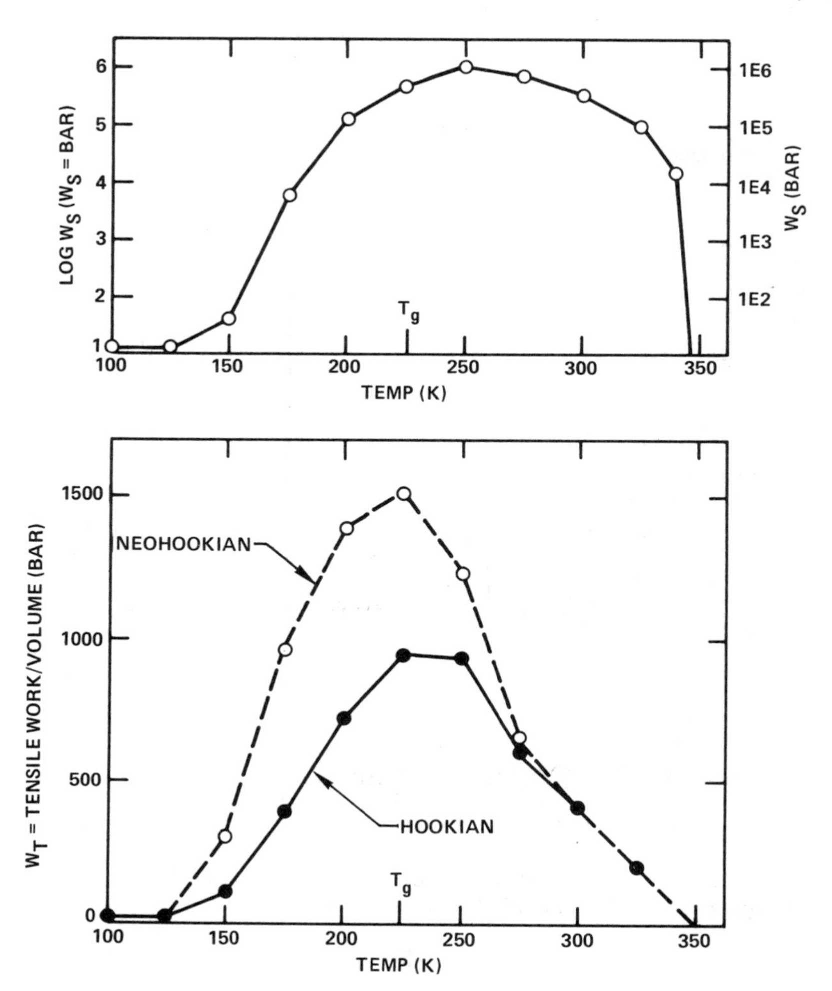

FIG. 6-7 Calculated shear (upper view) and tensile (lower view) works of deformation per unit volume.

conservative estimate of mechanical response and is therefore a preferable mechanical response for use in fracture mechanics and stress analysis.

The dramatic effects of light crosslinking, with $M_c = M_e = 34200$ g/mol, as the only change in the example of Table 6-5 is shown in the calculated shear stress strain curves of Fig. 6-8. In this case the entanglements cannot slip and dissipate mechanical energy so that molecular orientation dominates the high strain response to produce "strain hardening," as opposed to the typical "stress softening" response of uncrosslinked polymer. The same light crosslinking effect is shown in the calculated tensile stress-strain curves of Fig. 6-9. The effects of the molecular orientation are present in the tensile response in a much less pronounced fashion than in shear. The curves of W_S and W_T versus tem-

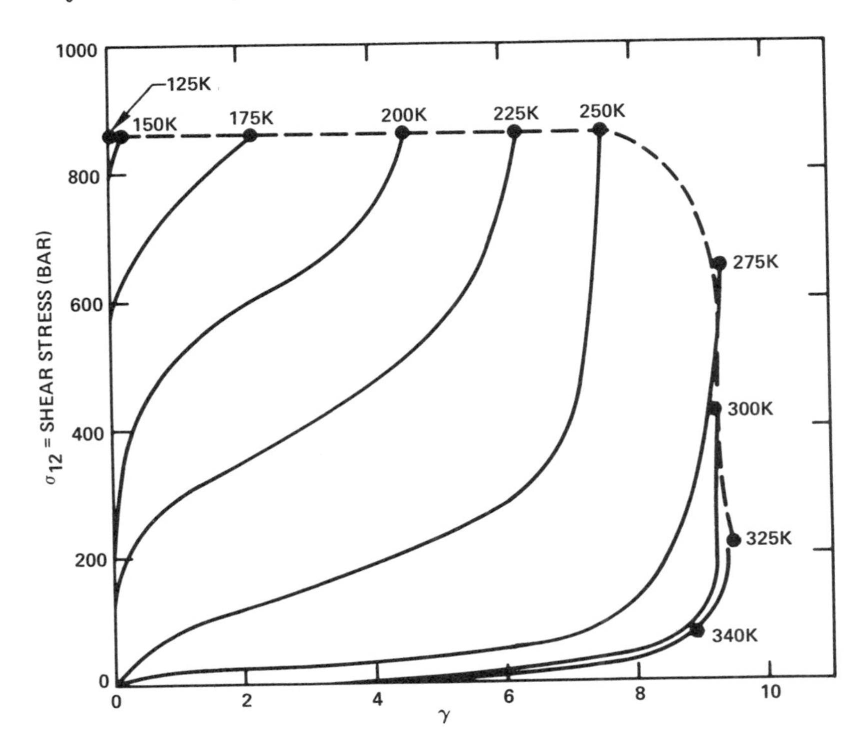

FIG. 6-8 Calculated shear stress vs strain response for equimolar isoamyl-neopentyl acrylate copolymer (M_n = 1.03E6 g/mol, T_g = 230K) with light crosslinking (M_c = 3.42E4 g/mol).

perature in Fig. 6-10 show that light crosslinking brings the magnitude of W_S and W_T much closer together over the whole range of temperature from T_b to T_m.

Several elastic-plastic analog curves of the calculated stress-strain responses of lightly crosslinked acrylate copolymer are shown in the dashed curves of Fig. 6-11. The upper curve of Fig. 6-8 is fit by the criteria of Eq. (6-12) and the lower curve by the criteria of Eq. (6-16).

The ease by which time dependent mechanical response can be computed is shown in Table 6-6. The data of this table are acquired by holding T = 296K constant and varying the loading time t over a range of log t which extends from brittle response at low values of t = 1E-14 s to flow response at high values of t > 2E2 s.

The examples shown by Fig. 6-5 through Fig. 6-10 are of importance in the area of composite technology dealing with adhesives, coatings, and sealants. In specialized composite tape constructions where aligned glass fibers are imbedded in the pressure sensitive adhesive the importance of entanglement slippage to preserve high shear bond strength and fracture toughness is well

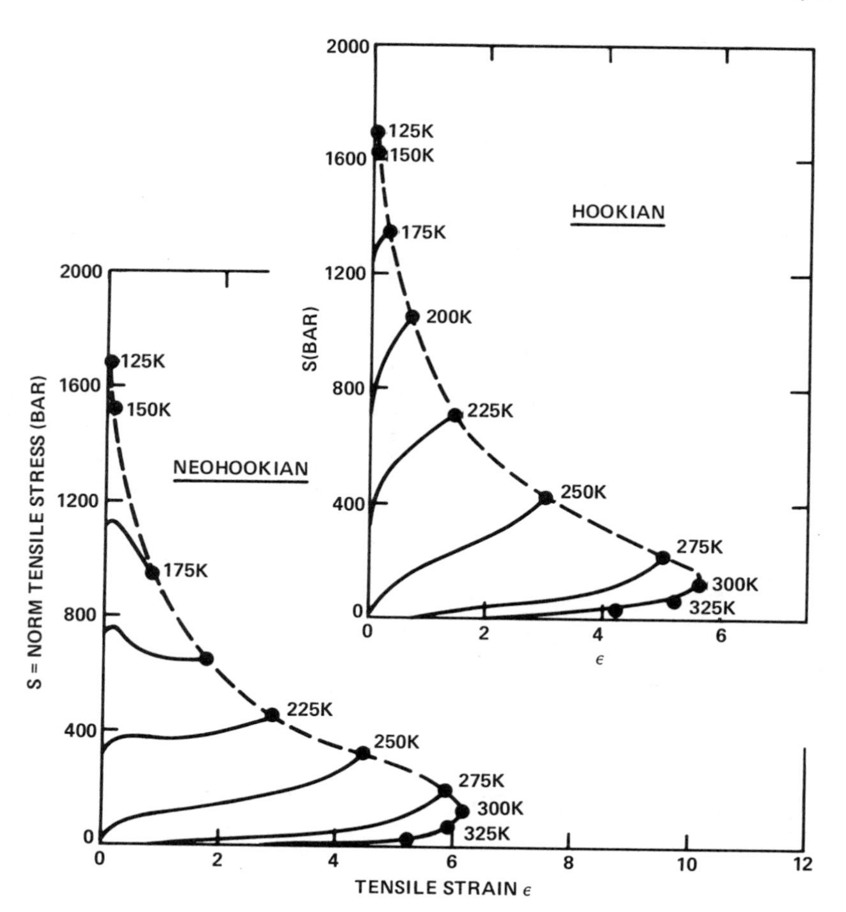

FIG. 6-9 Calculated tensile stress vs strain response for equimolar isoamyl-neopentyl copolymer (M_n = 1.06E6 g/mol, T_g = 230K) and light crosslinking (M_c = 3.42E4 g/mol).

understood. This subject will be subsequently examined relative to the peel mechanics model.

To further demonstrate the versatility of the chemical to mechanical property model we can computationally simulate the plasticizing effects of water on the mechanical response of the cured epoxy thermoset in Table 5-22 and Fig. 5-14 of Chapter 5. The physical properties and cure curve for this non-stoichiometric TGMDA/DDS thermoset are typical of the commercial important epoxy resins. The review of Delmonte (see Ref. 6, Chap 9) details the many studies of both moisture and combined moisture and temperature effects on graphite-epoxy composites and also a group of pure epoxy matrix materials. The earlier discussion of Table 1-6 and Figs. 1-11 through 1-14 also are relevent to this computational analysis of moisture effects on mechanical response.

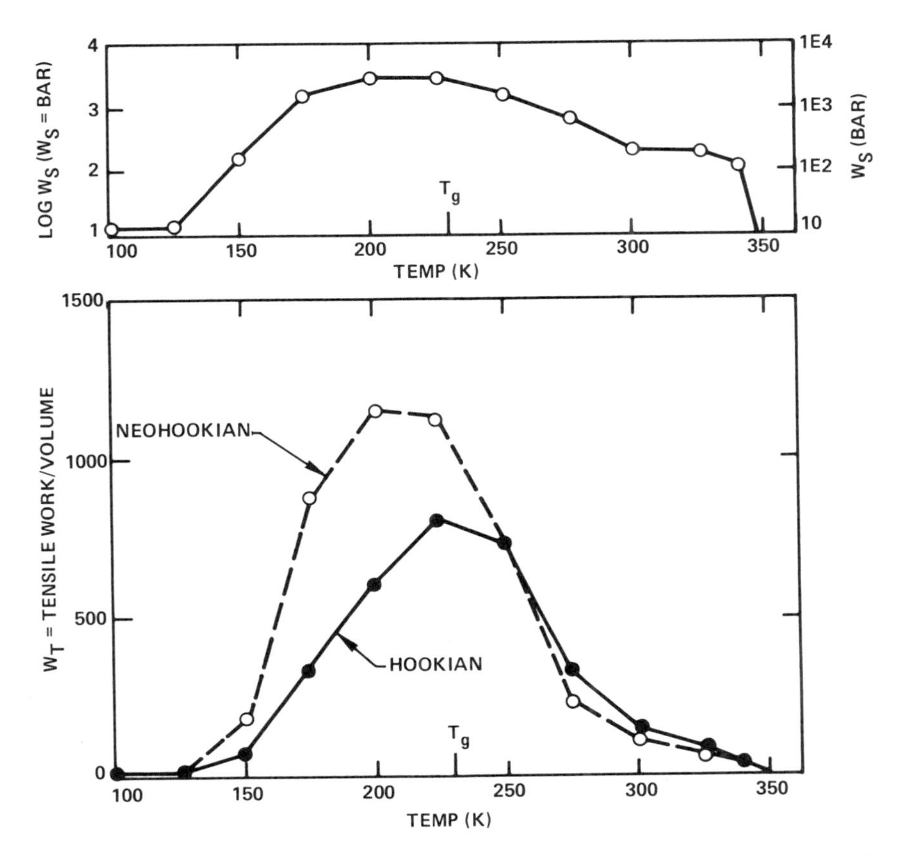

FIG. 6-10 Calculated shear (upper view) and tensile (lower view) works of deformation per unit volume.

6-4. Moisture Degradation Effects

When fiber reinforced composites are developed with moisture sensitive properties at the fiber-matrix interface or in the rheological response of the matrix it follows that moisture degradation becomes a subject of major importance.[6] As pointed out in recent review of this subject by Delmonte (Ref. 6, Chap. 9) the study of environmental aging generally commences with a study of the effects of liquid water and water vapor exposure upon the critical properties of composite response. In general, the interrelated effects of moisture, temperature, and mechanical stress are combined in studies of hydrothermal degradation of mechanical response. The reader is referred to this general review of environmental effects by Delmonte[6] which provides a detailed discussion and reference to the many recent studies of this subject over the last decade.

In this section we will more narrowly review the special studies conducted by Kaelble and coworkers[13-33] which are now incorporated into the com-

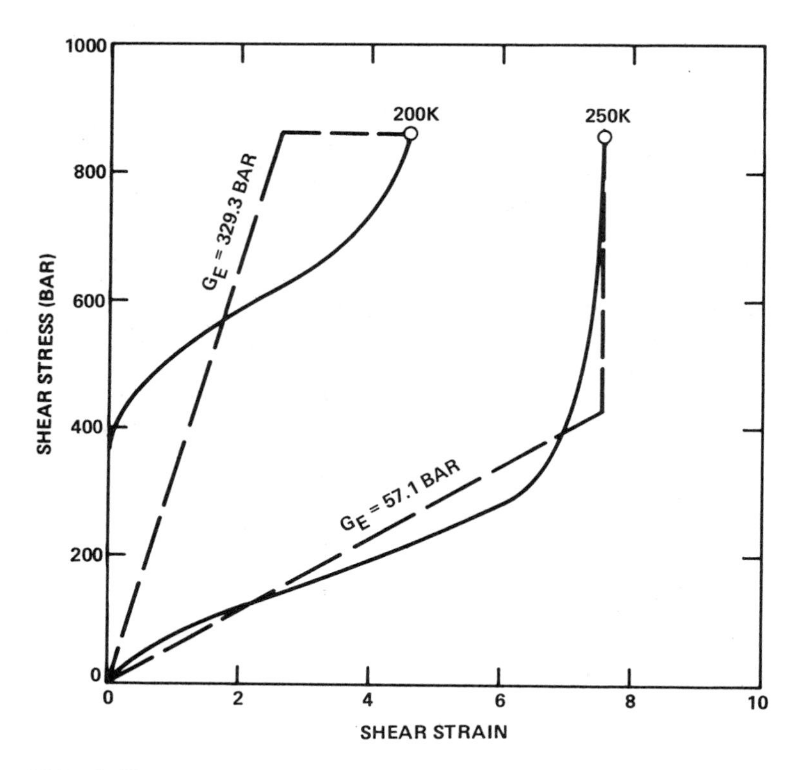

FIG. 6-11 Calculated shear stress vs strain (solid curves) response (see Fig. 6-8) and elastic-plastic analogs (dashed curves) for lightly crosslinked equimolar isoamyl-neopentyl acrylate copolymer.

TABLE 6-6 Calculated Effects of Isochronal Loading Time t Upon the Shear and Tensile Failure Properties of Equimolar Isoamyl-Neopentyl Acrylate Linear Polymer (M_n = 1.06 E6) at T = 296K with Hookian Response

	Shear			Tension		
t (S)	σ_b (bar)	γ_b	W_S (bar)	S_b (bar)	ε_b	W_T (bar)
2E2	9.6	55	243	2.82	5.8	8.19
1E2	52.4	334	8.3E3	7.37	13.2	48.7
1E1	224	1200	1.35E5	19.3	22.2	215
1	461	2156	5.2E5	34.5	25.8	444
1E-2	860	2310	1.13E6	107	15.1	806
1E-4	860	582	3.08E5	278	5.2	814
1E-5	860	186	9.88E4	441	3.2	791
1E-6	860	58	3.1E4	582	1.96	726
1E-8	860	5.9	3.1E3	1010	0.70	491
1E-10	860	0.58	312	1449	0.18	196
1E-12	860	0.067	32	1664	0.033	35.9
1E-14	860	0.034	13.5	1686	0.020	17.2

puter models presented here. An early study of polyimide resin by Kaelble and Cirlin[13] showed that prior exposure of the uncured polyamic acid based polyimide resin to moisture altered both the subsequent curing kinetics and the thermomechanical response of the cured polyimide. An early report by Kaelble[14] reported the analysis of the moisture sensitivity of the surface and fiber-matirx interface properties of graphite and boron fibers as well as epoxy and polyimide. Several studies by Kaelble, Dynes and Maus[15,16] defined surface energy analysis and surface chemical treatments for controlling the moisture sensitivity of graphite fibers. These early studies utilized the single fiber as a Wilhelmy plate for quantitative measurement of contact angle. The method of surface energy analysis for environmental sensitivity of fiber-matrix bonding was introduced by Kaelble[17] into the Griffith fracture mechanics model for environmental degradation. This combined surface energetics and fracture mechanics analysis is outlined in Chapter 1, and now finds a wide range of applications as described in structural composites, biomaterials, and printing processes as described in reviews by Kaelble.[18,19]

An analysis of bioadhesion and biocompatibility by Kaelble and Moacanin[20] provides extensive surface energy data on 190 reference surfaces and applies the modified Griffith surface energy analysis to define biocompatibility. A number of studies of moisture degradation in graphite-epoxy composites by Kaelble and coworkers[21-25] have tested various predictions of the interface degradation model with good results. These studies of interface degradation were extended by the development of nondestructive evaluation (NDE) methods for detecting moisture and moisture degradation.[26-33] These studies showed that both ultrasonic (2.25 MHz) sound velocity and acoustic energy absorption measurements are sensitive to moisture degradation.[26,27]

Moisture diffusion analysis was shown to provide for direct nondestructive analysis not only of the extent but also the discrete mechanisms of moisture degradation.[28-31] The primary effect of hydrothermal aging was shown to increase the moisture diffusion coefficients along the fiber axis of uniaxial reinforced graphite-epoxy composites. This result is consistent with the appearance of debonds or micro-cracks which initiate at the fiber-matrix interface but fail to interconnect between adjacent fibers to alter the transverse moisture diffusion coefficients. These detailed experimental studies of moisture degradation have culminated in comprehensive physiochemical characterizations which quantify material compositions and processing properties which are critical to successful manufacturing of the cured composite.[32,33] The end point of this decade of study of moisture degradation is now being usefully applied in the CAD/CAM models currently under extensive development for automation of composite manufacturing. Many of these current activities are reviewed in a recent symposium on the chemorheology of thermosetting polymers.[34]

The polymer chemistry to mechanical property program just introduced in the previous section permits a direct method for analyzing the effects of moisture degradation. Moisture can be treated in the computer as a polymer diluent and also as a polymer plasticizer which lowers the matrix glass temperature T_g. The plasticizing effects of water on cured epoxy matrix resins is documented in detailed studies of moisture content versus T_g.[35,36] The studies of De Iasi and Whitside[36] of moisture effects on the T_g of six cured epoxy resins used in high performance composites are summarized in the circled data points of Fig. 6-12. The six dry resin display values of T_g from 463K to 498K. These T_g values evidently depend upon specific details of chemical composition and degree of cure as discussed in Chapter 5. As shown in Fig. 6-12 the addition of moisture has a relative uniform effect of lowering T_g for all six resins to an extent which depends upon moisture content. As indicated by the solid curves of Fig. 6-12, it is evident that the plasticizing effect of moisture in lowering T_g can be readily analyzed by the chemical to mechanical property model. The computations for moisture effects on T_g are summarized in Tables 6-7 through 6-9 which uses a lower estimate of $\Sigma H = 20$ for the rotational degrees of freedom of the water molecule. A second estimate of $\Sigma H = 28$ for water produces the computed values of T_g summarized in Table 6-10 through Table 6-12.

All of the diluent properties of water are commonly known except its T_g, degrees of rotational freedom ΣH at T_g and the solvent-polymer interaction

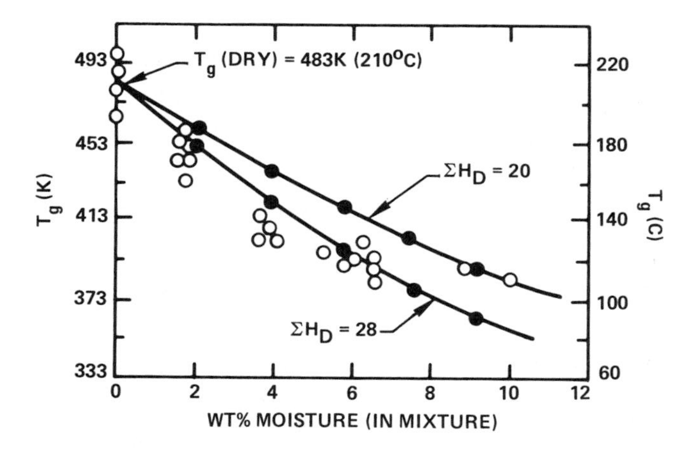

FIG. 6-12 Calculated (•) and experiment (○) effects of moisture on T_g of six cured epoxy resins (3501-5, 3501-6, 5208, 934, 3502, and NMD 2373); (for data see Refs. 6, 36).

TABLE 6-7 Estimated Effects of Low Moisture (0-2 Wt%) on Cured Epoxy Thermoset

```
POLYMER-DILUENT=TGMDA/DDS EPOXY+0% H20
POLYMER PROPERTIES:
AA,AB,AC,AD,AE= 495000  1  319  1.16  483
AF,AG,AH,AI,AJ= 10  943  9  12 6  1
AK,AL,AM,AN,AP= 1094  1709  5168  1  .32
DILUENT PROPERTIES:
AQ,AR,AS,AT,AU= 13  0  20  1  137
AV,AW,AX,AY,AZ= 10  13  11  23.2  1
TEST CONDITIONS:
BA,BB,BC,BD,BE= 1  1  300  25  15
FRACTION NEOHOOKIAN TENSILE RESPONSE= 0
PRESS ENTER TO CONTINJE

INTERMEDIATE RESULTS:
BF,BG,BH,BI,BJ= 3.81052  1  0  483  .158045
BK,BL,BM,BN= 1  0  137  .0393688
TG,UR= 483  158045
BO,BQ,BR= 1  0  158.76
BS,BT= .993095  .979119
BU,BV,BX.BY,BZ= 53163.7  53163.7  9.0124  1.30231  35.8285
SB,TL,TM,NH= 1660.63  11.5743  585.517  0
T,SM,SS= 300  1806 56  1660.63
SHEAR AND TENSION ANALYSIS:
INPUT NUMBER OF STRESS INCREMENTS? _

POLYMER-DILUENT=TGMDA/DDS EPOXY+2% H20
POLYMER PROPERTIES:
AA,AB,AC,AD,AE= 495000  1  319  1.16  483
AF,AG,AH,AI,AJ= 10  943  9  12.6  1
AK,AL,AM,AN,AP= 1094  1709  5168  1  .32
DILUENT PROPERTIES:
AQ,AR,AS,AT,AU= 13  550  20  1  137
AV,AW,AX,AY,AZ= 10  13  11  23.2  1
TEST CONDITIONS:
BA,BB,BC,BD,BE= 1  1  300  25  15
FRACTION NEOHOOKIAN TENSILE RESPONSE= 0
PRESS ENTER TO CONTINUE

INTERMEDIATE RESULTS:
BF,BG,BH,BI,BJ= 3.81052  1  0  483  .158045
BK,BL,BM,BN= 1  0  137  .0393688
TG,UR= 458.498  .149641
BO,BQ,BR= .977326  .022674        156.065
BS,BT= .993095  .978635
BU,BV,BX,BY,BZ= 52261.1  51958.2  8.36124  1.22215  33.2358
SB,TL,TM,NH= 1632.44  11.5956  561.582  0
T,SM,SS= 300  1748.06  1632.44
SHEAR AND TENSION ANALYSIS:
INPUT NUMBER OF STRESS INCREMENTS? _
```

TABLE 6-8 Estimated Effects of Medium Moisture (4-6 Wt%) on Cured Epoxy Thermoset

```
POLYMER-DILUENT=TGMDA/DDS EPOXY+4% H20
POLYMER PROPERTIES:
AA,AB,AC,AD,AE= 495000  1  319  1.16  483
AF,AG,AH,AI,AJ= 10  943  9  12.6  1
AK,AL,AM,AN,AP= 1094  1709  5168  1  .32
DILUENT PROPERTIES:
AQ,AR,AS,AT,AU= 18  1100  20  1  137
AV,AW,AX,AY.AZ= 10  18  11  23.2  1
TEST CONDITIONS:
BA,BB,BC,BD,BE= 1  1  300  25  15
FRACTION NEOHOOKIAN TENSILE RESPONSE= 0
PRESS ENTER TO CONTINUE


INTERMEDIATE RESULTS:
BF,BG,BH,BI,BJ= 3.81052  1  0  483  .158045
BK,BL,BM,BN= 1  0  137  .0393688
TG,UR= 437.238  .142349
BO,BQ,BR= .955658  .0443425        153.979
BS,BT= .993095  .97815
BU,BV,BX,BY,BZ= 51562.6  50806.3  7.79674  1.15249  30.9882
SB,TL,TM,NH= 1610.62  11.6135  540 799  0
T,SM,SS= 300  1691.61  1610.62
SHEAR AND TENSION ANALYSIS:
INPUT NUMBER OF STRESS INCREMENTS? _
```

```
POLYMER-DILUENT=TGMDA/DDS EPOXY+6% H20
POLYMER PROPERTIES:
AA,AB,AC,AD,AE= 495000  1  319  1.16  483
AF,AG,AH,AI,AJ= 10  943  9  12.6  1
AK,AL,AM,AN,AP= 1094  1709  5168  1  .32
DILUENT PROPERTIES:
AQ,AR,AS,AT,AU= 18  1650  20  1  137
AV,AW,AX,AY,AZ= 10  18  11  23.2  1
TEST CONDITIONS:
BA,BB,BC,BD,BE= 1  1  300  25  15
FRACTION NEOHOOKIAN TENSILE RESPONSE= 0
PRESS ENTER TO CONTINUE


INTERMEDIATE RESULTS:
BF,BG,BH,BI,BJ= 3.81052  1  0  483  .158045
BK,BL,BM,BN= 1  0  137  .0393688
TG,UR= 418.614  .135961
BO,BQ,BR= .934929  .0650711        152.432
BS,BT= .993095  .977666
BU,BV,BX,BY,BZ= 51044.5  49704.3  7.30274  1.09137  29.0213
SB,TL,TM,NH= 1594.43  11.6284  522.577  0
T,SM,SS= 300  1637.07  1594.43
SHEAR AND TENSION ANALYSIS:
INPUT NUMBER OF STRESS INCREMENTS? _
```

TABLE 6-9 Estimated Effects of High Moisture (8-10 Wt%) on Cured Epoxy Thermoset

```
POLYMER-DILUENT=TMDA/DDS EPOXY+8% H2O
POLYMER PROPERTIES:
AA,AB,AC,AD,AE= 495000  1  319  1.16  483
AF,AG,AH,AI,AJ= 10  943  9  12.6  1
AK,AL,AM,AN,AP= 1094  1709  5168  1  .32
DILUENT PROPERTIES:
AQ,AR,AS,AT,AU= 18  2200  20  1  137
AV,AW,AX,AY,AZ= 10  18  11  23.2  1
TEST CONDITIONS:
BA,BB,BC,BD,BE= 1  1  300  25  15
FRACTION NEOHOOKIAN TENSILE RESPONSE= 0
PRESS ENTER TO CONTINUE


INTERMEDIATE RESULTS:
BF,BG,BH,BI,BJ= 3.81052  1  0  483   158045
BK,BL,BM,BN= 1  0  137  .0393688
TG,UR= 402.166  .130319
BO,BQ,BR= .915001  .0849195       151.361
BS,BT= .993095  .977182
BU,BV,BX,BY,BZ= 50685.9  48649  6.86686  1.0373  27.2858
SB,TL,TM,NH= 1583.23  11.6409  506.465  0
T,SM,SS= 300  1584.3  1583.23
SHEAR AND TENSION ANALYSIS:
INPUT NUMBER OF STRESS INCREMENTS? _


POLYMER-DILUENT=TGMDA/DDS EPOXY+10% H2O
POLYMER PROPERTIES:
AA,AB,AC,AD,AE= 495000  1  319  1.16  483
AF,AG,AH,AI,AJ= 10  943  9  12.6  1
AK,AL,AM,AN,AP= 1094  1709  5168  1  .32
DILUENT PROPERTIES:
AQ,AR,AS,AT,AU= 18  2750  20  1  137
AV,AW,AX,AY,AZ= 10  18  11  23.2  1
TEST CONDITIONS:
BA,BB,BC,BD,BE= 1  1  300  25  15
FRACTION NEOHOOKIAN TENSILE RESPONSE= 0
PRESS ENTER TO CONTINUE


INTERMEDIATE RESULTS:
BF,BG,BH,BI,BJ= 3.81052  1  0  483  .158045
BK,BL,BM,BN= 1  0  137  .0393688
TG,UR= 387.534  .125301
BO,BQ,BR= .896057  .103943       150.711
BS,BT= .993095  .976697
BU,BV,BX,BY,BZ= 50468.4  47637.7  6.47946  .989189  25.7433
SB,TL,TM,NH= 1576.44  11.6512  492.112  0
T,SM,SS= 300  1533.21  1533.21
SHEAR AND TENSION ANALYSIS:
INPUT NUMBER OF STRESS INCREMENTS? _
```

TABLE 6-10 Second Estimated Effects of Low Moisture (0-2 Wt%) on Cured Epoxy Thermoset

```
POLYMER-DILUENT=TGMDA DDS EPOXY+0% H2O
POLYMER PROPERTIES
AA,AB,AC,AD,AE= 495000  1  319  1.16  483
AF,AG,AH,AI,AJ= 10  943  9  12.6  1
AK,AL,AM,AN,AP= 1094  1709  5168  1  .32
DILUENT PROPERTIES:
AQ,AR,AS,AT,AU= 18  0  28  1  137
AV,AW,AX,AY,AZ= 10  18  11  23.2  1
TEST CONDITIONS:
BA,BB,BC,BD,BE= 1  1  220  40  15
FRACTION NEOHOOKIAN TENSILE RESPONSE= 0
PRESS ENTER TO CONTINUE


INTERMEDIATE RESULTS:
BF,BG,BH,BI,BJ= 5.33473  1  0  483  .158045
BK,BL,BM,BN= 1  0  137  .0281206
TG,UR= 483  .158045
BO,BQ,BR= 1  0  158.76
BS,BT= .993095  .979119
BU,BV,BX,BY,BZ= 53163.7  53163.7  9.0124  1.30231  35.8285
SB,TL,TM,NH= 1660.63  11.5743  585.517  0
T,SM,SS= 220  2312.74  1660.63
SHEAR AND TENSION ANALYSIS:
INPUT NUMBER OF STRESS INCREMENTS? _


POLYMER-DILUENT=TGMDA/DDS EPOXY+2% H2O
POLYMER PROPERTIES:
AA,AB,AC,AD,AE= 495000  1  319  1.16  483
AF,AG,AH,AI,AJ= 10  943  9  12.6  1
AK,AL,AM,AN,AP= 1094  1739  5168  1  .32
DILUENT PROPERTIES:
AQ,AR,AS,AT,AU= 18  550  29  1  137
AV,AW,AX,AY,AZ= 10  18  11  23.2  1
TEST CONDITIONS:
BA,BB,BC,BD,BE= 1  1  223  40  15
FRACTION NEOHOOKIAN TENSILE RESPONSE= 0
PRESS ENTER TO CONTINUE


INTERMEDIATE RESULTS:
BF,BG,BH,BI,BJ= 5.33473  1  0  483  .158045
BK,BL,BM,BN= 1  0  137  .0281206
TG,UR= 449.643  .145519
BO,BQ,BR= .977326  .022674       156.065
BS,BT= .993095  .978635
BU,BV,BX,BY,BZ= 52261.1  51958.2  8.19975  1.19854  32.5939
SB,TL,TM,NH= 1632.44  11.6126  553.179  0
T,SM,SS= 220  2289.59  1632.44
SHEAR AND TENSION ANALYSIS:
INPUT NUMBER OF STRESS INCREMENTS? _
```

TABLE 6-11 Second Estimated Effects of Medium Moisture (4-6 Wt%) on Cured Epoxy Thermoset

```
POLYMER-DILUENT=TGMDA/DDS EPOXY+4% H2O
POLYMER PROPERTIES:
AA,AB,AC,AD,AE= 495000  1  319  1.16  483
AF,AG,AH,AI,AJ= 10  943  9  12.5  1
AK,AL,AM,AN,AP= 1094  1709  5168  1  .32
DILUENT PROPERTIES:
AQ,AR,AS,AT,AU= 18  1100  28  1  137
AV,AW,AX,AY,AZ= 10  19  11  23.2  1
TEST CONDITIONS:
BA,BB,BC,BD,BE= 1  1  220  40  15
FRACTION NEOHOOKIAN TENSILE RESPONSE= 0
PRESS ENTER TO CONTINUE


INTERMEDIATE RESULTS:
BF,BG,BH,BI,BJ= 5.33473  1  0  483  .158045
BK,BL,BM,BN= 1  0  137  .0281206
TG,UR= 422.152  .135196
BO,BQ,BR= .955658  .0443425        153.979
BS,BT= .993095  .97815
BU,BV,BX,BY,BZ= 51562.6  50306.3  7.52773  1.11272  29.919
SB,TL,TM,NH= 1610.62  11.644  526.535  0
T,SM,SS= 220  2267.33  1610.62
SHEAR AND TENSION ANALYSIS:
INPUT NUMBER OF STRESS INCREMENTS? _


POLYMER-DILUENT=TGMDA/DDS EPOXY+6% H2O
POLYMER PROPERTIES:
AA,AB,AC,AD,AE= 495000  1  319  1.16  483
AF,AG,AH,AI,AJ= 10  943  9  12.6  1
AK,AL,AM,AN,AP= 1094  1709  5168  1  .32
DILUENT PROPERTIES:
AQ,AR,AS,AT,AU= 18  1650  28  1  137
AV,AW,AX,AY,AZ= 10  19  11  23.2  1
TEST CONDITIONS:
BA,BB,BC,BD,BE= 1  1  220  40  15
FRACTION NEOHOOKIAN TENSILE RESPONSE= 0
PRESS ENTER TO CONTINUE


INTERMEDIATE RESULTS:
BF,BG,BH,BI,BJ= 5.33473  1  0  483  .152045
BK,BL,BM,BN= 1  0  137  .0281206
TG,UR= 399.105  .126542
BO,BQ,BR= .934929  .0650711        152.432
BS,BT= .993095  .977666
BU,BV,BX,BY,BZ= 51044.5  49704.3  6.9624  1.0405  27.6687
SB,TL,TM,NH= 1594.43  11.6699  504.193  0
T,SM,SS= 220  2245.84  1594.43
SHEAR AND TENSION ANALYSIS:
INPUT NUMBER OF STRESS INCREMENTS? _
```

TABLE 6-12 Second Estimated Effects of High Moisture (8-10 Wt%) on Cured Epoxy Thermoset

```
POLYMER-DILUENT=TGMDA/DDS EPOXY+8% H2O
POLYMER PROPERTIES:
AA,AB,AC,AD,AE= 495000  1  319  1.16  483
AF,AG,AH,AI,AJ= 10  943  9  12.6  1
AK,AL,AM,AN,AP= 1094  1709  5168  1  .32
DILUENT PROPERTIES:
AQ,AR,AS,AT,AU= 18  2200  28  1  137
AV,AW,AX,AY,AZ= 10  18  11  23.2  1
TEST CONDITIONS:
BA,BB,BC,BD,BE= 1  1  220  40  15
FRACTION NEOHOOKIAN TENSILE RESPONSE= 0


INTERMEDIATE RESULTS:
BF,BG,BH,BI,BJ= 5.33473  1  0  483  .158045
BK,BL,BM,BN= 1  0  137  .0281206
TG,UR= 379.504  .119182
BO,BQ,BR= .915081  .0849195        151.361
BS,BT= .993095  .977182
BU,BV,BX,BY,BZ= 50685.9  48649  6.47992  .978844  25.7482
SB,TL,TM,NH= 1583.23  11.6913  485.179  0
T,SM,SS= 220  2225  1583.23
SHEAR AND TENSION ANALYSIS:
INPUT NUMBER OF STRESS INCREMENTS? _


POLYMER-DILUENT=TGMDA/DDS EPOXY+10% H2O
POLYMER PROPERTIES:
AA,AB,AC,AD,AE= 495000  1  319  1.16  483
AF,AG,AH,AI,AJ= 10  943  9  12.6  1
AK,AL,AM,AN,AP= 1094  1709  5168  1  .32
DILUENT PROPERTIES:
AQ,AR,AS,AT,AU= 18  2750  28  1  137
AV,AW,AX,AY,AZ= 10  18  11  23.2  1
TEST CONDITIONS:
BA,BB,BC,BD,BE= 1  1  220  40  15
FRACTION NEOHOOKIAN TENSILE RESPONSE= 0
PRESS ENTER TO CONTINUE


INTERMEDIATE RESULTS:
BF,BG,BH,BI,BJ= 5.33473  1  0  483  .158045
BK,BL,BM,BN= 1  0  137  .0281206
TG,UR= 362.632  .112846
BO,BQ,BR= .896057  .103943        150.711
BS,BT= .993095  .976697
BU,BV,BX,BY,BZ= 50468.4  47637.7  6.0631  .925551  24.0891
SB,TL,TM,NH= 1576.44  11.7089  468.793  0
T,SM,SS= 220  2204.71  1576.44
SHEAR AND TENSION ANALYSIS:
INPUT NUMBER OF STRESS INCREMENTS? _
```

parameter AP = ϕ as defined by Eq. (6-1). The experimentally determined glass temperature of water is reported as T_g = 137 ± 2K by Fedors[37] and this value is consistent with other hydroxy compounds. The solvent-polymer interaction parameter Φ for water in epoxy resin can be estimated from solubility parameter data reviewed by Kaelble.[32] In terms of the definition of Eq. (6-1) water is a dominantly polar liquid with a dispersion fraction $d_D \simeq 0.20$ and a polar fraction $P_D \simeq 0.80$ for the diluent solubility parameter δ_D. Epoxy resin is dominated by the dispersion fraction $d_P \simeq 0.80$ and a smaller polar fraction $p_P \simeq 0.20$ to the polymer solubility parameter δ_P. There is no evidence for strong interactions such as hydrogen bonding between water and epoxy so $\Delta_{PD} \simeq 0$ in Eq. (6-1). Combining the above terms in Eq. (6-1) provides an estimate for the water-epoxy interaction parameter of ϕ = 0.32 which reflects a low interaction efficiency between the polar diluent and dominantly nonpolar epoxy polymer.

The higher number of degrees of freedom H assigned to water is typical of other small molecules. In Table 4-1 are listed the degrees of freedom for chemical repeat units in linear polymer chains. The values of H in Table 4-1 are roughly multiplied by three if that unit is an independent small molecule with additional rotational and translational degrees of freedom. This result is embodied in the derivation of Eq. (4-5) as discussed by Kaelble[11] and accounts for the notable molecular weight dependence of T_g as shown in Figs. 5-9 and 5-10 where the extra motion of the chain ends becomes dominant at low molecular weight.

The surface energy effects of water exposure are incorporated in Eq. (1-13) through Eq. (1-17) and demonstrated relative to detailed studies of adhesive bond strength degradation. This same surface energy model has been applied to graphite-epoxy composites to predict the degradation of interlaminar shear strength.[21-24] The general analysis of moisture effects on the matrix response of epoxy resin extend over the wide range of temperatures of aerospace applications, temperatures from 163 to 448K (-110°C to 175°C), variable moisture (space vacuum to ambient moisture condensation) and variable ultraviolet radiation (dark to unfiltered sunlight) and represent the exposure ranges of space vehicles such as the Space Shuttler Orbiter.[6,33] Direct testing of hydrothermal effects over this range of variables involves major research test programs. Computer simulation involves essentially no cost and provides insights of material response of a unique type.

The predicted effects of moisture with the higher estimated degrees of freedom ΣH_D = 28 as shown in the lower curve of Fig. 6-12 is incorporated in this computational analysis. The epoxy matrix is defined by the nonstoichiometric molecular structure of Fig. 5-4 with two moles of TGMDA and one mole of DDS. The physical properties and cure path of this typical epoxy resin are generated by computer models as reported in Table 5-7 and Table 5-22. Only the experimental average value of T_g = 483K as shown for experimental data in Fig. 6-12 is substituted for the calculated T_g = 530K reported in Table 5-22 for the model cured epoxy.

As mentioned previously the effects of low intermediate and high moisture levels of water on the complete set of intermediate physical properties of the cured epoxy are shown for water with degrees of freedom $\Sigma H = 20$ in Table 6-7 through Table 6-9. The calculated values of T_g are shown in the upper curve of Fig. 6-12. Equivalent calculations are shown in Table 6-10 through Table 6-12 for water with degrees of freedom $\Sigma H = 28$ and the computed plasticization of T_g for this case is seen to closely model the experimental data a low and intermediate moisture content. The calculations of Table 6-10 through Table 6-12 were extended to a complete analysis of shear and tensile mechanicl response from 220K to 585K and loading time t_1 = 1.0 s. The results of these calculations are tabulated in the detailed data summaries of Table 6-13 through Table 6-19. These tables present calculated values of strength, extensibility, fracture energy and also the four elastic-plastic analog stress-strain responses described in earlier discussions and defined in Eq. (6-12) through Eq. (6-19). The large temperature increments of 40K permit a casual survey over a broad temperature range which is the objective of this initial computation.

Graphical plots of the data of Table 6-13 for dry epoxy, Table 6-15 for 4 wt% moisture, and Table 6-18 for 10 wt% moisture are presented in Fig. 6-13 through Fig. 6-16. The analog elastic shear G_E ansd tensile E_E modulus curves of Fig. 6-13 show the notable stress softening influence of water plasticization over the whole range of temperature from 220K to 580K. As shown by the curves of Fig. 6-13 it requires the full 10 wt% moisture to lower the values of both G_E and E_E at the low temperature of 220K and 4 wt% moisture influences these modulii only above 260K. The modulus of Fig. 6-13 are plotted on the logarithmic scale and it should be recalled represent the averaged slope of the stress-strain curve at all stresses up to the failure stress. Many experimental studies as reported by Delmonte[6] show the notable stress softening effects of moisture on the modulus of epoxy resins in composites.

The calculated effects of moisture on shear stength and extensibility are shown in the curves of Fig. 6-14. The corresponding curves for tensile strength and extensibility correspond closely to those shown for shear. The upper curves of Fig. 6-14 show that shear strength begins to diminish slightly above ambient temperature (300K) while shear extensibility for moisture plasticized epoxy increases from brittle levels at lower temperature (200K to 260K). This somewhat unexpected result, that strength and extensibility are not closely

TABLE 6-13 Calculated Mechanical Response of Cured TGMDA/DDS Epoxy (0 Wt% H_2O, $T_g = 483K$, $t = 1$ s)

Shear

T (K)	σ_b (bar)	γ_b	W_S (bar)	W_E (bar)	W_p (bar)	G_E (bar)	γ_E
220	1.66E3	2.24E-2	1.85E1	1.85E1	0	7.39E4	2.34E-2
260	1.66E3	2.24E-2	1.85E1	1.85E1	0	7.39E4	2.34E-2
300	1.66E3	3.22E-2	3.42E1	1.92E1	1.50E1	7.17E4	2.32E-2
340	1.55E3	1.31E-1	1.77E2	2.81E1	1.49E2	4.30E4	3.61E-2
380	1.30E3	2.30E-1	2.56E2	4.29E1	2.13E2	1.97E4	6.59E-2
420	1.05E3	3.61E-1	3.02E2	7.55E1	2.26E2	7.27E3	1.44E-1
460	7.94E2	5.94E-1	3.25E2	1.46E2	1.78E2	2.15E3	3.70E-1
500	5.41E2	9.27E-1	2.01E2	2.01E2	0	4.67E2	9.27E-1
540	2.88E2	1.08	7.80E1	7.80E1	0	1.35E2	1.08
580	3.49E1	4.84E-1	7.46	7.46	0	6.36E1	4.84E-1
585	0	0	0	0	0	0	0

Tension

T (K)	S_b (bar)	ε_b	W_T (bar)	W_E (bar)	W_p (bar)	E_E (bar)	ε_E
220	3.27E3	1.47E-2	2.40E1	2.40E1	0	2.22E5	1.47E-2
260	3.27E3	1.47E-2	2.40E1	2.40E1	0	2.22E5	1.47E-2
300	3.26E3	1.94E-2	3.89E1	2.44E1	1.45E1	2.18E5	1.49E-2
340	2.94E3	5.64E-2	1.40E2	2.56E1	1.14E2	1.69E5	1.74E-2
380	2.37E3	9.66E-2	2.01E2	2.80E1	1.73E2	1.00E5	2.36E-2
420	1.80E3	1.61E-1	2.48E2	4.24E1	2.05E2	3.85E4	4.70E-2
460	1.25E3	2.71E-1	2.55E2	8.29E1	1.73E2	9.42E3	1.33E-1
500	7.20E2	5.03E-1	1.64E2	1.64E2	0	1.30E3	5.03E-1
540	3.53E2	6.30E-1	6.39E1	6.39E1	0	3.22E2	6.30E-1
580	5.49E1	2.72E-1	6.76	6.76	0	1.83E2	2.71E-1
585	0	0	0	0	0	0	0

TABLE 6-14 Calculated Mechanical Response of 2 Wt% Moisture in Cured Epoxy (T_g = 449.6K, t = 1 s)

Shear

T (K)	σ_b (bar)	γ_b	W_S (bar)	W_E (bar)	W_p (bar)	G_E (bar)	γ_E
220	1.63E3	2.23E-2	1.82E1	1.82E1	0	7.26E4	2.24E-2
260	1.63E3	2.23E-2	1.82E1	1.82E1	0	7.26E4	2.24E-2
300	1.63E3	8.79E-2	1.21E2	2.26E1	9.03E1	5.89E4	2.77E-2
340	1.46E3	1.85E-1	2.29E2	4.24E1	1.87E1	2.53E4	5.79E-2
380	1.19E3	3.43E-1	3.30E2	7.76E1	2.53E2	9.12E3	1.30E-1
420	9.15E2	3.36E-1	1.90E2	1.18E2	7.18E1	3.56E3	2.57E-1
460	6.40E2	8.26E-1	2.25E2	2.25E2	0	6.61E2	8.26E-1
500	3.65E2	1.18	9.28E1	9.28E1	0	1.34E2	1.18
540	9.05E1	8.49E-1	2.76E1	2.76E1	0	7.66E1	8.49E-1
553	0	0	0	0	0	0	0

Tension

T (K)	S_b (bar)	ε_b	W_T (bar)	W_E (bar)	W_p (bar)	E_E (bar)	ε_E
220	3.22E3	1.47E-2	2.36E1	2.36E1	0	2.18E5	1.47E-2
260	3.22E3	1.47E-2	2.36E1	2.36E1	0	2.18E5	1.47E-2
300	3.15E3	3.69E-2	9.16E1	2.46E1	6.70E1	2.02E5	1.56E-2
340	2.71E3	8.29E-2	1.91E2	3.36E1	1.58E2	1.10E5	2.47E-2
380	2.08E3	1.45E-1	2.54E2	4.71E1	2.08E2	4.59E4	4.53E-2
420	1.53E3	1.95E-1	2.04E2	9.44E1	1.10E2	1.24E4	1.23E-1
460	8.72E2	4.69E-1	2.12E2	1.96E2	1.61E1	1.94E3	4.50E-1
500	4.27E2	7.13E-1	8.40E1	8.40E1	0	3.30E2	7.13E-1
540	1.23E2	4.74E-1	2.30E1	2.30E1	0	2.05E2	4.74E-1
553	0	0	0	0	0	0	0

TABLE 6-15 Calculated Mechanical Response of 4 Wt% Moisture in Cured Epoxy (T_g = 422K, t = 1 s)

Shear

T (K)	σ_b (bar)	γ_b	W_S (bar)	W_E (bar)	W_p (bar)	G_E (bar)	γ_E
220	1.61E3	2.24E-2	1.80E1	1.80E1	0	7.17E4	2.24E-2
260	1.61E3	2.24E-2	1.80E1	1.80E1	0	7.17E4	2.24E-2
300	1.61E3	1.72E-1	2.36E2	4.08E1	1.95E2	3.18E4	5.06E-2
340	1.38E3	3.20E-1	3.65E2	7.75E1	2.87E2	1.23E4	1.12E-1
380	1.08E3	5.21E-1	4.24E2	1.40E2	2.84E2	4.19E3	2.59E-1
420	7.88E2	7.41E-1	3.06E2	2.78E2	2.75E1	1.12E3	7.06E-1
460	4.92E2	1.15	1.04E2	1.04E2	0	1.58E2	1.15
500	1.96E2	1.09	5.75E1	5.75E1	0	9.72E1	1.09
527	0	0	0	0	0	0	0

Tension

T (K)	S_b (bar)	ε_b	W_T (bar)	W_E (bar)	W_p (bar)	E_E (bar)	ε_E
220	3.17E3	1.47E-2	2.33E1	2.33E1	0	2.15E5	1.47E-2
260	3.17E3	1.47E-2	2.33E1	2.33E1	0	2.15E5	1.47E-2
300	2.99E3	7.61E-2	1.95E2	3.26E1	1.63E2	1.37E5	2.18E-2
340	2.43E3	1.35E-1	2.84E2	4.43E1	2.40E2	6.67E4	3.65E-2
380	1.78E3	2.20E-1	3.14E2	7.61E1	2.38E2	2.07E4	8.57E-2
420	1.13E3	3.96E-1	2.85E2	1.62E2	1.23E2	3.94E3	2.87E-1
460	5.77E2	7.04E-1	9.07E1	9.07E1	0	3.66E2	7.04E-1
500	2.39E2	6.39E-1	4.99E1	4.99E1	0	2.44E2	6.39E-1
527	0	0	0	0	0	0	0

TABLE 6-16 Calculated Mechanical Response of 6 Wt% Moisture in Cured Epoxy (T_g = 399K, t = 1 s)

Shear

T (K)	σ_b (bar)	γ_b	W_S (bar)	W_E (bar)	W_p (bar)	G_E (bar)	γ_E
220	1.59E3	2.24E-2	1.78E1	1.78E1	0	7.09E4	2.24E-2
260	1.59E3	3.32E-2	3.28E1	1.85E1	1.44E1	6.88E4	2.32E-2
300	1.59E3	1.15E-1	1.34E2	4.85E1	8.57E1	2.62E4	6.08E-2
340	1.30E3	4.35E-1	4.45E2	1.20E2	3.25E2	7.04E3	1.84E-1
380	9.81E2	6.71E-1	4.33E2	2.26E2	2.06E2	2.13E3	4.61E-1
420	6.65E2	1.02	2.11E2	2.11E2	0	4.04E2	1.02
460	3.49E2	1.13	6.77E1	6.77E1	0	1.07E2	1.13
500	3.31E1	5.56E-1	7.89	7.89	0	5.11E1	5.56E-1
504	0	0	0	0	0	0	0

Tension

T (K)	S_b (bar)	ε_b	W_T (bar)	W_E (bar)	W_p (bar)	E_E (bar)	ε_E
220	3.14E3	1.47E-2	2.30E1	2.30E1	0	2.13E5	1.47E-2
260	3.13E3	1.94E-2	3.74E1	2.34E1	1.40E1	2.09E5	1.50E-2
300	2.98E3	7.17E-2	1.65E2	4.89E1	1.15E2	9.06E4	3.28E-2
340	2.20E3	1.81E-1	3.31E2	6.60E1	2.64E2	3.66E4	6.01E-2
380	1.50E3	3.06E-1	3.32E2	1.28E2	2.03E2	8.80E3	1.71E-1
420	8.29E2	6.05E-1	2.05E2	2.06E2	0	1.12E3	6.05E-1
460	4.03E2	7.35E-1	8.15E1	8.15E1	0	3.02E2	7.35E-1
500	5.06E1	3.09E-1	6.96	6.96	0	1.46E2	3.09E-1
504	0	0	0	0	0	0	0

TABLE 6-17 Calculated Mechanical Response of 8 Wt% Moisture in Cured Epoxy (T_g = 380 K, t = 1 s)

Shear

T (K)	σ_b (bar)	γ_b	W_S (bar)	W_E (bar)	W_p (bar)	G_E (bar)	γ_E
220	1.58E3	2.24E-2	1.77E1	1.77E1	0	7.04E4	2.24E-2
260	1.58E3	7.64E-2	9.99E1	2.11E1	7.88E1	5.94E4	2.66E-2
300	1.55E3	3.49E-1	4.42E2	1.01E2	3.41E2	1.20E4	1.30E-1
340	1.22E3	3.79E-1	3.21E2	1.40E2	1.81E2	5.32E3	2.29E-1
380	8.82E2	4.59E-1	1.56E2	1.56E2	0	1.48E3	4.59E-1
420	5.47E2	1.19	8.62E1	8.62E1	0	1.22E2	1.19
460	2.11E2	1.14	5.80E1	5.80E1	0	9.00E1	1.14
485	0	0	0	0	0	0	0

Tension

T (K)	S_b (bar)	ε_b	W_T (bar)	W_E (bar)	W_p (bar)	E_E (bar)	ε_E
220	3.12E3	1.47E-2	2.29E1	2.29E1	0	2.11E5	1.47E-2
260	3.06E3	3.42E-2	8.09E1	2.37E1	5.72E1	1.98E5	1.55E-2
300	2.71E3	1.45E-1	3.33E2	5.99E1	2.74E2	6.15E4	4.41E-2
340	2.05E3	1.88E-1	2.84E2	1.01E2	1.82E2	2.07E4	9.89E-2
380	1.38E3	2.81E-1	1.70E2	1.70E2	0	4.29E3	2.81E-1
420	6.14E2	7.81E-1	1.08E2	1.08E2	0	3.52E2	7.81E-1
460	2.52E2	6.76E-1	5.09E1	5.09E1	0	2.23E2	6.76E-1
485	0	0	0	0	0	0	0

related properties, is reemphasized in the curves of calculated fracture toughness for shear W_S and tension W_T shown in Fig. 6-15. The magnitude of W_S and W_T measures the energy per unit volume to deform the unnotched specimen.

As shown in Fig. 6-15 the shape of the W_S and W_T curves are quite similar for three moisture contents. The chief effect of moisture appears in shifting these curves to lower temperature as is predicted by the plasticizer effect of water on the T_g of the epoxy. The curves of Fig. 6-15 also show that increasing moisture content substantially increases the calculated fracture toughness at ambient temperature around 300K. This result is not obvious and is of great practical importance in graphite-epoxy composites which are notably sensitive to impact damage as discussed by Delmonte.[6]

TABLE 6-18 Calculated Mechanical Response of 10 Wt% Moisture in Cured Epoxy (T_g = 363K, t = 1 s)

Shear

T (K)	σ_b (bar)	γ_b	W_S (bar)	W_E (bar)	W_p (bar)	G_E (bar)	γ_E
220	1.58E3	2.24E-2	1.76E1	1.76E1	0	7.01E4	2.24E-2
260	1.58E3	1.02E-1	1.33E2	2.85E1	1.05E2	4.37E4	3.61E-2
300	1.50E3	4.33E-1	5.11E2	1.37E2	3.74E2	8.16E3	1.83E-1
340	1.14E3	6.78E-1	5.14E2	2.60E2	2.54E2	2.51E3	4.55E-1
380	7.87E2	1.05	2.98E2	2.98E2	0	5.38E2	1.05
420	4.32E2	1.24	8.89E1	8.89E1	0	1.16E2	1.24
460	7.79E1	9.04E-1	2.41E1	2.41E1	0	5.89E1	9.04E-1
468	0	0	0	0	0	0	0

Tension

T (K)	S_b (bar)	ε_b	W_T (bar)	W_E (bar)	W_p (bar)	E_E (bar)	ε_E
220	3.11E3	1.47E-2	2.28E1	2.28E1	0	2.10E5	1.47E-2
260	3.05E3	4.92E-2	1.19E2	2.93E1	8.94E1	1.54E5	1.95E-2
300	2.54E3	1.79E-1	3.79E2	7.62E1	3.03E2	4.22E4	6.01E-2
340	1.74E3	3.15E-1	4.10E2	1.36E2	2.74E2	1.10E4	1.57E-1
380	9.93E2	5.85E-1	2.48E2	2.48E2	0	1.45E3	5.85E-1
420	4.87E2	7.74E-1	8.53E1	8.53E1	0	2.84E2	7.74E-1
460	1.03E2	5.08E-1	2.01E1	2.01E1	0	1.56E2	5.09E-1
468	0	0	0	0	0	0	0

A cross plot of calculated strength versus extensibility provides so-called "failure envelopes" for shear and tensile response shown in Fig. 6-16. These curves show the interesting result that increased moisture enlarges the failure envelopes of the epoxy resin both in shear and tension. The result is, of course, consistent with the calculated increase in maximum fracture toughness shown for moisture plasticized epoxy in the curves of Fig. 6-15.

The main objective of this section is to show that the polymer chemistry to mechanical property model can treat water in epoxy resin as a molecular composite response. As indicated in Table 6-7 through Table 6-12 different estimates of the molecular properties of water can produce minimum and maximum

TABLE 6-19 Relations Between English and SI Units in the Composite Fracture Energy and Strength Model

Input Variable	English Units	SI Units
D, V	2E-4 in, 0.5	5.08E-6m, 0.5
E, S	1E7 psi, 4E5 psi	6.89E10 N/m^2, 2.76E9N/m^2
G, L	5E5 psi, 5E3 psi	3.44E9N/m^2, 3.44E7N/m^2
Y, YY	1E6 psi, 1E4 psi	6.89E9N/m^2, 6.89E7N/m^2
LB, LF	5E3 psi, 5E2 psi	3.44E7N/m^2, 3.44E6N/m^2
<u>Output (1)</u>		
Inter-fiber Spacing	(in.) = 1.7E-4	(m) = 4.30E-6
Shear Stress Conc.	$(in.)^{-1}$ = 4360	$(m)^{-1}$ = 1.71E5
Max. F-M Bond St.	(psi) = 8.7E5	(N/m^2) = 6.01E8
F-M Debond Length	(in.) = 3.7E-3	(m) = 9.61E-4
Inter. Shear St.	(psi) = 5000	(N/m^2) = 3.44E7
Comp. Tens. Mod.	(psi) = 5.5E6	(N/m^2) = 3.79E10
Comp. Tens. St.	(psi) = 1.04E5	(N/m^2) = 7.18E8
Crit. Crack Length	(in.) = 7.5E-3	(m) = 1.92E-3
<u>Output (2)</u>		
Unflawed St. (min)	(psi) = 6.11E4	(N/m^2) = 4.2E8
Unflawed St. (max)	(psi) = 1.76E4	(N/m^2) = 1.22E9
Crit. Stress Int (min)	$(lb^2/in.^3)^{1/2}$ = 9406	$(N^2/m^3)^{1/2}$ = 3.27E7
Crit. Stress Int (max)	$(lb^2/in.^3)^{1/2}$ = 2.7E4	$(N^2/m^3)^{1/2}$ = 9.47E7
<u>Output (3)</u>		
F-M Bond Stress	(psi) = 5000	(N/m^2) = 3.44E7
Fiber (W_{Fb}/A)	(lb/in.) = 1.67	(N/m) = 2.95E3
Matrix (W_{Sb}/A)	(lb/in.) = 14.4	(N/m) = 2.53E4
Frict. (W_{Fb}/A)	(lb/in.) = 118	(N/m) = 2.08E5
Tot. (W_b/A)	(lb/in.) = 134	(N/m) = 2.36E5
Crit. Length (L_c)	(in.) = 7.54E-3	(m) = 1.92E-3

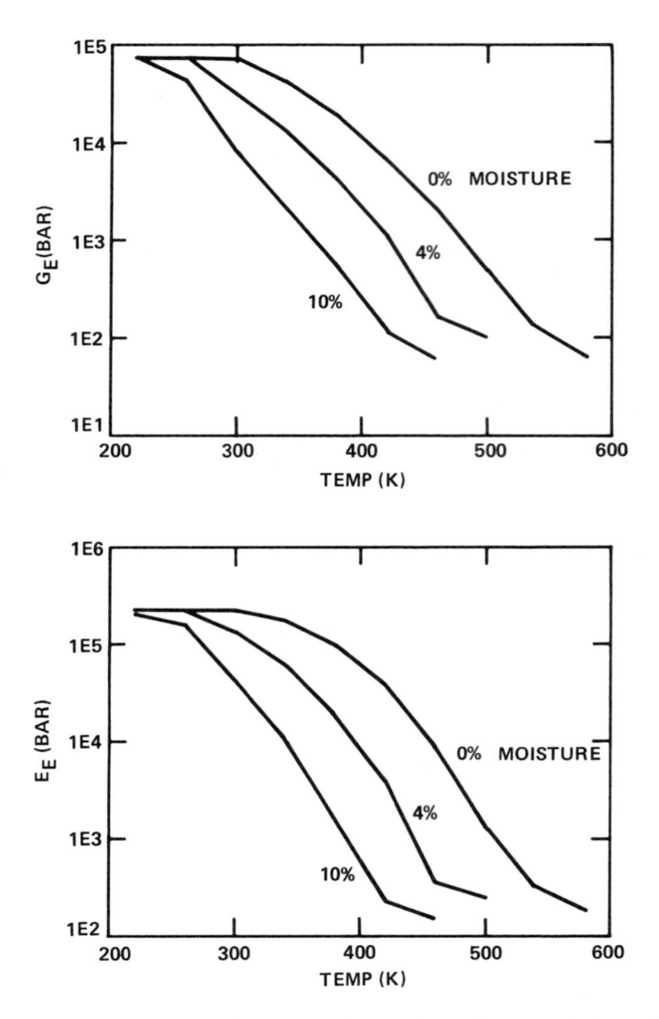

FIG. 6-13 Calculated engineering shear modulus (upper) and tensile modulus (lower curves) for cured epoxy with varied wt% moisture.

estimates of the moisture degradation of T_g. At the same time the computation defines the effects of moisture on 28 other intermediate physical properties, defined in Table 6-4, which influence mechanical response. Finally the computation produces the summary of thermomechanical response properties over a wide range of moisture contents. Experimental verification of these combined physical and mechanical effects of moisture degradation are well documented in a number of studies of Kaelble and co-workers[13-33] for uniaxial reinforced graphite-epoxy composites and many studies of cross ply composites as reviewed by Delmonte.[6]

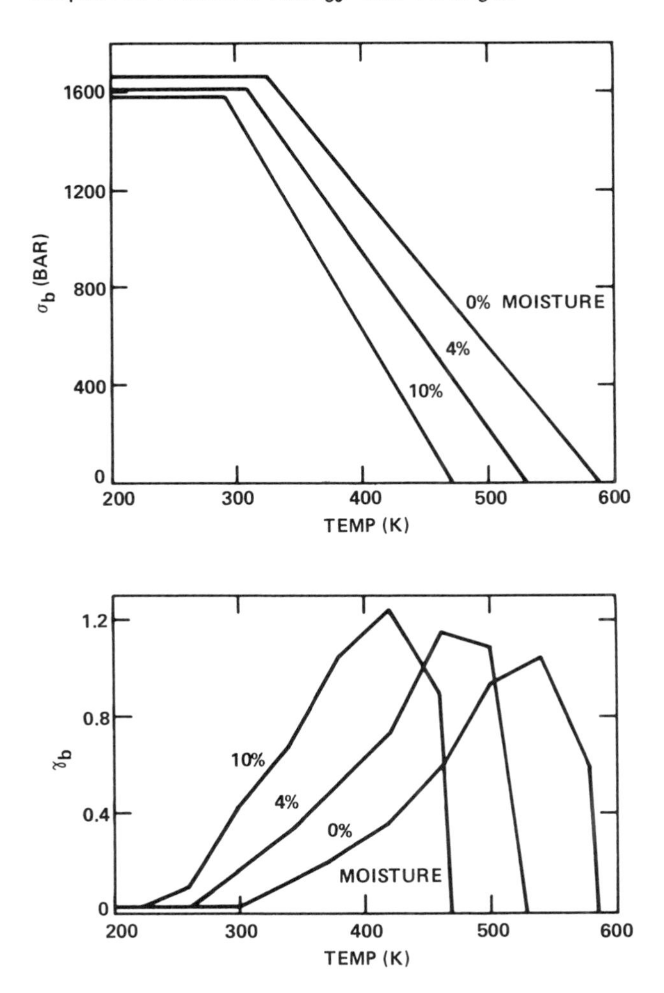

FIG. 6-14 Calculated shear strength (upper) and shear extensibility (lower curves) for cured epoxy with varied wt% moisture.

6-5. Composite Fracture Energy and Strength

In most common cases of composite laminate design the layers of reinforcement are arranged to produce transverse isotropy and the composite is used in plate form with plane stress loading. For this common case of composite design the elastic constants as schematically described in Fig. 6-17 reduce to four as expressed by the following standard relations:[39]

$$\begin{bmatrix} \varepsilon_{11} \\ \varepsilon_{22} \\ \varepsilon_{12} \end{bmatrix} = \begin{vmatrix} S_{11} & S_{12} & 0 \\ S_{12} & S_{22} & 0 \\ 0 & 0 & S_{66} \end{vmatrix} \begin{bmatrix} \sigma_{11} \\ \sigma_{22} \\ \sigma_{12} \end{bmatrix} \qquad (6\text{-}20)$$

$$S_{11} = \frac{1}{E_{11}} \tag{6-21}$$

$$S_{22} = \frac{1}{E_{22}} \tag{6-22}$$

$$S_{12} = \frac{-\mu_{12}}{E_{11}} = \frac{-\mu_{21}}{E_{22}} \tag{6-23}$$

$$S_{66} = \frac{1}{G_{12}} \tag{6-24}$$

The variation of elastic properties with the angle θ between fibers and applied stress for the uniaxial reinforced composite of Fig. 6-17 is given by the following relation:[39]

$$\frac{1}{E_\theta} = S_{11} \cos^4 \theta + S_{22} \sin^4 \theta + (S_{66} - 2S_{12}) \cos^2 \theta \sin^2 \theta \tag{6-25}$$

the maximum strain energy criteria of composite strength σ_θ at an angle θ

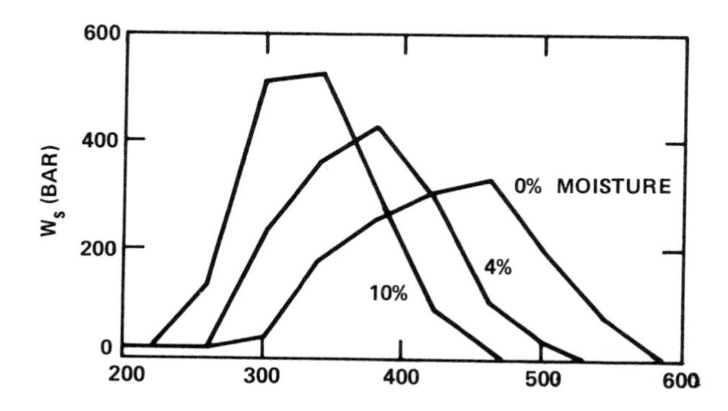

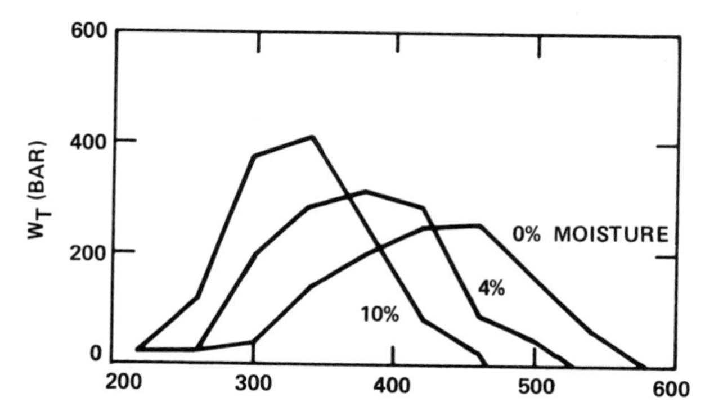

FIG. 6-15 Calculated specific fracture energy in shear (upper) and tension (lower curves) for cured epoxy with varied wt% moisture.

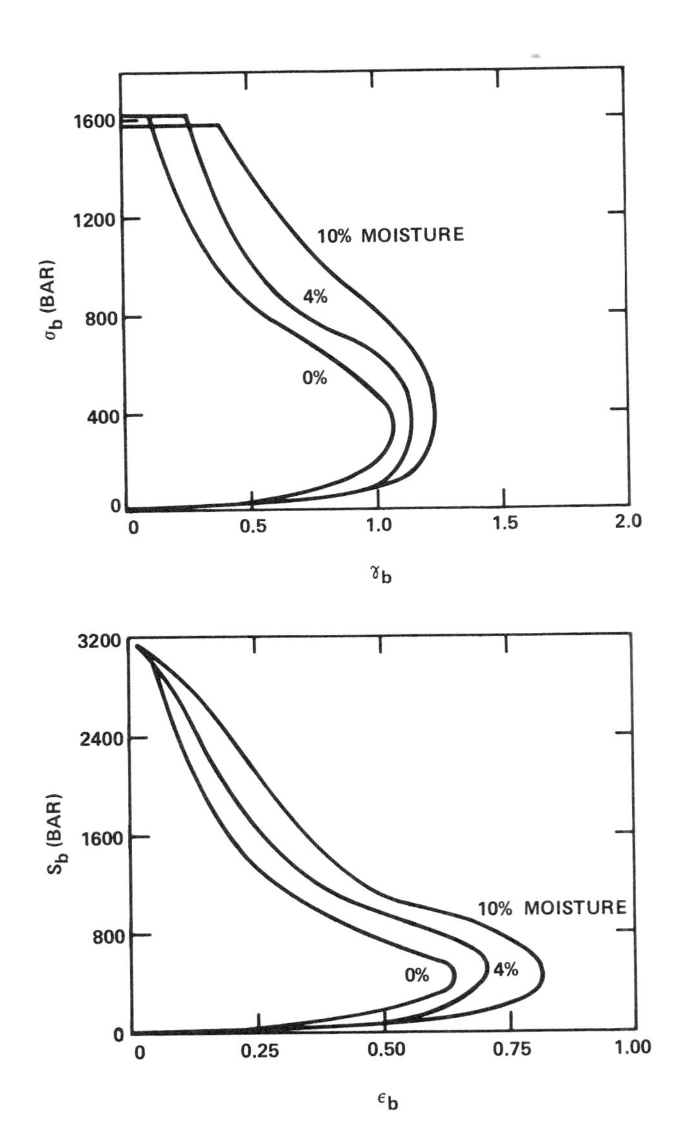

FIG. 6-16 Calculated failure envelopes in shear (upper) and tension (lower curves) for cured epoxy with varied wt% moisture.

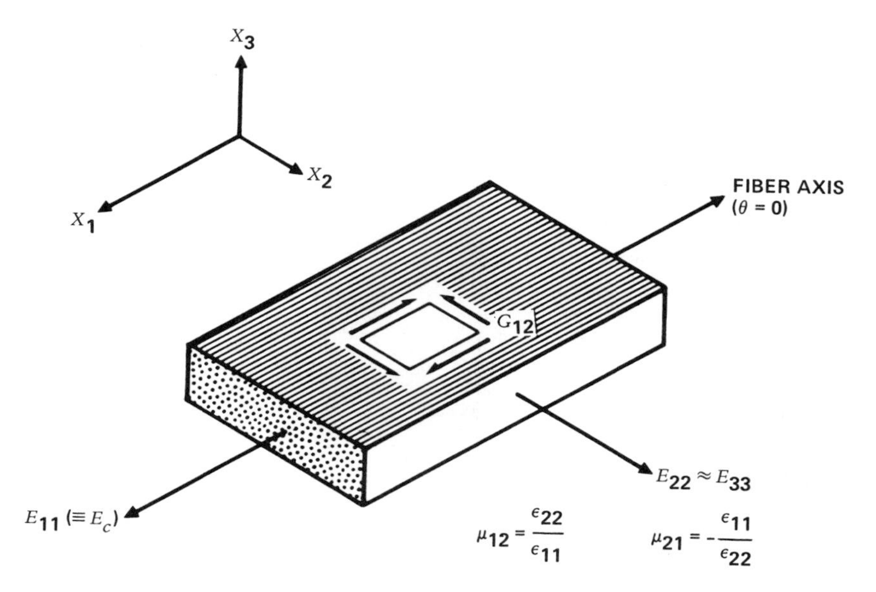

FIG. 6-17 Unidirectional reinforced composite.

between fibers and applied stress is defined by the following relation:[39]

$$\frac{1}{\sigma_\theta^2} = \frac{\cos^4 \theta}{\sigma_c^2} + \frac{\sin^4 \theta}{\sigma_t^2} + \left(\frac{1}{\tau_c^2} - \frac{1}{\sigma_c^2}\right) \sin^2 \theta \cos^2 \theta \qquad (6\text{-}26)$$

where σ_c is the composite tensile strength parallel to the fibers $\theta = 0$, σ_t is the transverse tensile strength perpendicular to the fibers ($\theta = \pi/2$) and τ_c is the in-plane shear strength.

The tensile strength and fracture energy for fracture transverse to the fiber axis $\theta = \pi/2$ has been analyzed by a number of micro-mechanic models. The major emphasis in the theories of Kelly, Cottrell and Cooper[40-43] is the frictional work W_F expended in extracting the fiber from the matrix. Outwater and Murphy[44] consider the deformational work W_D developed through tensile strain of the reinforcing fiber in the region of debonding. Linear elastic analysis for the shear stress distributions around bonded fibers in a fiber pull-out geometry is provided by Greszczuk[45] and Lawrence.[46] An elastic-plastic analysis of an elastic fiber in a ductile matrix is provided by Lin, Salinas and Ito.[47] More recently Kaelble[2] and Wells and Beaumont[48] have included the work of shear deformation W_S stored in the matrix and released during fiber debonding.

The microfracture mechanics model of Kaelble[2] considers the case of uniform circular fibers in square or hexagonal packing as shown in Fig. 6-18.

LATTICE TYPE	SQUARE	HEXAGONAL
UNIT GEOMETRY		
FIBERS/UNIT CELL	2.0	3.0
FIBER VOLUME FRACTION (v)	$2\pi(r_0/L)^2$	$1.1548\pi(r_0/L)^2$
UNIT CELL AREA (A)	$2\pi r_0^2/V$	$3\pi r_0^2/V$
$a = (r_1 - r_0)$	$r_0[(\pi/V)^{1/2}-2]$	$r_0[1.074(\pi/V)^{1/2}-2]$
v AT $(r_1 - r_0) = 0$	0.785	0.906

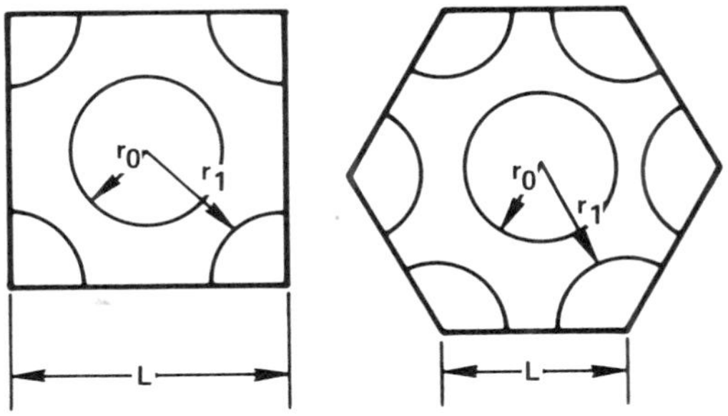

FIG. 6-18 Packing geometries for regular uniaxial fiber arrays.

Also illustrated in Fig. 6-18 are the significant parameters relating to fiber packing in a square or hexagonal unit cell geometry. For the square fiber packing with the fiber volume fraction V in the range $0 \leqslant V \leqslant 0.785$ inspection of Fig. 6-18 defines the following volume fraction functions:[2]

$$f_1(V) = \left(\ln\left(\frac{r_1}{r_0}\right)\right)^{1/2} = \left(\ln\left((\pi/V)^{1/2} - 1\right)\right)^{1/2} \quad (6\text{-}27)$$

$$f_2(V) = V \quad (6\text{-}28)$$

$$f_3(V) = r_0/(r_0 + r_1) = (V/\pi)^{1/2} \quad (6\text{-}29)$$

and for hexagonal packing where $0 \leqslant V \leqslant 0.906$ as second set of volume fraction functions are defined by Fig. 6-18 as follows:[2]

$$f_1(V) = \left(\ln\left(\frac{r_1}{r_0}\right)\right)^{1/2} = \left(\ln\left(1.074(\pi/V)^{1/2} - 1\right)\right)^{1/2} \quad (6\text{-}30)$$

$$f_2(V) = 2V/3 \quad (6\text{-}31)$$

$$f_3(V) = 2r_0/3(r_0 + r_1) = \frac{2}{3(1.074)}\left(\frac{V}{\pi}\right)^{1/2} \quad (6\text{-}32)$$

The shear stress distribution λ_s at the bonded fiber-matrix interface can be derived from linear elastic arguments and provides the following relations:[2]

$$\lambda_s = \frac{\alpha P \exp(\alpha x)}{2\pi r_0} = \lambda_0 \exp(\alpha x) \quad (6\text{-}33)$$

where x is a distance along the fiber length where x = 0 at the point of disbonding and P is the applied force on the fiber. The stress concentration factor α of this shear stress distribution has units of reciprocal length as defined by the following relation:(2)

$$\alpha = (2G_m/E_f)^{1/2}/r_0 \; f_1(V) \quad (6\text{-}34)$$

where G_m is the matrix shear modulus and E_f the fiber tensile modulus. The maximum length L_b for fiber-matrix debonding is expressed by the following relation:(2)

$$L_b = \frac{r_0}{\lambda_f} \; \frac{\sigma_b}{2} - \lambda_0(E/2G)^{1/2} \; f_1(V) \;\geq 0 \quad (6\text{-}35)$$

and L_b is directly related to the critical fiber length L_c defined in the models of Kelly, Cottrell and Cooper by the following simple relation:(2)

$$L_c = 2L_b \quad (6\text{-}36)$$

The maximum values for W_S, W_B, and W_F per unit cell area A are given by the following relations:(2)

$$W_{Sb}/A = (\lambda_0 E_f/2G_m)^{1/2} \; L_b \; f_1(V) \; f_2(V) \quad (6\text{-}37)$$

$$W_{Bb}/A = (\sigma_b{}^2/12E_f) \; L_b \; f_2(V) \quad (6\text{-}38)$$

$$W_{Fb}/A = \lambda_F \; L_F^2 \; f_2(V)/2r_o \quad (6\text{-}39)$$

where the fiber pull-out length L_F lies in the range $0 \leq L_F \leq L_b$ since the locus of fiber failure in the debond length is undefined. The total fracture energy per unit area W_b/A is then the sum of the above parts as follows:(2)

$$\frac{W_b}{A} = \frac{W_{Sb} + W_{Bb} + W_{Fb}}{A}$$

$$= \lambda_o \left(\frac{E_f}{G_m}\right)^{1/2} L_b \; f_1(V) \; f_2(V) + \left(\frac{\sigma_b{}^2}{12E_f}\right) L_b \; f_2(V) + \lambda_F L_F^2 \; \frac{f_2(V)}{2r_o} \quad (6\text{-}40)$$

The above solution has been incorporated in the computer model for Listing No. 5 of the Appendix. In addition, this model includes the standard relations for parallel interaction of fiber and matrix in terms of composite tensile modulus E_c in the fiber axis ($\theta = 0$):[39]

$$E_c = V\, E_f + (1 - V)\, E_m \tag{6-41}$$

and the composite tensile strength S_c in the fiber axis as follows:[2]

$$\sigma_c = S_f\, f_3(V) + (1 - V)\, S_m \tag{6-42}$$

where S_f and S_m are the respective fiber and matrix tensile strengths. The simple additive stiffness relations for E_c assumes the fiber and matrix act independently while the relations for composite strength S_c and the specific fracture energy W_b/A are completely specified by the micromechanics. The prominent micromechanics parameters are the shear stress concentration factor α and the maximum fiber debond length L_b in the above relations.

Other properties of the uniaxial composite can be also estimated such as the composite shear modulus G_c from the standard rule of mixtures relation:[39]

$$G_c = G_m\, \frac{(G_f + G_m) + (G_f - G_m)V}{(G_f + G_m) - (G_f - G_m)V} \tag{6-43}$$

where G_f and G_m are the respective shear modulus of fiber and matrix. The interlaminar shear strength λ_c of the uniaxial composite can be defined in terms of the unit cell models of Fig. 6-18 by the following relations:[24]

$$\lambda_c = f_I\, \lambda_I + (1 - f_I)\lambda_m \tag{6-44}$$

$$f_I = \frac{\pi r_0}{\pi r_0 + a} \tag{6-45}$$

where λ_I and λ_m are the respective shear fracture stresses for the fiber-matrix interface (I) and matrix (M) cohesive failure and f_I as defined by Fig. 6-18 is the area fraction of interface as defined by fiber radius r_0 and interfiber spacing a. The effect of moisture degradation of interlaminar shear strength has been analyzed by Kaelble in terms of the surface energetics model for interface crack propagation as discussed in Eqs. (1-13) to (1-17). When applied to uniaxial reinforced composites this model of interface degradation provides the following relevant relations:[24]

$$\frac{\lambda_{c\infty}}{\lambda_{c0}} = (1 - f_I) + f_I\, \frac{\sigma_c(H_2O)}{\sigma_c(\text{air})} \tag{6-46}$$

where the Griffith critical stresses $\sigma_c(\text{air})$ and $\sigma_c(H_2O)$ represent the respective initial dry air immersion state and the equilibrium moisture immersion state where for most composites:[22-24]

$$\frac{\sigma_c(H_2O)}{\sigma_c(\text{air})} \simeq 0.5 - 0.70 \tag{6-47}$$

due to polar interactions at the fiber-matrix interface which are degraded by moisture absorption.

The extension of the above arguments of micro-fracture to macroscopic strength of crossply composites appears in the general observation of Wells and Beaumont[49] in a recent comprehensive review of 16 studies on this subject. This review shows that the strength of crossply composite laminates with manufactured defects such as circular holes, central slits, and inclined slits of widely varied size can be fit to "master curves" of tensile strength σ_b versus crack length c as shown in the solid curve of Fig. 6-20. The crack length c is defined as one-half the diameter of a circular hole or one-half the projected crack length normal to the loading direction of perpendicular or tilted slits. This data survey includes carbon, graphite, glass, and polymer (aramide) reinforcements in epoxy and polyester matrix phases. Both linearly aligned and woven reinforcement is included and the only apparent restriction is that the crossply laminate contain one ply with the aligned fiber axis parallel to the loading direction (ply fiber angle $\theta = 0$). The general observation is that failure generally involves splitting parallel to the fibers which is consistent with the mechanism of fiber shear debonding described in Fig. 6-19.

The general conclusion from the review of Wells and Beaumont[49] is that composites with no stress concentrating notch or a notch shorter than a critical crack length a, as shown in Fig. 6-20 display no tendency to develop a larger crack during loading and fail by a microcracking and crazing process which extends through the entire composite and defines a characteristic unnotched tensile strength σ_u. This strength is independent of initial crack length c when $c \leqslant a$ and is described as the microfracture or yielding mode of fracture.

For larger initial cracks where $c > a$ the crack-tip stress concentration is large enough to localize failure and micro-damage at the crack tip in a so-called damage "process zone." Stresses outside the process zone remain elastic with a magnitude which is proportional to the square root of the reciprocal crack length $1/\sqrt{c}$. Failure in this second process is by the single crack and the composite macroscopic strength σ_b becomes proportional to crack length by the relation shown in Fig. 6-20 which is $\sigma_b \propto (1/c)^{1/2}$.

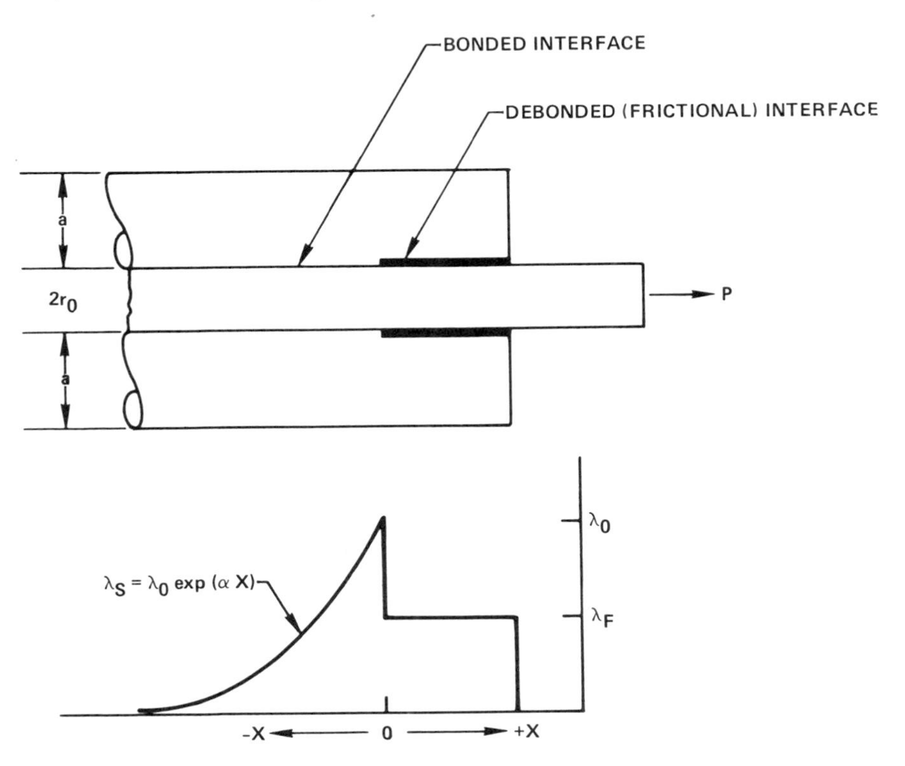

FIG. 6-19 Frictional (λ_F) and bonded (λ_S) interfacial shear stresses during fiber pull-out.

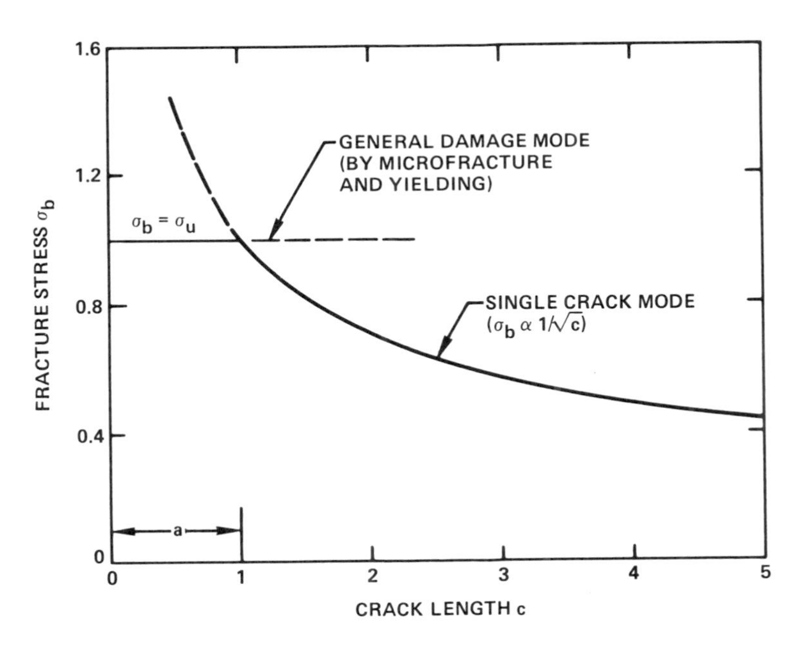

FIG. 6-20 Schematic showing the observed variation in failure mode and fracture stress σ_b with crack length c in damage tolerant composites.

As pointed out by Wells and Beaumont[49] and the more detailed analysis of this phenomena by Nuismer and Whitney[50] the measured strength of a finite width specimen with crack length c can be corrected to a specimen of infinite width (representing a large plate loading) by the following standard relations:[49,50]

$$\sigma_{b\infty} = \frac{Y}{\sqrt{\pi}} \sigma_b \text{ (exp.)} \tag{6-48}$$

$$Y = 1.77 + 0.227 \left(\frac{2c}{W}\right) - 0.510 \left(\frac{2c}{W}\right)^2 + 2.70 \left(\frac{2c}{W}\right)^3 \tag{6-49}$$

where the central slit is of length 2c and the specimen width is W and $\sigma_{b\infty}$ is the strength of the infinite plate.

All of the micromechanics arguments introduced in this section can now be readily incorporated in a modified Griffith-Dugdale model for crack propagation. We can first write a conventional statement for Griffith crack propagation in plane stress:[17]

$$\sigma_b \leqslant \sigma_c = \left(\frac{2E\gamma_G}{\pi c_1}\right)^{1/2} = \frac{K_{1C}}{(\pi c_1)^{1/2}} \tag{6-50}$$

where σ_c is the critical stress for crack propagation and E is the plate modulus. The Griffith fracture energy $2\gamma_G$ for creation of two free surfaces by crack propagation will now be equated to the specific fracture work W_b/A so that:

$$2\gamma_G = \left(\frac{W_b}{A}\right) = \frac{W_{Sb} + W_{Bb} + W_{Fb}}{A} \tag{6-51}$$

The critical stress intensity K_{1C} is thus defined as follows:

$$K_{1C} = (E_c{*}W_b/A)^{1/2} \tag{6-52}$$

where E_c is the composite tensile modulus.

A second lower estimate of the estimate of the Griffith fracture energy is available by subtracting W_{Fb} from the definition of Eq. (6-51) as follows:

$$(2\gamma_G)_{min} = \frac{W_{Sb} + W_{Bb}}{A} \tag{6-53}$$

The second lower estimate of the critical stress intensity is as follows:

$$(K_{1C})_{min} = \left[E_c \frac{(W_{Sb} + W_{Bb})}{A}\right]^{1/2} \tag{6-54}$$

These second minimum estimates for $(2\gamma_G)_{min}$ and $(K_{1C})_{min}$ describe the special case described by theory where fiber-matrix debonding occurs over the fiber length L_b but the fiber breaks at the crack tip such that the effective frictional pull out length $L_F = 0$. In general, studies show that fracture energy data scatter so as to be statistically distributed between the higher estimates of Eq. (6-51) and Eq. (6-52) and the minimum estimates of Eq. (6-53) and Eq. (6-54). The effective crack length c, in the modified form of the Griffith equation has two parts as follows:

$$C_1 = L_c F + (1 - F)\, c \qquad (6\text{-}55)$$

where $L_c = 2L_b$ is the critical fiber length that determines the plastic zone size, c is the crack length and F is a factor which defines the uniformity of micro stresses by the following relation:

$$F = \exp - \left(c/L_c\right)^g \qquad (6\text{-}56)$$

where

$g = 3$ for spherical cavity growth
$g = 2$ for prolate spheroidal craze growth
$g = 1$ for linear crack growth.

When $c < L_c$ it follows that $F \simeq 1.0$ and the critical stress σ_c is insensitive to crack length as described by the Dugdale model. When $c > L_c$ it follows that $F \simeq 0$ and $c_1 \simeq c$ in which case the normal Griffith condition that $\sigma_c \propto \sqrt{1/c_1}$ is obtained. The combination of these conditions is presented in the solid curve of Fig. 6-20 where the critical crack length for ductile response is given as $a = 2L_b = L_c$.

In this revised Griffith-Dugdale model all of the detailed design features of the microfracture mechanics model are incorporated. The effects of environmental aging appear both in their impact upon the matrix mechanical response properties and also upon the fiber matrix shear failure stress λ_I as influenced by the effects of moisture upon the interface work of adhesion. The microfracture mechanics model discussed here intrinsically defines the effects of fiber volume fraction V and fiber packing upon both specific fracture energy (W_b/A) and also upon the strength σ_b of composite plates with preformed cracks of length c through the modified Griffith-Dugdale crack propagation model presented above.

The composite fracture energy and strength model presented as Listing No. 5 in the Appendix incorporates the relations developed in this section for the case of hexagonal fiber packing as shown in Fig. 6-18. This model accepts any self consistent physical units in order to be more useful to both material scientists and engineers. Table 6-19 compares inputs and outputs in standard English and SI units and Table 6-20 presents the complete input and output report for this sample calculation.

The upper part of Table 6-20 presents the ten input variables and their assigned values. The first output segment of Table 6-20 presents a group of intermediate composite properties calculated by the computer model. The second output segment reports the fracture mechanics analysis and calculates a table of composite tensile strengths as a function of flaw size and crack length relative to the critical length L_c. The third output segment calculates the specific fracture energies for the input variables followed by a table of specific fracture energies over the range of fiber-matrix stresses varying from zero to the maximum stress λ_0 where $L_b = 0$ in Eq. (6-35).

In the model calculation of Table 6-20 the fiber-matrix interface stress $\lambda_0 = 5000$ psi is proportionately large compared to the frictional stress $\lambda_F = 500$ psi. The calculated curves of composite strength σ_b versus crack length presented in Table 6-20 are plotted in the solid curves of Fig. 6-21. The upper solid curves in Fig. 6-21 describe the transition from the general damage mode (see Fig. 6-20) to the single crack mode of failure at a critical crack length $C = L_c = 0.075$ in. The lower solid curves of Fig. 6-21 present the same data in bilogarithmic curves of σ_b versus C which define linear curves for ease of interpretation.

A second estimate of composite fracture energy and strength is presented in Table 6-21. All the input variables of Table 6-20 are retained except the interface is made a regular elastic-plastic response with $\lambda_0 = \lambda_F = 5000$ psi. The data summary of Table 6-21 shows all the property changes calculated by this proportionate increase in frictional stress. Intercomparing the first output segment in upper Table 6-20 and Table 6-21 one notes the only intermediate property change is the tenfold reduction in debond length L_b and critical crack length in the second estimate of Table 6-21. The dramatic effect on estimated composite strength versus crack length is illustrated by the dashed curves of Fig. 6-21 where the transition from the general damage mode to the single mode occurs at a smaller critical crack length $C = L_c = 0.0075$ in. The lower curves of Fig. 6-21 show this increase in λ_F as merely shifting the curves of log G_b versus log c a factor of 10 to lower values of log c.

It is left to the reader to explore the many relevant combinations of inputs to this composite fracture energy and strength model. The inputs can come from either experiments such as the detailed characterization of microfrac-

TABLE 6-20 First Estimate of Composite Fracture Energy and Strength (English Units, LB = 5000 psi, LF = 500 psi)

```
FIBER DIAMETER(D), VOLUME FRACTION(V)= 2E-04  .5
FIBER TENSILE MODULUS(E), STRENGTH(S)= 1E+07  400000
MATRIX SHEAR MODULUS(G), STRENGTH(L)= 500000  5000
MATRIX TENSILE(Y), STRENGTH(YY)= 1E+06  10000
F-M BOND STRENGTH(LB), FRICT. STRENGTH(LF)= 5000  500

INTER-FIBER SPACING(R1)= 1.63212E-04
SHEAR STRESS CONC.(A)= 4360.29
MAX. F-M BOND STRENGTH(LM)= 87205.7
F-M DEBOND LENGTH(BL)= .0377066
INTERLAM. SHEAR STRENGTH(IL)= 5000
COMPOSITE TENSILE MODULUS= 5.5E+06
COMPOSITE CONTINUUM TENSILE STRENGTH= 104055
CRITICAL CRACK LENGTH= .0754131
TO CONTINUE PRESS ENTER? _

          FRACTURE MECHANICS ANALYSIS
UNFLAWED STRENGTH (MIN.)= 61114.4 (MAX.)= 176749
CRIT. STRESS INTENSITY (MIN.)= 29746.9 (MAX.)= 86031.3
FLAW SIZE        CRACK LENGTH     MIN  STRENGTH    MAX  STRENGTH
 .0753959         4.71332E-03       61121.3         176769
 .0752844         9.42664E-03       61166.6         176900
 .0745363          .0188533         61472.8         177786
 .0709825          .0377065         62992.8         182182
 .0754131          .0754131         61114.4         176749
 .150801           .150826          43218           124991
 .301652           .301652          30557.2         88374.6
 .603304           .603304          21607.2         62490.3
 1.20661           1.20661          15278.6         44187.3
 2.41322           2.41322          10803.6         31245.2
TO CONTINUE PRESS ENTER? _

          FRACTURE WORK PER UNIT CROSSECTION AREA
F-M BOND  FIBER     MATRIX    FRICT     TOTAL     CRIT.FIBER
STRESS    WORK      WORK      WORK      WORK      LENGTH
 5000     16.7585   144.129   1184.82   1345.71   .0754131
PRESS ENTER TO CONTINUE?
 0        17.7778   0         1333.33   1351.11   .08
 9689.53  15.8025   263.374   1053.5    1332.67   .0711111
 19379.1  13.8272   460.905   806.584   1281.32   .0622222
 29068.6  11.8519   592.593   592.593   1197.04   .0533333
 38758.1  9.87654   658.436   411.523   1079.84   .0444444
 48447.6  7.90124   658.436   263.375   929.712   .0355556
 58137.2  5.92592   592.593   148.148   746.667   .0266667
 67826.7  3.95062   460.905   65.8436   530.7     .0177778
 77516.2  1.97531   263.374   16.4609   281.811   8.88889E-03
 87205.7  0         0         0         0         0
PRESS ENTER TO CONTINUE? _
```

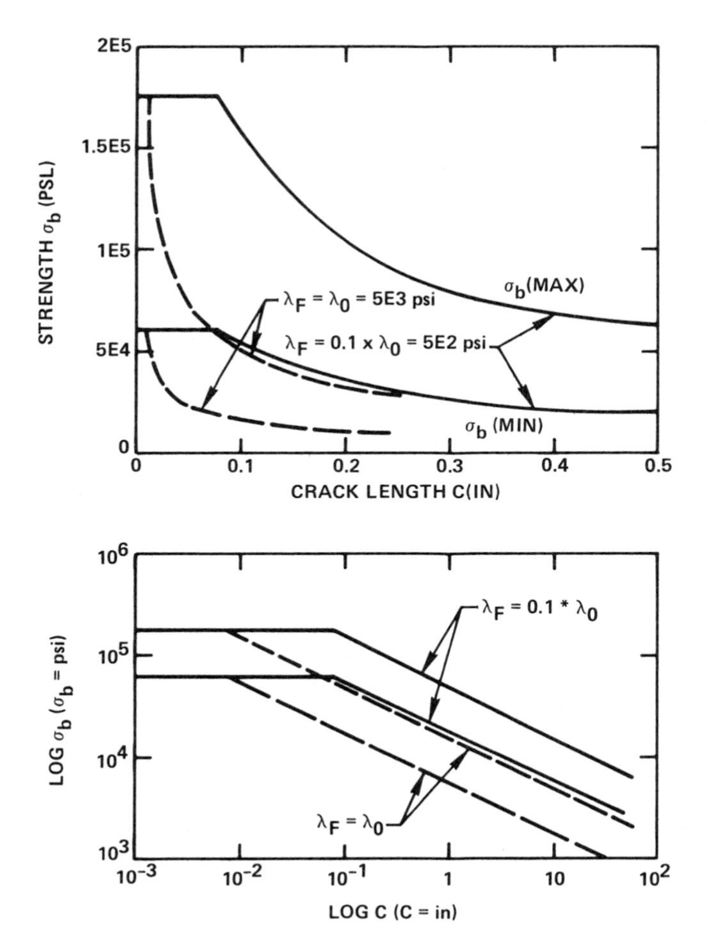

FIG. 6-21 Calculated curves of composite strength maximum σ_b(max) and minimum σ_b(min) vs crack length c.

ture mechanics in single fiber tests or from model calculations in which direct effects of chemical, physical or environmental factors are generated by computer simulation as already discussed. The presently developed methods of single fiber testing and fiber-matrix debond testing discussed in Chapter 2 which include detailed statistical analysis present an important new range of inputs to testing and improving the properties analyzed in this discussion. A recent technical report of McMahon and Ling[51] reviews the recent experimental methods and analysis for single fiber strength and interface bond strength measurements which incorporate detailed statistical and stress optical analysis. The correlation of fiber matrix interactions as discussed by McMahon and Long[51] and strength and fracture energy as discussed by Wells and Beaumont[49] represents one of the central issues in current composites research and development. The

TABLE 6-21 Second Estimate of Composite Fracture Energy and Strength (English Units, LB = 5000 psi, LF = 5000 psi)

FIBER DIAMETER(D), VOLUME FRACTION(V)= 2E-04 .5
FIBER TENSILE MODULUS(E), STRENGTH(S)= 1E+07 400000
MATRIX SHEAR MODULUS(G), STRENGTH(L)= 500000 5000
MATRIX TENSILE(Y), STRENGTH(YY)= 1E+06 10000
F-M BOND STRENGTH(LB), FRICT. STRENGTH(LF)= 5000 5000

INTER-FIBER SPACING(R1)= 1.69212E-04
SHEAR STRESS CONC.(A)= 4360.29
MAX. F-M BOND STRENGTH(LM)= 87205.7
F-M DEBOND LENGTH(BL)= 3.77066E-03
INTERLAM. SHEAR STRENGTH(IL)= 5000
COMPOSITE TENSILE MODULUS= 5.5E+06
COMPOSITE CONTINUUM TENSILE STRENGTH= 104055
CRITICAL CRACK LENGTH= 7.54131E-03
TO CONTINUE PRESS ENTER? _

FRACTURE MECHANICS ANALYSIS
UNFLAWED STRENGTH (MIN.)= 61114.4 (MAX.)= 176249
CRIT. STRESS INTENSITY (MIN.)= 9406.81 (MAX.)= 27205.5

FLAW SIZE	CRACK LENGTH	MIN STRENGTH	MAX. STRENGTH
7.53959E-03	4.71332E-04	61121.4	176770
7.52844E-03	9.42664E-04	61166.6	176900
7.45363E-03	1.88533E-03	61472.8	177786
7.09825E-03	3.77065E-03	62992.8	182182
7.54131E-03	7.54131E-03	61114.4	176749
.0150801	.0150826	43218	124991
.0301652	.0301652	30557.2	88374.7
.0603304	.0603304	21607.2	62490.4
.120661	.120661	15278.6	44187.4
.241322	.241322	10803.6	31245.2

TO CONTINUE PRESS ENTER? _

FRACTURE WORK PER UNIT CROSSECTION AREA

F-M BOND STRESS	FIBER WORK	MATRIX WORK	FRICT. WORK	TOTAL WORK	CRIT.FIBER LENGTH
5000	1.67585	14.4129	118.482	134.571	7.54131E-03
PRESS ENTER TO CONTINUE?					
0	1.77778	0	133.333	135.111	8E-03
9689.53	1.58025	26.3375	105.35	133.267	7.11111E-03
19379.1	1.38272	46.0905	80.6584	128.132	6.22222E-03
29068.6	1.18519	59.2593	59.2593	119.704	5.33333E-03
38758.1	.987654	65.8436	41.1523	107.984	4.44444E-03
48447.6	.790124	65.8436	26.3375	92.9712	3.55556E-03
58137.2	.592593	59.2592	14.8148	74.6667	2.66667E-03
67826.7	.395062	46.0905	6.58437	53.07	1.77778E-03
77516.2	.197531	26.3374	1.64609	28.1811	8.88889E-04
87205.7	0	0	0	0	0

PRESS ENTER TO CONTINUE? _

elastic-plastic analog properties of the polymer chemistry to mechanical property discussed earlier can provide input data to the microfracture model discussed in this section.

6-6. Peel Mechanics

The peel mechanics of laminates is an important subject in both fabrication and service evaluation of composite materials. The subject of peel mechanics has been of long term interest to this author.(1-4) The peel mechanics model developed by Kaelble is unique in that the effects of all-angle peeling, with peel angle $w = o$ for simple shear to peel angle $w = \pi$ rad = 180 deg, are treated as a continuous range of action for the external peeling forces. The peel mechanics model illustrates many important phenomena in the formation and fracture of laminates where cleavage (mode I in fracture mechanics) interacts and competes with simple shear (Mode II in fracture mechanics). Thus the peel mechanics model introduces the fracture mechanics of beams and plates by application of the theory of bending of cantilever beams on a viscoelastic foundation. The schematic illustrations of Fig. 6-22 describe the cross section of the peel geometry where an adhesive layer of thickness a separates a flexible adherend of thickness 2h from the rigid substrate. The cleavage and shear stress distributions are plotted in the lower left portion of Fig. 6-22 where the stresses are zero for positive X and display a highly attenuated sinusoidal response in cleavage with alternating regions of tension at the boundary and compression stress interior from the boundary. The shear stress distribution is a simple exponential decay with monotonically decreasing shear stress as distance increases into the bond interior from the boundary. The predicted forms of these cleavage and shear stress distributions has been confirmed by direct stress analysis of pressure sensitive tapes.(52) The fracture criteria for this peel model are illustrated by the maximum stress criteria shown in lower right Fig. 6-22. The bond will fail in cleavage if the boundary cleavage stress reaches a critical value described by the stress $\sigma_0(w = \pi)$ and by shear under a maximum shear stress $\pm \lambda_0$ $(w = o)$.

The new design property introduced in the bending theory of peel mechanics is the cross section moment of inertia I of the flexible adherend as defined by the following standard relation:(3-5)

$$I = \frac{2b\ h^3}{3} \tag{6-57}$$

where b is the bond width and 2h is the flexible adherend thickness. As shown in Fig. 6-22 the peel force P acts on the noncurved portion of the flexible

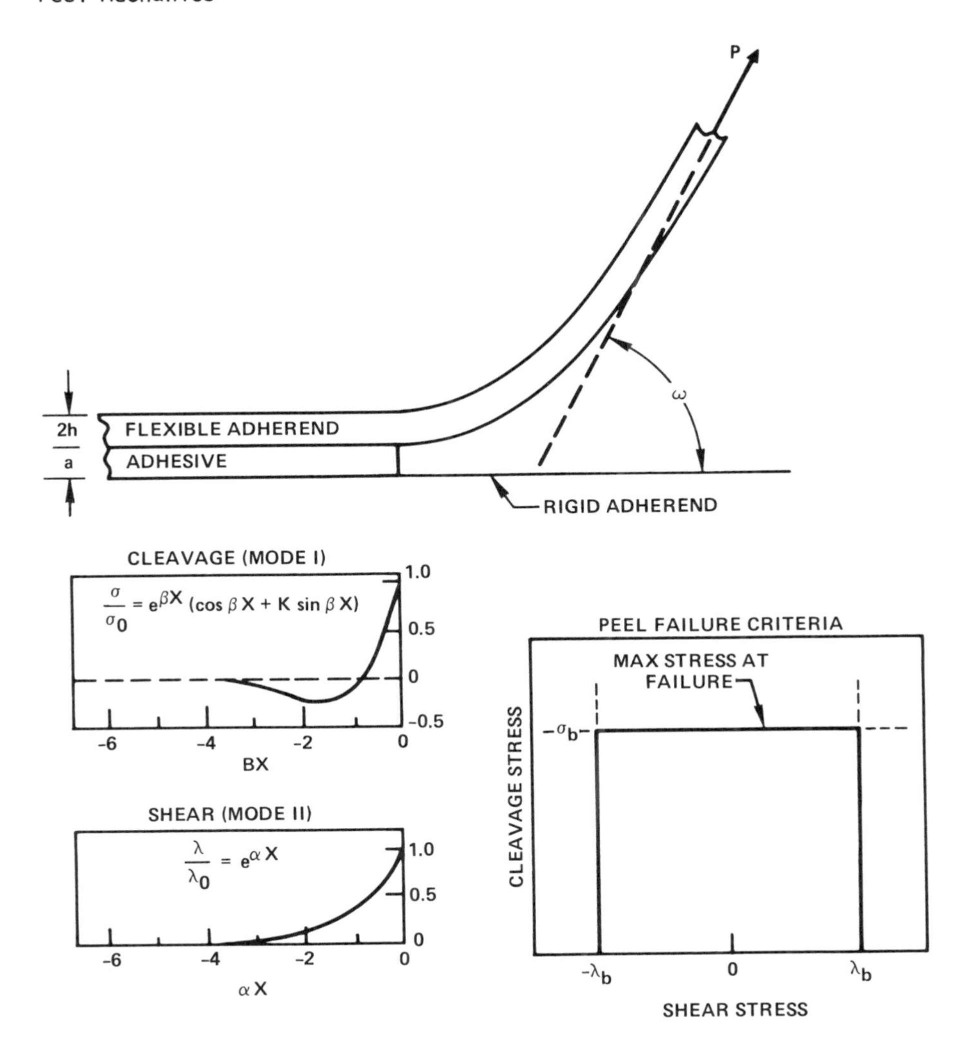

FIG. 6-22 Peel mechanics (upper and left views) and failure criteria.

adherend at a constant angle w. The peel model describes stress concentations α for shear and β for cleavage by the following relations:(3-5)

$$\alpha = \left(\frac{G}{2Eha}\right)^{1/2} \tag{6-58}$$

$$\beta = \left(\frac{3Y}{8Eh^3a}\right)^{1/4} \tag{6-59}$$

where E is the flexible adherend tensile modulus and G and Y are the respective adhesive shear and tensile modulii. The moment of force M_π and radius of curvature R_π at a peel angle $w = \pi$ rad are defined by the following relations:(3-5)

$$M_\pi = \frac{b\sigma_0}{2\beta^2} \quad (6\text{-}60)$$

$$R_\pi = \frac{EI}{M_\pi} = \frac{2EI\beta^2}{b\sigma_o} = \frac{1}{\sigma_o}\left(\frac{2YEh^3}{3a}\right)^{1/2} \quad (6\text{-}61)$$

The peeling forces $P(w = 0)$ and $P(w = \pi)$ for zero and 180 degree peel angle are defined as follows:(3-5)

$$P(w = 0) = \frac{\lambda_b b}{\alpha} \quad (6\text{-}62)$$

$$P(w = \pi) = \frac{\sigma_b^2 b\ a}{4Y} \quad (6\text{-}63)$$

where λ_b and σ_b are the critical maximum shear and tensile stresses.

The iterative procedure for the solution of the parameter K in the cleavage stress distribution (see Fig. 6-22) is the critical step in the computer analysis of peel mechanics. The cleavage stress distribution is defined as follows:

$$\sigma_x = \sigma_o \exp(\beta X)\left[\cos(\beta X) + K \sin(\beta X)\right] \quad (6\text{-}64)$$

To numerically solve for K we initially set a high peel angle $w = \pi$ rad where the following relations hold:

$$\sigma_o = \sigma_b \quad (6\text{-}65)$$

$$AA = \frac{\sin(w)}{1 - \cos(w)} \quad (6\text{-}66)$$

$$T = \frac{\sigma_o * AA}{2\beta\sigma_b R_\pi} \quad (6\text{-}67)$$

$$K = \frac{(1 + 4T)^{1/2} - 1}{2T} \quad (6\text{-}68)$$

$$\lambda_o = \frac{\sigma_b\alpha(1 - K)}{2\beta \tan(w)} \quad (6\text{-}69)$$

If the boundary shear stress λ_o is less than λ_b then the value of K is defined by Eq. (6-67) and peel failure is by the cleavage. Conversely if λ_o in Eq. (6-68) exceeds the shear stress λ_b the peel model adopts a new criteria as follows:

$$\lambda_x = \lambda_o \exp(\alpha x) \quad (6\text{-}70)$$

where now we set:

$$\lambda_o = \lambda_b \tag{6-71}$$

and the boundary cleavage stress is defined as follows:

$$\sigma_o = \frac{2\lambda_b\beta \tan (w)}{\alpha (1 - K)} \tag{6-72}$$

and two test conditions are defined which are:

$$N = K \tag{6-73}$$

$$T = \frac{\sigma_o * AA}{2\beta\sigma_b R_\pi} \tag{6-74}$$

if T = 0 then:

$$K = 1.0 \tag{6-75}$$

if T < 0 then

$$K = \frac{1 - (1 - 4T)^{1/2}}{2T} \tag{6-76}$$

if T > 0 then

$$K = \frac{(1 + 4T)^{1/2} - 1}{2T} \tag{6-77}$$

and the difference D is defined as the absolute difference

$$D = |K - N| \tag{6-78}$$

in Eq. (6-78) if D => 0.0005 then the current value of K is returned to Eq. (6-72) to calculate a new σ_o and to Eq. (6-73) to reset N. Equations (6-74) through (6-78) are solved until D < 0.0005 to provide resolved values of λ_o and K.

The normal P_N and tangential P_T components of the peel force P are defined as follows:

$$P_N = \frac{\sigma_o b (1 - K)}{2\beta} \tag{6-79}$$

$$P_T = \frac{\lambda_o b}{\alpha} \tag{6-80}$$

$$P = (P_N{}^2 + P_T{}^2)^{1/2} \tag{6-81}$$

The specific work of peel per unit area unbonded is defined by the work of bonding translation $(W/A)_B$ and the work of deformation of the flexible member $(W/A)_D$ as follows:

$$(W/A)_B = (P/b)(1 - \cos(w)) \qquad (6\text{-}82)$$

$$(W/A)_D = \frac{(P/b)^2}{4hE} \qquad (6\text{-}83)$$

$$(W/A)_T = (W_B + W_D)/A \qquad (6\text{-}84)$$

where $(W/A)_T$ is the total work of peel. After solving the above relations the value of the peel angle is reduced and the calculation returns to Eq. (6-65) to provide a new set of values for the peeling process.

The above peel model is developed in the computer program Listing No. 6 in the Appendix. This peel mechanics model accepts any self consistent physical units in order to extend the utility of the calculation. In Table 6-22 is listed comparative inputs and outputs in standard English and SI units while Table 6-23 provides the complete input and output report for this sample calculation.

The upper part of Table 6-22 defines the eight input variables and their assigned values in English units. As shown in upper Table 6-23 the model reports the cleavage and shear stress concentrations, the adherend radius of curvature R at peel angle 180 deg, and the peeling force at w = 180 deg and in shear where w = 0 deg. The input variables of Table 6-22 are selected to illustrate the important case where peeling is dominated by cleavage failure at high peel angles, $w > 28$ deg, and displays shear failure at low peel angles where $w < 23$ deg.

The lower portion of Table 6-23 provides a table of peel angle versus peel work (W_T/A), peel force P, cleavage stress coefficient K, and the boundary cleavage stress σ_0 and shear stress λ_0. As indicated in the lower Table 6-23 when the tensile stress $\sigma_0 = \sigma_b = 2E4$ psi failure is by the cleavage mechanism. At low peel angles the shear stress rises to its maximum value $\lambda_0 = \lambda_b$ and failure is by the shear mechanism with tensile stress values $\sigma_0 < \sigma_b$. The computations of Table 6-23 provide the lower of the three peel force P versus peel angle w curves in Fig. 6-23.

In Table 6-24 a second estimate of laminate peel properties is calculated in which the flexible adherend (ribbon) modulus is increased by a factor of ten to become E = 1E5 psi. The effect of this change in E is shown to

TABLE 6-22 Relations Between English and SI Units in the Peel Mechanics Model

Input Variable	English Units	SI Units
H, A	1E-3 in., 8E-3 in.	2.54E-5m, 2.03E-4m
B, E	1.0 in., 1E4 psi	2.54E-2m, 6.89E7 N/m^2
Y, SA	5E4 psi, 2E4 psi	3.45E8m, 1.38E8N/m^2
G, LA	1.67E4 psi, 6.67E3 psi	1.15E8N/m^2, 4.60E7 N/m^2
Output (1)		
Cleavage Stress Conc.	(in.$^{-1}$) = 6.95E2	(m^{-1}) = 2.74E4
Shear Stress Conc.	(in.$^{-1}$) = 3.23E2	(m^{-1}) = 1.27E4
180 Deg. Radius of Curv.	(in.) = 3.22E-4	(m) = 8.209E-6
0 Deg. Peel Force	(lb) = 20.6	(N) = 91.8
180 Deg. Peel Force	(lb) = 16.0	(N) = 71.2
Output (2)		
Peel Angle	(deg) = 177.9	(deg) = 177.9
Peel Work	(lb/in.) = 35.2	(N/m) = 6.16E3
Peel Force	(lb) = 14.8	(N) = 65.8
K	.963	.963
Tensile Stress	(psi) = 2E4	(N/m^2) = 1.38E8
Shear Stress	(psi) = -4.79E3	(N/m^2) = -3.30E7

TABLE 6-23 First Estimate of Laminate Peel and Shear Properties (Flexible Adherend Tensile Modulus = 1E4 psi, English Units)

```
RIBBON HALF THICKNESS(H), ADHESIVE THICKNESS(A)? .001,.008
BOND WIDTH(B), RIBBON   TENSILE MODULUS(E)? 1,1E4
ADHESIVE TENSILE MODULUS(Y),STRENGTH(SA)? 5E4,2E4
ADHESIVE SHEAR MODULUS(G),STRENGTH(LA)? 1.67E4,6.67E3
CLEAVAGE STRESS CONC.(BA)= 695.789
SHEAR STRESS CONC.(GA)= 323.071
180 DEG. RAD. OF CURV.(R)= 3.22749E-04
0 DEG. PEEL FORCE(PS)= 20.6456
180 DEG. PEEL FORCE(PC)= 16
```

PEEL ANGLE	PEEL WORK	PEEL FORCE	K	TENSILE STRESS	SHEAR STRESS
177.943	35.1768	14.8403	.962936	20000	-4791.38
147.943	17.0924	8.31579	.692902	20000	-2276.95
117.944	11.5577	7.82883	.567962	20000	-1064.12
87.9443	8.67872	7.53096	.476341	20000	87.2691
77.9462	8.06533	8.11386	.447893	20000	547.412
67.9481	7.65196	9.00554	.419243	20000	1092.33
57.95	7.53574	10.35	.389619	20000	1774.4
47.9519	7.96186	12.424	.358076	20000	2688.27
37.9538	9.59627	15.8136	.323288	20000	4028.42
32.9542	11.4109	18.3874	.304058	20000	4984.66
27.9546	14.6422	21.9796	.283097	20000	6272.44
22.955	14.3437	22.4215	.279083	16881.5	6670
17.9553	12.8328	21.7033	.280759	12948.7	6670
12.9557	11.7599	21.1855	.281979	9210.26	6670
7.9561	11.0648	20.8462	.282874	5598.38	6670

TABLE 6-24 Second Estimate of Laminate Peel and Shear Properties (Flexible Adherend Tensile Modulus = 5E4 psi, English Units)

```
RIBBON HALF THICKNESS(H), ADHESIVE THICKNESS(A)?  001, .308
BOND WIDTH(B), RIBBON   TENSILE MODULUS(E)? 1,5E4
ADHESIVE TENSILE MODULUS(Y),STRENGTH(SA)? 5E4,3E4
ADHESIVE SHEAR MODULUS(G),STRENGTH(LA)? 1.67E4,6.57E3
CLEAVAGE STRESS CONC.(BA)= 465.303
SHEAR STRESS CONC.(GA)= 144.482
180 DEG. RAD. OF CURV.(R)= 7.21688E-04
0 DEG. PEEL FORCE(PS)= 46.165
180 DEG. PEEL FORCE(PC)= 16
```

PEEL ANGLE	PEEL WORK	PEEL FORCE	K	TENSILE STRESS	SHEAR STRESS
177.943	31.5538	15.2039	.974606	20000	-2195.27
147.943	18.7644	9.89172	.755712	20000	-1211.26
117.944	13.3668	8.83588	.6368	20000	-598.236
87.9443	9.94176	9.81231	.543725	20000	50.8506
77.9462	9.02199	10.6822	.513914	20000	322.301
67.9481	8.19742	11.9768	.483483	20000	649.682
57.95	7.49303	13.9051	.451604	20000	1066.11
47.9519	6.99257	16.8666	.417214	20000	1632.14
37.9538	6.94726	21.7079	.378777	20000	2473.07
32.9542	7.30912	25.3919	.357368	20000	3078.38
27.9546	8.22931	30.5457	.333734	20000	3898.35
22.955	10.307	38.1692	.307336	20000	5078.06
17.9553	14.1381	48.5277	.280889	19354.7	6670
12.9557	12.4262	47.3715	.282006	13769.4	6670
7.9561	11.3131	46.6142	.282796	8377	6670

decrease both the cleavage and shear stress concentrations from the first estimate of Table 6-23. The adherend radius of curvature R at w = 180 deg is increased as is the peel force under shear failure at w = 0 deg. The calculation displays an initially surprising result that the peel force at w = 180 deg is the same for the first and second estimates of Table 6-23 and Table 6-24. The lower portion of Table 6-24 and the middle curve of Fig. 6-23 shows in detail the complex changes in peel properties produced by the modulus increase of the flexible adherend.

A third estimate of peel properties shown in Table 6-25 describes a further increase in flexible adherend modulus to E = 1E6 psi with other factors held constant. This single change in one input variable creates a group of changes in calculated peel respones yet leaves the peel force P at w = 180 deg unchanged as shown by the upper curve of Fig. 6-23. This group of three estimates in peel force versus peel angle summarized in the curves of Fig. 6-23 demonstrates the unique advantage of a computerized model in rapidly tracing the complex effects on laminate peel performance produced by a single input variable. The reason why peel force at peel angle w = 180 degree does not change is

TABLE 6-25 Third Estimate of Laminate Peel and Shear Properties (Flexible Adherend Tensile Modulus = 2.5E5 psi, English Units)

```
RIBBON HALF THICKNESS(H), ADHESIVE THICKNESS(A)? .001, .003
BOND WIDTH(B), RIBBON   TENSILE MODULUS(E)? 1,2.5E5
ADHESIVE TENSILE MODULUS(Y),STRENGTH(SA)? 5E4,2E4
ADHESIVE SHEAR MODULUS(G),STRENGTH(LA)? 1.67E4,6.67E3
CLEAVAGE STRESS CONC.(BA)= 311.166
SHEAR STRESS CONC.(GA)= 64.6142
180 DEG. RAD  OF CURV.(R)= 1.61374E-03
0 DEG. PEEL FORCE(PS)= 103.228
180 DEG. PEEL FORCE(PC)= 16
```

PEEL ANGLE	PEEL WORK	PEEL FORCE	K	TENSILE STRESS	SHEAR STRESS
177.943	31.1472	15.4591	.982733	20000	-998.235
147.943	21.2076	11.4085	.811585	20000	-624.753
117.944	15.3552	10.7349	.703537	20000	-326.554
87.9443	12.1653	12.4569	.612631	20000	28.8703
77.9462	11.0442	13.7214	.582451	20000	185.146
67.9481	9.96251	15.5635	.551143	20000	377.556
57.95	8.91483	18.2822	.517826	20000	626.859
47.9519	7.91709	22.4474	.481313	20000	971.426
37.9538	7.04729	29.2705	.439834	20000	1491.29
32.9542	6.73657	34.4802	.416371	20000	1869.45
27.9546	6.62322	41.7936	.390373	20000	2385.37
22.955	6.94244	52.6562	.360978	20000	3132.91
17.9553	8.34284	70.178	.326816	20000	4313.65
12.9557	13.0984	102.425	.285449	20000	6449.67
7.9561	11.8676	104.232	.282823	12523.8	6670

clarified by Eq. (6-63) which shows that the flexible member modulus E does not appear as a variable parameter.

While the tensile modulus of the flexible member does not influence the estimate of peel force at peel angle w = 180 deg the curves of Fig. 6-23 show this variable has an important influence at all lower peel angles. At low peel angles where shear failure predominates the maximum effect of variable E is displayed by the wide separation of peel forces in the curves of Fig. 6-23.

The peel mechanics model can be utilized to simulate a variety of laminate structures. For example, the flexible member can represent a fiber reinforced composite. The adhesive layer can represent an uncured epoxy or an epoxy resin at various stages of cure as discussed in Chapter 5. The polymeric constituents of either the flexible member or adhesive layer can be defined by the viscoelastic or analog elastic-plastic properties defined by the polymer chemistry to mechanical properties model of Chapter 6-3.

A recent summary review and rederivation of the peel mechanics model presented here is provided by C.A. Mylonas[53] and extensive experimental

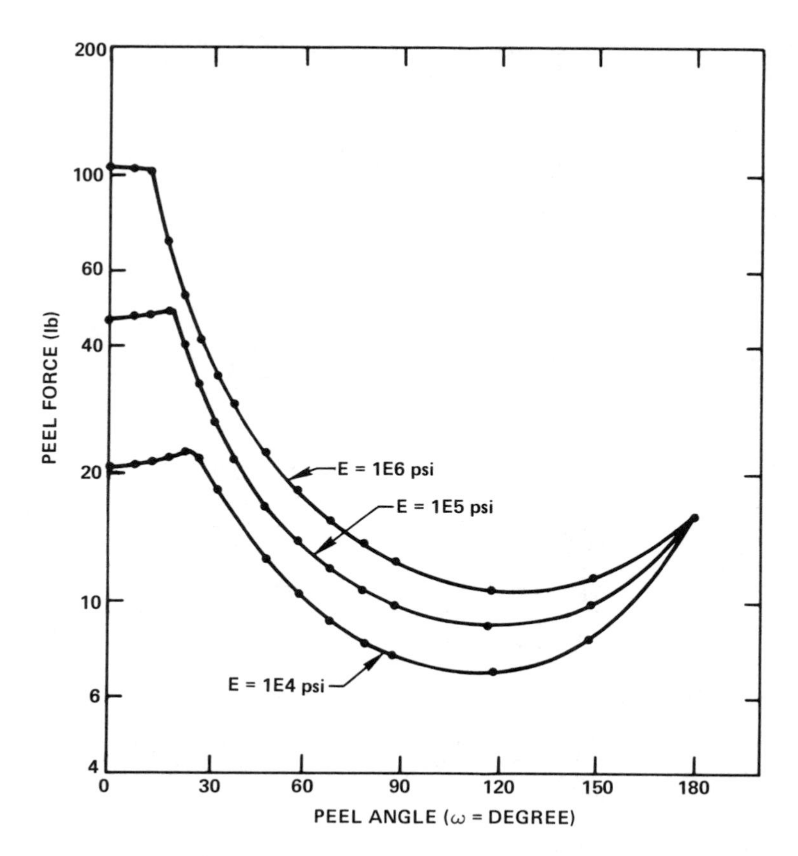

FIG. 6-23 Calculated curves of peel force P vs peel angle W for three values of flexible adherend tensile modulus E.

verification provided in papers by Kaelble.(1-4) Recent theory and analysis of peel mechanics is also provided in other recent papers which discuss this important and interesting subject.(54-57)

References

1. D.H. Kaelble, "Physical Chemistry of Adhesion," Wiley-Interscience, New York, (1971), Chaps. 7-12.

2. D.H. Kaelble, "Theory and Analysis of Fracture Energy in Fiber Reinforced Composites," J. Adhesion, 5, (1973), p. 245.

3. D.H. Kaelble, "Theory and Analysis of Peel Adhesion: Mechanisms and Mechanics," Trans. Soc. Rheol., 3, (1959), p. 161.

4. D.H. Kaelble, "Theory and Analysis of Peel Adhesion: Bond Stresses and Distributions," Ibid., 4, (1960), p. 45.

5. D.H. Kaelble, "Theory and Analysis of Peel Ahesion," Adhesion and Cohesion (Editor, P. Weiss), Elsevier, Amsterdam, (1962), p. 75.

6. J. Delmonte, "Technology of Carbon and Graphite Fiber Composites," Van Nostrand, New York, (1981).

7. D.H. Kaelble and E.H. Cirlin, J. Poly. Sci., Part C, 35, (1971), p. 79.

8. R.F. Boyer, Rubber Chem. & Technol. 36 (1963), p. 1303; J. Poly Sci.: Symposium No. 50, (1975), p. 189.

9. N.G. McCrum, B.E. Read, and G. Williams, Anelastic and Dielectric Effects in Polymeric Solids, Wiley, New York (1967).

10. I.M. Ward, "Mechanical Properties of Solid Polymers," Wiley-Interscience, New York, (1971), pp. 35-40.

11. D.S. Dugdale, J. Mech. Phys. Solids, 8, (1960) p. 100.

12. L.J. Hart-Smith, "Single-Bonded Single-Lap Joints," NASA-Langeley Report CR-112236, January 1973.

13. D.H. Kaelble and E.H. Cirlin, "Chemo-Rheology of Curing of Structural Adhesives II: Polyimide Systems," J. Poly, Sci., Part C, 35, (1971), p. 101.

14. D.H. Kaelble, "Adsorption and Interdiffusion Mechanisms of Bonding in Polymer Composites," Proc. 23rd Int. Cong. on Pure and Applied Chem., 8, Butterworths, London (1971) pp. 265-302.

15. P.J. Dynes and D.H. Kaelble, "Surface Energy Analysis of Carbon Fibers and Films," J. Adhesion, 6, (1974) p. 195.

16. D.H. Kaelble, P.J. Dynes, and L. Maus, "Surface Energy Analysis of Treated Graphite Fibers," Ibid., 6, (1974), p. 239.

17. D.H. Kaelble, "Relationship Between Fracture Mechanics and Surface Energetic Failure Criteria," J. Appl. Poly. Sci., 18, (1974), p. 1869.

18. D.H. Kaelble, "Surface Energetics Criteria of Bonding and Fracture," SAMPE Quarterly, 7(3), (1976) p. 30.

19. D.H. Kaelble, "Interface Degradation Processes and Durability," Poly Eng. and Sci., 17(7), (1977), p. 474.

20. D.H. Kaelble and J. Moacanin, "A Surface Energy Analysis of Bioadhesion," Polymer, 18(5), (1977), p. 475.

21. D.H. Kaelble, P.J. Dynes and E.A. Cirlin, "Interfacial Bonding and Environmental Stability in Polymer Matrix Composites," J. Adhesion, 6, (1974), p. 23.

22. D.H. Kaelble, P.J. Dynes, L.W. Crane and L. Maus, "Interfacial Mechanisms of Moisture Degradation in Graphite-Epoxy Composites," Ibid., 7, (1974), p. 25.

23. D.H. Kaelble, P.J. Dynes, L.W. Crane and L. Maus, "Kinetics of Environmental Degradation in Graphite-Epoxy Laminates," ASTM Spec. Tech. Pub. ASTM STP 580, (1975), pp. 247-262.

24. D.H. Kaelble, P.J. Dynes, and L. Maus, "Hydrothermal Aging of Composite Materials Part 1: Interfacial Aspects," J. Adhesion, 8, (1976), p. 121.

25. D.H. Kaelble and P.J. Dynes, "Hydrothermal Aging of Composite Materials Part II: Matrix Aspects," Ibid., 8, (1976), p. 195.

26. D.H. Kaelble and P.J. Dynes, "Nondestructive Test for Shear Strength Degradation of a Graphite-Epoxy Composite," ASTM Spec. Tech. Pub. STP 617, (1977), pp. 190-200.

27. D.H. Kaelble and P.J. Dynes, "Methods for Detecting Moisture Degradation in Graphite-Epoxy Composites," Materials Evaluation, 34(4), (1977), p. 103.

28. C.L. Leung and D.H. Kaelble, "Moisture Diffusion and Microdamage in Composites," Resins for Aerospace, (Ed. C.A. May), ACS Symposium Series No. 132, (1980), pp. 419-434.

29. D.H. Kaelble, "Experimental Analysis of Hydrothermal Aging in Fiber Reinforced Composites," Ibid., (1980), pp. 385-417.

30. C.L. Leung and D.H. Kaelble, "Moisture Diffusion Analysis of Composite Microdamage," Advanced Composites and Applications, U.S. Nat. Bu. Stds. Spec. Tech. Pub. No. 563, (1979), pp. 32-39.

31. C.L. Leung, P.J. Dynes, and D.H. Kaelble, "Moisture Diffusion Analysis of Microstructure Degradation in Graphite-Epoxy Composites," ASTM Spec. Tech. Pub. STP 696 (1979), pp. 298-315.

32. P.J. Dynes and D.H. Kaelble, "Physiochemical Analysis of Graphite-Epoxy Composite Systems, ASTM Spec. Tech. Pub. STP 674, (1979), p. 577.

33. D.H. Kaelble and P.J. Dynes, "Preventative Nondestructive Evaluation (PNDE) of Graphite Epoxy Composites," Ceramic Eng. and Sci. Proc., 1(7-8A), (1980), pp. 458-472.

34. Proc. Symposium on Chemorheology of Thermosetting Resin, ACS Org. Coatings and App. Poly. Sci. Proc., 47 (1982).

35. C.E. Browning and J.T. Hartness, "Effects of Moisture on Properties of High Performance Structural Resins and Composites," ASTM Spec. Tech. Pub. No. STP 546, (1974), p. 284.

36. R. DeIasi and J.B. Whiteside, "Effect of Moisture on Epoxy Resins and Composites," ASTM Spec. Tech. Pub. STP No. 658, (1978), p. 2.

37. R.F. Fedors, J. Poly. Sci., Polymer Letters Ed. 17, (1979), p. 719.

38. D.H. Kaelble, Physical Chemistry of Adhesion, Wiley-Interscience, New York (1971), Chap. 3

39. R.M. Jones, Mechanics of Composite Materials, McGraw-Hill, New York, (1975).

40. A.H. Cottrell, Proc. Roy. Soc., A282, (1964), p. 2.

41. G.A. Cooper and A. Kelly, J. Mech. and Phy. of Solids, 15, (1967), p. 279.

42. G.A. Cooper and A. Kelly, "Interfaces in Composites, " ASTM Spec. Tech. Pub. STP 452, (1969), pp. 90-106.

43. A. Kelly, Proc. Roy. Soc., A319, (1970), p. 95.

44. J.O. Outwater and J.C. Murphy, 26th Annual Tech. Conf., Reinforced Plastics/Composites Div., Soc. of Plastics Industry, (1969), Paper 11-C.

45. L.B. Greszczuk, "Interfaces in Composites," ASTM Spec. Tech. Pub. STP 452, (1969), pp. 42-58.

46. P. Lawrence, J. Matls. Sci., 7, (1972), p. 1.

47. T.H. Lin, D. Salinas, and Y.M. Ito, Ibid. 6, (1972), p. 48.

48. J.K. Wells and P.W.R. Beaumont, J. Matls. Sci., 17 (1982), p. 397.

49. J.K. Wells and P.W.R. Beaumont, Scripta Metallurgica, 16, (1982), p. 99.

50. R.J. Nuismer and J.M. Whitney, "Fracture Mechanics of Composites," ASTM Spec. Tech. Pub. STP 593, (1975), pp. 117-142.

51. P.E. Mahon and L. Ying, "Effects of Fiber-Matrix Interaction on the Properties of Graphite-Epoxy Composites," NASA, Contract Report 3607, Contract NAS1-15749 September 1982.

52. D.H. Kaelble and C.L. Ho, "Biaxial Bond Stress Analysis in Peeling," Trans. Soc. Rheol., 18, (1974), p. 219.

53. C.A. Mylonas, "Proc. 4th Int. Cong. on Rheol.," Part 2, Interscience, New York, (1965), p. 423.

54. W.T. Chen and T.F. Flavin, "Mechanics of Film Adhesion: Elastic and Elastic-Plastic Behavior," IBM J. Res. and Dev., 16, (1972), p. 203.

55. S. Yamamoto, M. Hayashi, T. Inoue, "Viscoelastic Analysis on Peel Adhesion of Adhesive Tape by Matrix Method," J. Appl. Poly. Sci., 19, (1975), p. 2107

56. T. Igarashi, "Fracture Criteria on Peeling, in Adhesion & Adsorption of Polymers," (Ed. L-H. Lee), Polymer Science & Technology, 12B, Plenum Press, New York (1979), p. 421.

57. T. Tsuji, M. Mauoka, and K. Nakao, "Superposition of Peel Rate, Temperature and Molecular Weight for T-Peel Strength of Polyisobutylene," Ibid., p. 439.

Appendix

LISTING 1 Atomic to Molecular Properties Model

```
1990 CLS
2000 PRINT"ELEMENTAL BONDING PART 1-D.H.KAELBLE NOV.20,1981"
2010 DIM A$(69):DIM B(69,7)
2020 FOR R=0 TO 68
2030 READ A$(R)
2040 NEXT R
2050 DATA H,LI,BE,B,C,N,O,F,NA,MG,AL
2060 DATA SI,P,S,CL,K,CA,SC,TI,V,CR,MN
2070 DATA FE,CO,NI,CU,ZN,GA,GE,AS,SE,BR
2080 DATA RB,SR,Y,ZR,NB,MO,TC,RU,RH,PD,AG
2090 DATA CD,IN,SN,SB,TE,I,CS,BA,LA,HF,TA
2100 DATA W,RE,OS,IR,PT,AU,HG,TL,PB,BI
2110 DATA TH,U,PU,(N2)/2,(O2)/2
2120 FOR S=0 TO 68
2130 FOR T=0 TO 6
2140 READ B(S,T)
2150 NEXT T,S
2160 DATA 1,1.008,4.35,2.20,0.32,1,7
2170 DATA 3,6.941,1.11,0.98,1.23,1,12
2180 DATA 4,9.012,2.28,1.57,0.90,2,7
2190 DATA 5,10.81,2.53,2.04,0.82,3,5
2200 DATA 6,12.01,3.48,2.55,0.77,4,5
2210 DATA 7,14.01,1.61,3.04,0.75,3,2
2220 DATA 8,16.00,1.39,3.44,0.73,2,2
2230 DATA 9,19.00,1.53,3.98,0.72,1,2
2240 DATA 11,22.99,0.753,0.93,1.54,1,12
2250 DATA 12,24.31,0.971,1.31,1.36,2,12
2260 DATA 13,26.98,2.06,1.61,1.18,3,7
2280 DATA 14,28.09,1.77,1.90,1.11,4,7
2290 DATA 15,30.97,2.15,2.19,1.06,5,5
2300 DATA 16,32.06,2.13,2.58,1.02,6,2
2310 DATA 17,35.45,2.43,3.16,0.99,1,2
2320 DATA 19,39.09,0.552,0.82,2.03,1,12
2330 DATA 20,40.08,1.15,1.00,1.74,2,12
2340 DATA 21,44.96,2.58,1.36,1.44,3,9
2350 DATA 22,47.90,2.64,1.54,1.32,4,7
2360 DATA 23,50.94,3.36,1.63,1.22,5,7
2370 DATA 24,52.00,2.38,1.66,1.18,3,2
2380 DATA 25,54.94,1.43,1.55,1.17,2,2
2390 DATA 26,55.85,2.03,1.83,1.17,3,7
2400 DATA 27,58.93,2.20,1.88,1.16,2,7
2410 DATA 28,58.70,2.12,1.91,1.15,2,9
2420 DATA 29,63.55,1.72,1.90,1.17,2,9
2430 DATA 30,65.38,0.653,1.65,1.25,2,7
2440 DATA 31,69.72,1.36,1.81,1.26,3,7
2450 DATA 32,72.59,1.57,2.01,1.22,4,7
2460 DATA 33,74.92,1.34,2.18,1.20,3,5
2470 DATA 34,78.96,1.84,2.55,1.16,4,2
2480 DATA 35,79.90,1.93,2.96,1.14,1,2
2490 DATA 37,85.47,0.519,0.82,2.16,1,12
2500 DATA 38,87.62,1.05,0.95,1.91,2,12
2510 DATA 39,88.91,2.74,1.22,1.62,3,9
2520 DATA 40,91.22,3.45,1.33,1.45,4,7
2530 DATA 41,92.91,4.85,1.60,1.34,5,5
2540 DATA 42,95.94,4.30,2.16,1.30,6,2
2550 DATA 43,98,3.35,1.90,1.27,7,2
2560 DATA 44,101.07,3.35,2.20,1.25,3,5
2570 DATA 45,102.91,3.24,2.28,1.25,3,7
2580 DATA 46,106.4,1.93,2.20,1.28,2,9
2590 DATA 47,107.87,1.44,1.93,1.34,1,7
2600 DATA 48,112.41,0.552,1.69,1.48,2,9
2610 DATA 49,114.82,1.18,1.78,1.44,3,7
2620 DATA 50,118.69,1.43,1.96,1.41,4,7
2630 DATA 51,121.75,1.26,2.05,1.40,3,5
2640 DATA 52,127.60,1.38,2.10,1.36,4,5
2650 DATA 53,126.90,1.51,2.66,1.33,1,2
2660 DATA 55,132.91,0.448,0.79,2.35,1,12
2670 DATA 56,137.33,1.12,0.89,1.98,2,12
```

```
2680 DATA 57,138.91,2.48,1.10,1.69,3,9
2690 DATA 72,178.49,4.72,1.30,1.44,4,7
2700 DATA 73,180.95,5.56,1.50,1.34,5,5
2710 DATA 74,183.85,5.61,2.36,1.30,6,5
2720 DATA 75,186.21,3.97,1.90,1.28,7,5
2730 DATA 76,190.2,3.64,2.20,1.26,4,5
2740 DATA 77,192.22,3.48,2.20,1.27,4,9
2750 DATA 78,195.09,2.79,2.28,1.30,4,9
2760 DATA 79,196.97,1.86,2.54,1.34,3,7
2770 DATA 80,200.59,0.301,2.00,1.49,2,9
2780 DATA 81,204.37,0.866,2.04,1.48,1,9
2790 DATA 82,207.2,0.992,2.33,1.47,2,7
2800 DATA 83,209.0,1.03,2.02,1.46,3,5
2810 DATA 90,232.04,3.42,1.30,1.65,4,9
2820 DATA 92,238.03,3.56,1.38,1.42,6,7
2830 DATA 94,244,2.29,1.28,1.21,4,7
2840 DATA 7,14.01,4.73,3.04,0.55,3,2
2850 DATA 8,16.00,2.01,3.44,0.62,2,2
2860 CLS
3400 ME=0:NB=0
3410 INPUT "HOW MANY ELEMENTS";ME
3420 DIM W(ME,2)
3430 FOR I=0 TO ME-1
3440 INPUT "ELEMENT CODE NO.=";W(I,0)
3450 INPUT "MOLES OF ELEMENT=";W(I,1)
3460 NEXT I
3470 INPUT "NUMBER OF CHEMICAL BOND TYPES=";NB
3480 DIM R(NB,3)
3490 FOR J=0 TO NB-1
3500 INPUT"FOR A-B BOND, ELEMENT A CODE NO.=";R(J,0)
3510 INPUT "ELEMENT B CODE NO.=";R(J,1)
3520 INPUT "MOLES OF A-B BONDS=";R(J,2)
3530 NEXT J
3540 C=0
3545 CLS
3550 PRINT "ELEMENTARY PROPERTIES:"
3560 PRINT "Z, SY, W, D/1E5, X, R/1E-10, V, PH ="
3570 FOR RZ=0 TO ME-1: C=W(RZ,0)-1
3580 PRINT B(C,0);A$(C);B(C,1);B(C,2);B(C,3);B(C,4);B(C,5);B(C,6)
3590 NEXT RZ
3600 PRINT "TO CONTINUE PRESS ENTER": INPUT D$
3610 PRINT "CHEMICAL ANALYSIS:":PRINT
3620 PRINT "BONDING";TAB(16)"BOND";TAB(26)"% IONIC";TAB(36)"BOND";TAB(46)"MOLES"
3630 PRINT"ELEMENTS";TAB(16)"ENERGY";TAB(26)"ENERGY";TAB(36)"LENGTH"
3640 PRINT "A";TAB(6)"B";TAB(16)"(J/MOLE)";TAB(36)"(M*1E-10)"
3650 DA=0:DB=0:M=0:F=0:G=0:H=0:BL=0:VN=0:GA=0:NA=0:NT=0
3660 FOR RY=0 TO NB-1: DA=R(RY,0)-1: DB=R(RY,1)-1: M=R(RY,2)
3670 F=9.65E4*((B(DA,3)-B(DB,3))[2)
3680 G=5.0E4*(B(DA,2)+B(DB,2))+F
3690 H=100*F/G
3700 BL=(B(DA,4)+B(DB,4))-0.09*ABS(B(DA,3)-B(DB,3))
3710 VN=VN+M*BL*BL*BL
3712 GA=GA+(G*M): NA=NA+M
3714 PRINT  A$(DA);TAB(6) A$(DB);TAB(16) G;TAB(26) H;TAB(36) BL;TAB(46) M
3715 NEXT RY
3720 PRINT "TOTAL";TAB(16) GA;TAB(46) NA
3740 PRINT "TO CONTINUE PRESS ENTER":INPUT D$
3750 PRINT "PHYSICAL ANALYSIS:"
3760 PRINT"ELEMENTS","MOLES"
3770 Q=0:V=0:W=0
3780 FOR P=0 TO ME-1: Q=W(P,0)-1: V=W(P,1)
3800 PRINT A$(Q),V
3810 W=W+((B(Q,1))*V):NT=NT+V
3820 NEXT P
3830 SV=4.26E-1*NT*(VN/(NA*W))
3840 PRINT "MOLECULAR WT. (KG/MOLE)=";W/1000
3850 PRINT TAB(15)"(G/MOLE)=";W
3860 PRINT "SPECIFIC VOLUME (CC/G) :"
3870 PRINT"(Z=12)","(Z=8)","(Z=6)","(Z=4)"
3880 PRINT SV,1.088*SV,1.414*SV,2.18*SV
3890 PRINT"TO CONTINUE TYPE RUN AND PRESS ENTER"
```

LISTING 2 Polymer Chemistry to Physical Properties Model

```
5 CLS
10 PRINT"MONOMER-POLYMER PREDICTION PART-1,D.H.KAELBLE MAY,81"
13 DIM A$(33,2): DIM B(33,5)
16 FOR R=0 TO 32
19 FOR C=0 TO 1
22 READ A$(R,C)
25 NEXT C,R
31 DATA -CH2-,ETHYLENE
32 DATA -CH(CH3)-,PROPYLENE
33 DATA -C((CH3)2)-,ISOBUTYLENE
34 DATA -CH(C6H5)-,STYRENE
35 DATA -P-C6H4-,TEREPHTHALATE
36 DATA -M-C6H4-,ISOPHTHALATE
37 DATA -C(CH3)CH-,ISOPRENE
38 DATA -CHCH-,1-4-BUTADIENE
39 DATA -CH(CHCH2)-,1-2-BUTADIENE
40 DATA -CH(C6H11)-,VINYL CYCLOHEXANE
41 DATA -CH(C(O)OCH3)-,METHYLACRYLATE
42 DATA -C(CH3)(C(O)OCH3)-,METHYL METHACRYLATE
43 DATA -CH(CH3)O-,PROPYLENE OXIDE
44 DATA -C(O)O-,ETHYLENE ADIPATE
45 DATA -CH(OC(O)CH3)-,VINYL ACETATE
46 DATA -C(O)-,KETONE
47 DATA -CH(C(O)OH)-,ACRYLIC ACID
48 DATA -CH(OH)-,VINYL ALCOHOL
49 DATA -CH(OCH(O))-,VINYL FORMATE
50 DATA -O-,ETHER
51 DATA -NHC(O)-,AMIDE
52 DATA -NHC(O)O-,URETHANE
53 DATA -CH(CN)-,ACRYLONITRILE
54 DATA -CH(CL)-,VINYL CHLORIDE
55 DATA -C(CL)CH-,CHLOROPRENE
56 DATA -C(CL2)-,VINYLIDENE CHLORIDE
57 DATA -CF2-,TETRAFLUOROETHLENE
58 DATA -CH2CF2-, VINYLIDENE FLUORIDE
59 DATA -CF(CF3)-,PERFLUOROPROPYLENE
60 DATA -SI((CH3)2)O-,DIMETHYLSILOXANE
61 DATA -N((C(O))2)C6H2((C(O))2)N-,IMIDE
62 DATA -S-,SULFIDE
63 DATA -S((O)2)-,SULFONE
164 FOR R1=0 TO 32
167 FOR C1=0 TO 4
170 READ B(R1,C1)
173 NEXT C1,R1
181 DATA 4.14E3,8,2.22E-5,1.4E-2,1
182 DATA 1.28E4,11,4.44E-5,2.8E-2,1
183 DATA 1.19E4,14,6.66E-5,4.2E-2,1
184 DATA 3.01E4,15,1.11E-4,9.0E-2,1
185 DATA 2.38E4,5,8.86E-5,7.6E-2,4
186 DATA 2.58E4,10,8.86E-5,7.6E-2,3
187 DATA 1.15E4,11,5.92E-5,4.0E-2,2
188 DATA 7.49E3,8,3.70E-5,2.6E-2,2
189 DATA 1.29E4,11,5.90E-5,4.0E-2,1
190 DATA 2.56E4,21,1.48E-4,9.6E-2,1
191 DATA 2.81E4,23,7.57E-5,7.2E-2,1
192 DATA 4.6E4,26,9.79E-5,8.6E-2,1
193 DATA 1.39E4,17,5.54E-5,4.4E-2,2
194 DATA 1.41E4,12,3.32E-5,4.4E-2,2
195 DATA 3.17E4,23,7.57E-5,7.2E-2,1
196 DATA 7.32E3,6,2.22E-5,2.8E-2,1
197 DATA 3.51E4,20,5.64E-5,5.8E-2,1
198 DATA 2.66E4,14,4.90E-5,3.0E-2,1
199 DATA 2.86E4,20,5.54E-5,5.8E-2,1
200 DATA 6.82E3,6,1.06E-5,1.6E-2,1
201 DATA 4.44E4,13,3.79E-5,4.3E-2,2
202 DATA 2.63E4,19,4.89E-5,5.9E-2,3
203 DATA 2.41E4,8,4.89E-5,3.9E-2,1
204 DATA 1.75E4,8,4.07E-5,4.85E-2,1
205 DATA 1.26E4,8,5.55E-5,4.85E-2,2
```

```
206 DATA 1.13E4,8,5.92E-5,8.30E-2,1
207 DATA 4.81E3,8,3.48E-5,5.0E-2,1
208 DATA 1.48E4,16,5.70E-5,6.4E-2,2
209 DATA 1.84E4,16,6.96E-5,1.0E-1,1
210 DATA 1.72E4,30,8.62E-5,7.4E-2,2
211 DATA 1.10E5,62,2.01E-4,2.14E-1,7
212 DATA 8.26E3,8,2.65E-5,3.20E-2,1
213 DATA 4.54E4,23,4.04E-5,6.40E-2,1
230 MU=0:SU=0
520 INPUT "HOW MANY MAIN CHAIN UNITS?";MU
530 DIM W(MU,2)
540 FOR I=0 TO MU-1
550 INPUT "STRUCTURE UNIT NO.=";W(I,0)
560 INPUT "MOLES OF STRUCTURE UNIT=";W(I,1)
570 NEXT I
580 INPUT "HOW MANY SIDE GROUPS? (NONE=0)";SU
581 IF SU<1 GOTO 640
590 DIM U(SU,2)
600 FOR J=0 TO SU-1
610 INPUT "STRUCTURE UNIT NO.=";U(J,0)
620 INPUT "MOLES OF STRUCTURE UNIT=";U(J,1)
630 NEXT J
640 V=0:M=0:U=0:H=0:N=0
650 FOR RR=0 TO MU-1:C=W(RR,0)-1:D=W(RR,1)
660 U=U+B(C,0)*D
670 H=H+B(C,1)*D
680 V=V+B(C,2)*D
690 M=M+B(C,3)*D
700 N=N+B(C,4)*D
710 NEXT RR
711 IF SU<1 GOTO770
720 FOR SS=0 TO SU-1:E=U(SS,0)-1:F=U(SS,1)
730 U=U+B(E,0)*F
740 H=H+B(E,1)*F
750 V=V+B(E,2)*F
760 M=M+B(E,3)*F
765 NEXT SS
770 VV=0.69*V/M:DD=1.44*U/V
780 TG=(0.241*U/H)+25
790 IF N=<0 PRINT"NEED MAIN CHAIN UNIT": GOTO520
800 ME=5.4E4*(M/N)*SQR((V/N)-1.48E-5)
805 CLS
810 PRINT"I. MAIN CHAIN UNITS:"
820 PRINT "UNIT NO.";TAB( 9)"MOLES";TAB(15)"STRUCTURE";TAB(45)"POLYMER"
830 PRINT "  ";TAB( 9 ) "  ";TAB(15)"UNIT";TAB(45)"REFERENCE"
840 FOR II=0 TO MU-1:CC=W(II,0)-1
850 PRINT W(II,0); TAB( 9) W(II,1); TAB(15) A$(CC,0); TAB(45) A$(CC,1)
860 NEXT II
871 IF SU<1 GOTO 930
880 PRINT "ii. SIDE CHAIN UNITS:"
890 FOR JJ=0 TO SU-1: EE=U(JJ,0)-1
900 PRINT U(JJ,0); TAB( 9) U(JJ,1); TAB(15) A$(EE,0); TAB(45) A$(EE,1)
910 NEXT JJ
930 PRINT "GLASS SPEC. VOL.(M*M*M/KG)= ";VV;"(CC/G)=";1E3*VV
940 PRINT "GLASS C.E.D. (J/M*M*M)= ";DD;"(CAL/CC)=";2.39E-7*DD
950 PRINT "GLASS TEMP. (K)= ";TG;"(C)=";TG-273.2
960 PRINT "ENTANG. MW. (KG/MOLE)= ";ME;"(G/MOLE)=";1000*ME
970 PRINT "SUMMED VALUES: U,H,V,M,N"; U;H;V;M;N
980 PRINT "TO CONTINUE TYPE RUN AND PRESS ENTER"
```

LISTING 3 Polymerization and Crosslinking Model

```
5 CLS
10 PRINT"A AND B COREACTION-MOL. WT. DIST.-THERMAL TRANS.-D.H.KAELBLE-OCT. 27,19
82
20 PRINT"IF NONSTOICHIOMETRIC REACTION HAVE MOLES OF B IN EXCESS"
40 INPUT"MOLES OF TYPE A (MOLE)=";N1
50 INPUT"TYPE A FUNTIONALITY(=>2)=";F1
60 INPUT"MOL. WT. OF TYPE A (G/MOLE)=";M1
70 INPUT"MOLES OF TYPE B (MOLE)=";N2
80 INPUT"TYPE B FUNCTIONALITY (=>2)=";F2
90 INPUT"MOL. WT. OF TYPE B (G/MOLE)=";M2
92 R=(N1*F1)/(N2*F2)
94 IF R>1 GOTO 20
100 INPUT"FRACTION OF MOLECULES OF FUNCTIONALITY >2=";A
105 INPUT"NUMBER OF A AND B MAIN CHAIN ATOMS (A1,A2)=";A1,A2
110 INPUT"MOL. WT. BETWEEN ENTANGLEMENTS (G/MOLE)=";ME
115 INPUT"GLASS COORDINATION NUMBER (8<Z<10)=";ZG
120 INPUT"MONOMER AND LINEAR POLYMER GLASS TEMPERATURES (T1,T2) IN DEG. K=";T1,T
2
130 N=N1+N2:M=((N1*M1)+(N2*M2))/N:AM=((A1*N1)+(A2*N2))/N
140 F=((N1*F1)+(N2*F2))/N:AC=1/(F-1)
145 RI=(2*ZG/(ZG-2))*(T1/(T2-T1))
150 IF A<1 AND A>0 THEN GOTO 200
160 IF A=0 THEN GOTO 180
170 IF A=1 THEN GOTO 190
180 PC=SQR(AC/R):GOTO 220
190 PC=AC/R:GOTO 220
200 PC=((SQR((A*A*R*R)+(4*(1-A)*R/(F-1)))-A*R)/(2*(1-A)*R))
220 PRINT"GEL POINT (% A REACTED)=";100*PC
230 PRINT"GEL POINT (% B REACTED)=";100*R*PC
240 PRINT"INITIAL NUM. AVE. DEG. OF POLYMERIZATION=";RI
250 PRINT"TO ANALYSE POLYMERIZATION PRESS ENTER";:INPUT D$
260 PRINT"% A";TAB(10)"BRANCH";TAB(20)"NUM. AVE.";TAB(30)"WT. AVE.";TAB(40)"GLAS
S";TAB(50)"FLOW"
270 PRINT"REACTED";TAB(10)"COEF.";TAB(20)"MW(G/MOL)";TAB(30)"MW(G/MOL)";TAB(40)"
TEMP(K)";TAB(50)"TEMP(K)"
340 IF PC>1 THEN PP=0.999 : GOTO 350
345 PP=.999*PC
350 PI=0.0999*PP
360 FOR P1=0 TO PP STEP PI
370 AL=(A*R*P1)+(1-A)*R*P1*P1
380 MN=M/(1-(F*AL/2)):RJ=RI*MN/M
390 MW=M*(1+AL)/(1-((F-1)*AL))
400 T3=T2*RJ*(ZG-2)/(RJ*(ZG-2)+(2*ZG))
410 IF 2*ME=>MW THEN BS=0 : GOTO 430
420 BS=1-(2*ME/MW)
430 TL=(.8686*LOG(MW*AM/M))+(1.0423*BS*LOG(MW/(2*ME)))
440 TM=T3+51.6*TL/(17.4-ABS(TL))
445 IF TL>17.3 THEN TM=0
450 PRINT 100*P1;TAB(10)AL;TAB(20)MN;TAB(30)MW;TAB(40)T3;TAB(50)TM
460 NEXT P1
470 IF PC=>1 :P1=P1-0.0999*PP:GOTO 910
550 PRINT"TO ANALYSE CROSSLINKING PRESS ENTER";:INPUT D$
560 PRINT"%A";TAB(10)"BRANCH";TAB(20)"WT. FR.";TAB(30)"NUM. AVE.";TAB(40)"X-LINK
 MW";TAB(52)"GLASS"
565 PRINT"REACTED";TAB(10)"COEF.";TAB(20)"GEL";TAB(30)"MW(G/MOL)";TAB(40)"(G/MOL
)";TAB(52)"TEMP(K)"
570 Z=1/F:Z1=(1-Z):WS=0.99:WF=-0.099
580 FOR WS=0.99 TO 0 STEP WF
590 AL=(1-(WS[Z))/(1-(WS[Z1))
600 IF A<1 AND A>0 THEN GOTO 650
610 IF A=0 THEN GOTO 630
620 IF A=1.0 THEN GOTO 640
630 P1=SQR(AL/R):GOTO 660
640 P1=(AL/R):GOTO 660
650 P1=((SQR((A*A*R*R)+4*(1-A)*R*AL)-A*R)/(2*(1-A)*R))
```

```
660 WG=1-WS:C=(N*WG/2)*((AL/AC)-1)
670 MC=M*N/C
680 D=1-(F*AL/2)
690 IF D<1E-6 THEN D=1E-6
700 RR=RI/D:GM=M*RR/RI
710 T4=T2*RR*(ZG-2)/((RR*(ZG-2-(2*C/N)))+(2*ZG))
715 IF P1>1 THEN GOTO 740
720 PRINT 100*P1;TAB(10)AL;TAB(20)WG;TAB(30)GM;TAB(40)MC;TAB(52)T4
730 NEXT WS
735 GOTO 910
740 WS=WS-WF:WF=WF/10
745 WS=WS+WF
750 AL=(1-(WS[Z))/(1-(WS[Z1))
760 IF A<1 AND A>0 THEN GOTO 810
770 IF A=0 THEN GOTO 790
780 IF A=1 THEN GOTO 800
790 P1=SQR(AL/R):GOTO 815
800 P1=(AL/R):GOTO 815
810 P1=((SQR((A*A*R*R)+4*R*(1-A)*AL)-A*R)/(2*(1-A)*R))
815 IF P1<.999 GOTO 745
820 IF P1=>0.999 AND P1<=1.00 GOTO 840
830 IF P1>1.0 GOTO 740
840 WG=1-WS:C=(N*WG/2)*((AL/AC)-1)
850 MC=M*N/C:D=1-(F*AL/2)
870 IF D<1E-6 THEN D=1E-6
880 RR=RI/D:GM=M*RR/RI
890 T4=T2*RR*(ZG-2)/((RR*(ZG-2-(2*C/N)))+(2*ZG))
900 PRINT 100*P1;TAB(10)AL;TAB(20)WG;TAB(30)GM;TAB(40)MC;TAB(52)T4
910 PRINT"% B REACTED=";100*R*P1;"  TO CONTINUE TYPE RUN AND PRESS ENTER"
930 END
```

LISTING 4 Polymer Chemistry to Mechanical Properties Model

```
5 CLS
10 PRINT"POLYMER CHEMISTRY TO MECHANICAL PROPERTIES PROGRAM"
11 PRINT"-D. H. KAELBLE-JULY 20,1982
20 INPUT"POLYMER-DILUENT DESCRIPTION";AO$
30 PRINT"INPUT POLYMER PROPERTIES"
40 INPUT"AA,AB,AC,AD,AE";AA,AB,AC,AD,AE
50 INPUT"AF,AG,AH,AI,AJ";AF,AG,AH,AI,AJ
60 INPUT"AK,AL,AM,AN,AP";AK,AL,AM,AN,AP
70 PRINT"INPUT DILUENT PROPERTIES"
80 INPUT"AQ,AR,AS,AT,AU";AQ,AR,AS,AT,AU
90 INPUT"AV,AW,AX,AY,AZ";AV,AW,AX,AY,AZ
100 PRINT"INPUT TEST CONDITIONS"
110 INPUT"BA,BB,BC,BD,BE";BA,BB,BC,BD,BE
111 INPUT"FRACTION NEOHOOKIAN(=1.0) VS. HOOKIAN(=0.0) RESPONSE=";NH
112 CLS:PRINT"POLYMER-DILUENT=";AO$
114 PRINT"POLYMER PROPERTIES:"
116 PRINT"AA,AB,AC,AD,AE=";AA;AB;AC;AD;AE
118 PRINT"AF,AG,AH,AI,AJ=";AF;AG;AH;AI;AJ
120 PRINT"AK,AL,AM,AN,AP=";AK;AL;AM;AN;AP
122 PRINT"DILUENT PROPERTIES:"
124 PRINT"AQ,AR,AS,AT,AU=";AQ;AR;AS;AT;AU
126 PRINT"AV,AW,AX,AY,AZ=";AV;AW;AX;AY;AZ
128 PRINT"TEST CONDITIONS:"
130 PRINT"BA,BB,BC,BD,BE=";BA;BB;BC;BD;BE
131 PRINT"FRACTION NEOHOOKIAN TENSILE RESPONSE=";NH
132 PRINT"PRESS ENTER TO CONTINUE":INPUT D$
160 BF=(AK*AS)/(AC*AQ)
170 BG=AJ/BA
180 BH=LOG(BG)/2.3026
190 BI=AE-((51.6*BH)/(17.4-ABS(BH)))
200 BJ=4*AF*AG/(83.13*AH*AC)
210 BK=AZ/BA
220 BL=LOG(BK)/2.3026
230 BM=AU-(51.6*BL/(17.4-ABS(BL)))
240 BN=4*AV*AW/(83.13*AX*AS)
250 TG=((AA*AB*BI)+(BF*AQ*AR*BM))/((AA*AB)+(BF*AQ*AR))
260 UR=((AA*AB*BJ)+(BF*AQ*AR*BN))/((AA*AB)+(BF*AQ*AR))
300 BO=(AB*AA/AD)/((AB*AA/AD)+(AR*AQ/AT))
310 BQ=1-BO
320 BR=BO*BO*AI*AI+BQ*BQ*AY*AY+2*AP*BO*BQ*AI*AY
330 IF 2*AL=>AA THEN BS=0 :GOTO350
340 BS=1-(2*AL/AA)
350 IF 2*AM<=BO*AA GOTO 360
353 IF 2*AM>BO*AA PRINT"NO ENTANGLEMENTS-RAISE AA OR AB"
356 INPUT"AA,AB";AA,AB:GOTO 160
360 BT=1-((2*AM)/(BO*AA))
370 BU=(8*BR/0.02389)
380 BV=(8*BO*AI*AI/0.02389)
390 BX=BO*AD*83.13*TG/AM
400 BY=BX*SQR(2*AM/(BO*AA))
405 BZ=BO*BS*AD*83.13*TG*((1/AL)+(BT/AM))
410 SB=10.46*BR
420 TL=0.4343*(2*LOG(BV/BX)+4.8*LOG(BX/BY))
423 TI=0.4343*LOG(BB/AN)
425 TX=TL-TI:TY=ABS(TX)
427 IF TY>17 THEN TY=17
430 TM=TG+((51.6*TX)/(17.4-TY))
440 FOR TD=0 TO (BE-1) STEP 1
450 T=BC+(BD*TD)
460 SM=(TM-T)/UR
470 IF SM<=SB THEN SS=SM ELSE SS=SB
475 CLS: PRINT"INTERMEDIATE RESULTS:"
480 PRINT"BF,BG,BH,BI,BJ=";BF;BG;BH;BI;BJ
490 PRINT"BK,BL,BM,BN=";BK;BL;BM;BN
500 PRINT"TG,UR=";TG;UR
510 PRINT"BO,BQ,BR=";BO;BQ,BR
520 PRINT"BS,BT=";BS;BT
530 PRINT"BU,BV,BX,BY,BZ=";BU;BV;BX;BY;BZ
```

```
540 PRINT"SB,TL,TM,NH=";SB;TL;TM;NH
550 PRINT"T,SM,SS=";T;SM;SS
555 IF SS<=0.0 THEN PRINT "ABOVE ZERO STRENGTH TEMPERATURE-CHOOSE NEW TEST CONDI
TIONS:":GOTO 100
760 ZB=(BU+BZ):CK=0.4343*LOG(ZB)
770 ZC=(BV+BZ):CL=0.4343*LOG(ZC)
780 ZD=(BX+BZ):CM=0.4343*LOG(ZD)
790 ZE=(BY+BZ):CP=0.4343*LOG(ZE)
800 PRINT"SHEAR AND TENSION ANALYSIS:"
810 INPUT"INPUT NUMBER OF STRESS INCREMENTS";SZ
815 ZJ=0:WS=0:WT=0:ZR=0:SR=0:WB=0:TS=0:TE=0:WC=0
820 CA=0.4343*LOG(AN):CB=0.4343*LOG(BB)
821 PRINT "SHEAR";TAB(12)"SHEAR";TAB(24)"SHEAR";TAB(36)"TENSILE";TAB(48)"TENSILE
"
822 PRINT "MODULUS";TAB(12)"STRESS";TAB(24)"STRAIN";TAB(36)"STRESS";TAB(48)"STRA
IN"
823 PRINT "(BAR)";TAB(12)"(BAR)";TAB(36)"(BAR)"
824 PRINT TAB(12)"0";TAB(24)"0";TAB(36)"0";TAB(48)"0"
830 FOR TF=0 TO SZ STEP 1
840 SI=TF*SS/SZ
850 SJ=SI+(SS/SZ)
860 SG=SI+((0.7071*SS)/SZ)
870 CD=TG-(UR*SG)
880 CE=CA-((17.4*(T-CD))/(51.6+ABS(T-CD)))
890 CF=CE+0.8686*(LOG(BV)-LOG(BX))
900 CG=CF+4.8*0.4343*(LOG(BX)-LOG(BY))
910 CH=EXP(CE/0.4343)
920 CI=EXP(CF/0.4343)
930 CJ=EXP(CG/0.4343)
940 IF CB<CE THEN G=(BU+BZ) :GOTO 1050
950  IF CB<=CF THEN CR=CL-(CL-CM)*(CE-CB)/(CE-CF):G=EXP(CR/0.4343):GOTO 1050
960 IF CB<=CG THEN CS=CM-(CM-CP)*(CF-CB)/(CF-CG):G=EXP(CS/0.4343): GOTO 1050
965 IF CB>CG AND BZ=0 THEN G=BY:GOTO 1050
970 IF CB>CG THEN G=BZ: GOTO 1050
1050 ZI=ZJ
1060 ZJ=ZJ+((SJ-SI)/G)
1070 LI=(2*ZI/3)+1.0:LJ=(2*ZJ/3)+1.0
1080 IF BS=0 THEN  RI=1.0:RJ=1.0:GOTO 1130
1090 MX=83.13*TG*BO*AD/BZ
1100 LM=0.057*SQR(MX/AD)
1110 IF LM<=1.0 PRINT "INEXTENSIBLE-RAISE INPUT AL":INPUT "AL=";AL:GOTO 160
1115 IF LJ=>LM THEN PRINT "EXCEEDED NETWORK STRAIN LIMIT- DOUBLE MAXIMUM STRESS
INCREMENTS":GOTO 810
1120 RI=(LM*LM)/((LM*LM)-(LI*LI)):RJ=(LM*LM)/((LM*LM)-(LJ*LJ))
1130 SM=SI*RI:SN=SJ*RJ
1135 IF SN<SS THEN WS=WS+((SM+SN)*(ZJ-ZI)/2):GOTO 1200
1140 IF SN=>SS THEN ZR=ZJ-(ZJ-ZI)*(SN-SS)/(SN-SM):SR=SS:WB=WS+((SM+SR)*(ZR-ZI)*0
.5)
1200 ZA=(2*ZI/3):ZB=(2*ZJ/3)
1205 IF LI<=1.0 THEN QI=1.0:GOTO 1230
1210 QI=((LI*LI*LI)-1)/(3*((LI*LI*LI)-(LI*LI)))
1220 QJ=((LJ*LJ*LJ)-1)/(3*((LJ*LJ*LJ)-(LJ*LJ)))
1230 CT=(2*(1-NH)*SM)+(2*NH*QI*SM):CU=(2*(1-NH)*SN)+(2*NH*QJ*SN)
1240 FI=CT*LI:FJ=CU*LJ
1245 IF FI=>2*SS THEN 1300
1250 IF FJ<2*SS THEN WT=WT+((CT+CU)*(ZB-ZA)/2):GOTO 1300
1260 IF FJ=>2*SS THEN A=(CU-CT)/(LJ-LI):B=CT-(A*LI):C=-2*SS:D=SQR((B*B)-(4*A*C))
:LB=(D-B)/(2*A):TS=2*SS/LB:TE=LB-1:WC=WT+((CT+TS)*(TE-ZA)*0.5)
1300 IF SN<=SS AND FJ<=2*SS THEN PRINT G;TAB(12) SN;TAB(24) ZJ;TAB(36) CU;TAB(48
) ZB:GOTO 1440
1310 IF SN<=SS  AND FJ=>2*SS THEN PRINT G;TAB(12) SN;TAB(24) ZJ:GOTO 1440
1320 IF SN>SS AND FJ>(2*SS) THEN GOTO 1330 ELSE GOTO 1440
1330 PRINT"SHEAR FAILURE PROPERTIES:"
1335 PRINT"STRESS(BAR)=";SR;TAB(20)"STRAIN=";ZR;TAB(40)"ENERGY/VOL(BAR)=";WB
1337 IF WB>SR*ZR THEN GOTO 1380
1340 IF WB>SR*ZR/2 THEN DA=0.5*(SR*SR)/((SR*ZR)-WB):DB=SR*SR/(2*DA):DC=WB-DB:DD=
2*DB/SR
1350 IF WB<=SR*ZR/2 THEN DA=2*WB/(ZR*ZR):DB=WB:DC=O:DD=ZR
```

```
1360 PRINT"E-MODULUS(BAR)=";DA;TAB(30)"E-WORK(BAR)=";DB
1370 PRINT"E-STRAIN=";DD;TAB(30)"P-WORK(BAR)=";DC
1380 PRINT:PRINT"TENSILE FAILURE PROPERTIES:"
1390 PRINT"STRESS(BAR)=";TS;TAB(20)"STRAIN=";TE;TAB(40)"ENERGY/VOL(BAR)=";WC
1395 IF WC>TS*TE THEN GOTO 1450
1400 IF WC>0.5*TS*TE THEN DE=(0.5*TS*TS)/((TS*TE)-WC):DF=TS*TS/(2*DE):DG=WC-DF:D
H=2*DF/TS
1410 IF WC<=0.5*TS*TE THEN DE=2*WC/(TE*TE):DF=WC:DG=0:DH=TE
1420 PRINT"E-MODULUS(BAR)=";DE;TAB(30)"E-WORK(BAR)=";DF
1430 PRINT"E-STRAIN=";DH;TAB(30)"P-WORK(BAR)=";DG:GOTO 1450
1440 NEXT TF
1450 INPUT"PRESS ENTER TO CONTINUE:";D$
1460 NEXT TD
```

LISTING 5 Composite Fracture Energy and Strength

```
5 CLS
10 PRINT"COMPOSITE FRACTURE ENERGY AND STRENGTH-D.H.KAELBLE-JAN. 15,1980"
20 INPUT"FIBER DIAMETER(D), VOLUME FRACTION(V)="; D,V
30 INPUT"FIBER TENSILE MODULUS(E), STRENGTH(S)="; E,S
40 INPUT"MATRIX SHEAR MODULUS(G), STRENGTH(L)="; G,L
50 INPUT"MATRIX TENSILE MODULUS(Y), STRENGTH(YY)="; Y,YY
60 INPUT "F-M BOND STRENGTH(LB), FRICT. STRENGTH(LF)="; LB,LF
61 IF LB>L THEN LB=L: IF LF>L THEN LF=L:CLS
62 CLS:PRINT"FIBER DIAMETER(D), VOLUME FRACTION(V)=";D;V
63 PRINT"FIBER TENSILE MODULUS(E), STRENGTH(S)=";E;S
64 PRINT"MATRIX SHEAR MODULUS(G), STRENGTH(L)=";G;L
65 PRINT"MATRIX TENSILE(Y), STRENGTH(YY)=";Y;YY
66 PRINT"F-M BOND STRENGTH(LB), FRICT. STRENGTH(LF)=";LB;LF
67 PRINT
70 R=D/2: R1=R*((1.074*SQR(3.14159/V))-1):R2=LOG(R1/R)
80 A=SQR(2*G/E)/(R*SQR(R2))
90 PS=2*3.14159*R*LB/A: AA=3*3.14159*R*R/V
100 F1=SQR(R2): F2=2*V/3: F3=(2/3.222)*SQR(V/3.14159)
110 BL=(R/LF)*((S/2)-(LB*F1*SQR(E/(2*G))))
120 IF BL<0 THEN BL=0
130 WS=LB*BL*F1*F2*SQR(E/(2*G))
140 WB=S*S*BL*F2/(12*E)
150 WF=LF*BL*BL*F2/(2*R)
160 W=WS+WB+WF
170 PB=F3*S+(1-V)*YY: EY=V*E+(1-V)*Y
180 LM=A*R*S/2
182 FI=(3.1416*R)/((3.1416*R)+(R1-R))
183 IL=(FI*LB)+((1-FI)*L)
190 PRINT"INTER-FIBER SPACING(R1)="; R1
200 PRINT "SHEAR STRESS CONC.(A)="; A
210 PRINT"MAX. F-M BOND STRENGTH(LM)=";LM
215 PRINT"F-M DEBOND LENGTH(BL)=";BL
220 PRINT"INTERLAM. SHEAR STRENGTH(IL)=";IL
230 PRINT"COMPOSITE TENSILE MODULUS=";EY
232 PRINT"COMPOSITE CONTINUUM TENSILE STRENGTH=";PB
234 PRINT"CRITICAL CRACK LENGTH=";2*BL
236 PRINT"TO CONTINUE PRESS ENTER";:INPUT D$:CLS
241 PRINT TAB(10)"FRACTURE MECHANICS ANALYSIS"
242 LC=2*BL:K5=SQR(EY*(WS+WB)):K6=SQR(EY*W)
243 P5=K5/SQR(3.1416*LC):P6=K6/SQR(3.1416*LC)
244 PRINT"UNFLAWED STRENGTH (MIN.)=";P5;"(MAX.)=";P6
245 PRINT"CRIT. STRESS INTENSITY (MIN.)=";K5;"(MAX.)=";K6
246 PRINT"FLAW SIZE","CRACK LENGTH","MIN. STRENGTH","MAX. STRENGTH"
247 FOR XX=0 TO 9 STEP 1
248 CC=LC*EXP(-2.772589+(0.693147*XX))
249 RF=CC/LC:RG=-RF*RF*RF:F=EXP(RG)
250 C1=(LC*F)+((1-F)*CC)
251 P7=K5/SQR(3.1416*C1):P8=K6/SQR(3.1416*C1)
252 PRINT C1,CC,P7,P8
253 NEXT XX
254 PRINT"TO CONTINUE PRESS ENTER";:INPUT D$:CLS
260 PRINT TAB(10)"FRACTURE WORK PER UNIT CROSSECTION AREA"
270 PRINT TAB(0)"F-M BOND";TAB(10)"FIBER";TAB(20)"MATRIX";TAB(30)"FRICT.";TAB(40
)"TOTAL";TAB(50)"CRIT.FIBER"
280 PRINT TAB(0)"STRESS";TAB(10)"WORK";TAB(20)"WORK";TAB(30)"WORK";TAB(40)"WORK"
;TAB(50)"LENGTH"
290 PRINT TAB(0) LB; TAB(10) WB; TAB(20) WS; TAB(30) WF; TAB(40) WB+WS+WF; TAB(5
0) 2*BL
300 PRINT "PRESS ENTER TO CONTINUE";:INPUT D$
310 FOR FX=0 TO 9 STEP 1
312 LX=LM*FX/9
315 BL=(R/LF)*((S/2)-(LX*F1*SQR(E/(2*G)))): IF BL<0 THEN BL=0
320 WF=LF*BL*BL*F2/(2*R)
330 WS=LX*BL*F1*F2*SQR(E/(2*G))
340 WB=S*S*BL*F2/(12*E)
350 PRINT TAB(0) LX; TAB(10) WB; TAB(20) WS; TAB(30) WF; TAB(40) WB+WS+WF; TAB(5
0) 2*BL
360 NEXT FX
370 PRINT"PRESS ENTER TO CONTINUE";:INPUT D$
380 GOTO 5
```

LISTING 6 Peel Mechanics

```
5 CLS
10 PRINT" PEEL CALCULATION N0. 2  -D.H.KAELBLE- NOV 27, 1979"
20 INPUT"RIBBON HALF THICKNESS(H), ADHESIVE THICKNESS(A)";H,A
30 INPUT"BOND WIDTH(B), RIBBON   TENSILE MODULUS(E)";B,E
40 INPUT"ADHESIVE TENSILE MODULUS(Y),STRENGTH(SA)";Y,SA
50 INPUT"ADHESIVE SHEAR MODULUS(G),STRENGTH(LA)";G,LA
60 BB=3*Y/(8*E*H*H*H*A): BA=BB[(1/4)
70 GB=G/(2*E*H*A): GA=SQR(GB)
80 RA=SQR(2*Y*E*H*H*H/(3*A)): R=RA/SA
90 PS=LA*B/GA: PC=SA*SA*B*A/(4*Y)
100 PRINT "CLEAVAGE STRESS CONC.(BA)=";BA
110 PRINT "SHEAR STRESS CONC.(GA)=";GA
120 PRINT "180 DEG. RAD. OF CURV.(R)=";R
130 PRINT "0 DEG. PEEL FORCE(PS)=";PS
140 PRINT "180 DEG. PEEL FORCE(PC)=";PC
150 PRINT TAB(0)"PEEL";TAB(10)"PEEL";TAB(20)"PEEL";TAB(30)"K";TAB(40)"TENSILE";T
AB(50)"SHEAR"
160 PRINT TAB(0)"ANGLE";TAB(10)"WORK ";TAB(20)"FORCE";TAB(40)"STRESS";TAB(50)"ST
RESS"
175 PRINT "PRESS ENTER TO CONTINUE----READY";:INPUT D$
180 W=3.10569
190 S=SA: AA=SIN(W)/(1-COS(W))
200 T=S*AA/(2*BA*SA*R)
230 K=(SQR(1+4*T)-1)/(2*T)
240 L=(SA*GA*(1-K))/(2*BA*TAN(W))
250 IF L<=LA THEN GOTO 350 ELSE GOTO 260
260 L=LA
270 S=2*LA*BA*TAN(W)/(GA*(1-K))
280 N=K
290 T=S*AA/(2*BA*SA*R)
300 IF T=0 THEN K=1.0 ELSE GOTO 310
305 GOTO 330
310 IF T<0 THEN K=(1-SQR(1-(4*T)))/(2*T) ELSE GOTO 320
315 GOTO 330
320 IF T>0 THEN K=(SQR(1+4*T)-1)/(2*T)
330 D=ABS(K-N)
340 IF D>=0.0005 THEN GOTO 270 ELSE GOTO 350
350 I=B*S*(1-K)/(2*BA): J=B*L/GA
360 PM=B*A*PA*PA/(1-COS(W))
380 PF=SQR(I*I+J*J)
390 WW=(PF/B)*(1-COS(W))+(PF/B)*(PF/B)/(4*H*E)
400 PRINT TAB(0) W*180/3.1416; TAB(10) WW; TAB(20) PF; TAB(30) K; TAB(40) S; TAB
(50) L
410 IF W>1.5708 THEN W=W-0.52359 ELSE GOTO 430
420 GOTO 190
430 IF W>0.6981 THEN W=W-0.1745 ELSE GOTO 450
440 GOTO 190
450 IF W>0.1745 THEN W=W-0.08726 ELSE GOTO 470
460 GOTO 190
470 END
```

Index